COLLAPSING SPACE AND TIME:

Geographic aspects of
communications and information

COLLAPSING SPACE AND TIME:

Geographic aspects of communications and information

Edited by

STANLEY D. BRUNN
THOMAS R. LEINBACH

HarperCollins*Academic*

An imprint of HarperCollins*Publishers*

Published by
HarperCollins*Academic*
77–85 Fulham Palace Road
Hammersmith
London W6 8JB
UK

Allen County Public Library
Ft. Wayne, Indiana

First published in 1991

Library of Congress Cataloging in Publication Data

Brunn, Stanley D.
  Collapsing space and time: geographic aspects of communication and
  information / by Stanley D. Brunn and Thomas R. Leinbach.
      p.   cm.
Includes bibliographical references and index.
ISBN 0–04–910119–6 (HB): $75.00 — ISBN 0–04–910120–X (PB): $24.95
1. Telecommunication.   2. Space in economics.   3. Economic development.
I. Leinbach, Thomas R., 1941–   . II. Title.

HE7631.B785   1990   90–12815

384′.041—dc20   CIP

British Library Cataloguing in Publication Data

  Collapsing space and time: geographic aspects of communications and
  information.
1. Social structure. Influence of ideology.
I. Brunn, Stanley D. 1939–   II. Leinbach, Thomas R. *1941–*
303.3

ISBN 0–04–910119–6
ISBN 0–04–910120–X   pbk

Typeset in 10 on 12 point Bembo by Computape (Pickering) Ltd,
North Yorkshire
and printed in Great Britain by
Cambridge University Press

Contents

List of tables

List of figures

List of contributors

Ronald F. Abler
Professor of Geography at Pennsylvania State University and Executive Director, Association of American Geographers, Washington, DC.

Anne Lyew-Ayee
Assistant Professor in the Department of Geography at the University of the West Indies, Mona, Kingston.

William R. Code
Assistant Professor in the Department of Geography at the University of Western Ontario, Canada.

Kenneth E. Corey
Dean of the College of Social Sciences at Michigan State University.

P. W. Daniels
Professor of Geography at Portsmouth Polytechnic, England.

Pip Forer
Senior Lecturer in the Department of Geography at the University of Canterbury, Christchurch, New Zealand.

Ake Forsström
Professor, Kulturgeografiska Institutionem, Göteborgs Universitet, Göteborg, Sweden.

John R. Gold
Joint Head at the Centre for Geography in Higher Education at Oxford Polytechnic, England.

Peter Gould
Evan Pugh Professor of Geography in the Department of Geography at Pennsylvania State University.

Mark Hepworth
The Centre for Urban and Regional Studies at The University of Newcastle upon Tyne, England.

Briavel Holcomb
Associate Professor in the Department of Urban Studies at Rutgers University, New Jersey.

Erick Howenstine
Asistant Professor in the Department of Geography at the University of Northeastern Illinois, Chicago.

Donald G. Janelle
Professor of Geography at the University of Western Ontario, Canada.

Aharon Kellerman
Associate Professor in the Department of Geography at the University of Haifa, Mount Carmel, Israel.

John V. Langdale
Senior Lecturer in the Department of Geography at Macquarie University, Australia.

Nancy Davis Lewis
Associate Professor in Department of Geography and Center for Pacific Island Studies at the University of Hawaii, Honolulu, Hawaii.

Sten Lorentzon
Professor, Kulturgeografiska Institutionem, Göteborgs Universitet, Göteborg, Sweden.

Lori Van Dusen Makaida
PEACESAT Project Coordinator at the Social Science Research Institute, University of Hawaii, Honolulu, Hawaii.

Nigel Parrott
Research Officer in the Department of Conservation at the University of Wellington, New Zealand.

John Pickles
Associate Professor in the Department of Geography at the University of Kentucky.

Roger W. Stump
Department of Geography and Planning at the State University of New York, Albany.

Introduction

In the last decade of the twentieth century we are especially aware of the rapid changes occurring on the world's economic and political maps as the world regions become more interdependent. An essential thrust of these changes is captured by both the amount and speed by which information of many varieties and from local to global scales is produced, acquired, processed, transmitted, and dispersed. In essence communications on a greatly enlarged scale is truly shaping the way we think and behave as individuals, members of communities and organizations, and as states. Clearly these changes are also manifested in the spatial patterns in organization, behavior, interaction, and decision making. Several themes and unifying concepts form the essence of our volume.

Emergence of the services sector

One striking and apparent development of the last third of the twentieth century is the emergence and dominance of the services sector in developed and developing economies of the world. That emergence in the more developed countries is evident in both increases in the labor force employed in a variety of services, and by the growth of that sector in terms of total income generated. The growth of a formal sector of services in the developing world, while involving a smaller total labor force and accounting for less total income generated, nonetheless represents emerging areas of growth.

The services sector has always been important in economies, even those with strong orientations to primary or extractive goods, such as agriculture. Services included those in government, transportation, trade, and selected professions. Employment in these occupations was not large, but was significant in the kinds and variety of services they performed. Economies that experienced rapid growth in industry, whether heavy or light, also had important service components. The demand for services, not only transportation and trade, but also finance, wholesaling, and retailing, were vital to those economies strongly committed to providing industrial goods for producers and consumers. We have seen this commitment evident in developed economies during periods of rapid industrialization experienced in the last and this century. Many developing countries themselves, during the past several decades have also seen the growth of services, both of a formal and informal variety.

It has only been in the past three decades in the United States, Canada, Japan and much of Europe, where the growth of the services sector has been rapid. Those examining changes in the composition of the labor force have found the label 'tertiary sector' satisfactory in describing the more specialized kinds of services that are produced and in demand. Whereas once the term 'services sector' or tertiary sector defined all kinds of service, now we define quaternary and quinary sectors. These represent more specialized kinds of service. Quaternary services include those in the FIRE (finance, insurance, and real estate) category as well as those in retail trade. This sector also includes those employed in the information industries, whether broadcasting, telecommunications, television, or journalism. Quinary defines those employed in health, education, research, government, and recreation or leisure. Tertiary (in the five-fold classification) includes those employed in transportation and utilities.

Postindustrial economies and societies

A related term that appears in much of the literature on the changing nature of the services sector, especially in developed regions, is 'postindustrial.' While there is much debate over what exactly constitutes postindustrial, especially since services are 'produced' (and therefore some scholars would include in the secondary sector) and consumed, the label is one that appears in the social science and public policy literatures. The components of the postindustrial economy are those that are heavily represented by a labor force employed in a variety of services. The services are performed for those in the primary and secondary sectors and for those in the services sector. The labor skills may be specialized and require much formal training or little by way of skills. Labor costs for those in the services sector also range from very high to very low salaries and wages. Associated with changes in the labor composition 'mix' are social changes in society. The postindustrial society is often characterized as being more wealthy than the industrial sector, emphasizing leisure and recreation, more tied to consumer purchases than producer goods, and appealing to cultural achievements over material possessions. Diversity rather than homogeneity, environmental appreciation rather than exploitation, and human value and worth over humans as 'numbers' are additional characteristics. The salient features of postindustrial societies, as they appear in various world regions, remain a topic of much interest in the social and behavioral sciences.

New meanings to time and distance

Two additional components are important in understanding the dynamic changes occurring in postindustrial economies and societies. They are defining and interpreting the meanings of time and distance. These terms are of note not only because they represent concepts that are important in the organization and functioning of any society, but because they assume a different meaning in postindustrial worlds. Technological advances in the services sector, especially in transportation, communication and tele-communication, have ushered in a host of new meanings to time and space. The catholic term 'space adjusting' mechanisms defines those changes to an economy and society that are wrought by improvements in transportation and communication. It is not hard to envisage how the 'worlds' of an individual, a culture, and a nation-state can 'shrink' because of faster rail and air transportation, more extensive airline connections, multiple channel television, the diffusion of post offices and satellite dishes, and policies to ensure quick delivery of phone messages or mail delivery.

The impress of advances in transportation, communications, and tele-communications are evident in varying degrees in almost all countries and cultural regions. Even in once isolated regions or pockets of traditional agriculture or cultures, there are examples of small airports on grass or dirt runways, electric power lines, a village phone, irregular postal delivery systems, semi-regular bookmobile services to rural villages, and individuals with transistor radios, pocket radio and tape cassette recorders, and occa-sionally a battery operated television set. Isolation in areas within technolo-gically advanced societies has diminished by improved highways, faster road vehicles, satellite communications, cable television, electronic mail, supersonic transport, FAX and TELEX communications, and high speed phone circuits. Even if not all parts of a region have such benefits, many more are affected by their presence. What is occurring with the blends and mixes of advances in communications, information processing and collect-ing is a world where cores, peripheries, and semiperipheries assume differ-ent meanings than earlier when transportation and communication were slow and mail sent by rail, road, or ship. Even cartographers are experienc-ing problems in how best to represent worlds, continents, and countries, where relative space and distance are the norm. Conceptualizing these notions and presenting them in maps and graphics represents a major challenge for any geographer or cartographer. Governmental and non-governmental organizations including national and international busi-nnesses, are learning that while shrinking spaces represent parts of the world, other parts are static or may even be farther apart than previously. Distances 'shrink' with faster and improved connections between places. Distances mean little and direction means even less. Relative location is more important than absolute location in a tightly connected and integra-ted world. Absolute location, via, where you are, has much less meaning today. What is more important as markets, societies, cultures and govern-ments are becoming more connected is whether one is 'connected,' how far one is from other places in time not in absolute distance, and how much one is connected with other places. Also the costs and time of transacting information is more important in human decision making than absolute distances between places. Terms that geographers have used to describe these shrinking and collapsing spaces include time-space convergence, cost-space convergence, human extensibility, and regionally and globally integrated systems. Global village, world cities, world systems, and ecumenopolis are additional terms that appear in the literature. Those applying such concepts to regional and international scales recognize that not all places 'shrink' or 'collapse.' Some diverge, others remain static, and still others find themselves farther away from the centers of networking and innovations; that is, the 'gaps' between places connected and not connected tend to widen in time, cost, and relative distance. Or to express it in another way, the world's map is comprised of 'bridges' of communication and transportation, some locations having dense networks, others sparse, and still others that appear as blanks or 'gaps.'

Setting of volume with existing geographic literature

It is important to review briefly some of the pertinent geographic literature in order to place our volume within current research trends in human geography. While the geographic literature is rather modest, the number of articles, chapters and books is increasing annually. But there is also much valuable research being carried out by scholars in related disciplines, especially mass communications and telecommunications as well as regional economics and sociology. Much of that literature contains important concepts, theories and models that could be used in examining spatial processes and patterns as well as examining the changing nature of places and regions.

We identify a number of major themes in the extant literature. One addresses the role of evolving postindustrial societies and economies (for example, Bell 1973; Masuda 1981; and Kellerman 1985; Semple 1985; Kim 1987; Ó hUallacháin 1989). The geography of information, telecommunications, and information economies has been a major theme of a number of writers, including Abler 1971, 1974, 1975; Falk and Abler 1980; Holmes 1980; Bakis 1980, 1981; Hepworth 1987 and 1990; Lamberton and Mandeville 1982; Warntz 1973; Törnqvist 1974; Janelle 1968, 1975; Gottman 1983; Daniels 1985 and 1987; Salomon 1985; Kim 1987; Ó hUallacháin 1989. The role of communications and telecommunications in regional development has been studied by Roy et al. 1969; Ubershall 1969; Törnqvist 1970, 1973; Hagerstrand and Kuklinski 1971; Rogers 1969, 1976; Helleiner and Cruise 1982; Nicol 1983; Saunders et al. 1983; Wellenius 1984; Baum et al. 1985; Curien and Gensollen 1985; Goddard and Gillispie 1986; Gillispie and Hepworth 1986, and Gillispie and Willams 1988. A number of studies have focused specifically on Third World regions, including Lerner and Schram 1967; Battillon 1977; McAnany 1980; Thornqvist 1970, and Katz 1986.

There have also been studies focusing on specific countries and regions, both in the developed and developing world. A partial list would include studies on France (Bakis 1986); East Africa (Soja 1968); SubSaharan Africa (Guttman 1986); Singapore (Chen and Kuo 1985); Portugal (Gaspar et al.); Indonesia and Thailand (Chu *et al.* 1985); Papua New Guinea (Allen 1977; Renick 1984); Uganda (Obseschall 1969); United Kingdom (Openshaw and Goddart 1987); Australia (Langdale 1982); Australia and Singapore (Langdale 1984a, 1984b); New Zealand (Parrott and Forer 1986), and various Pacific Islands (Davey 1984; Karunaratne 1984; Kissling 1984; Langdale 1990).

Another cluster of studies has examined specific services, for example, television (Laboda 1984; Gould 1984; Thomas 1984; Gould and Lyew-Ayee 1985); radio (Innis 1953; Doucet 1983); telegraph service (Johnson 1968; Langdale 1979); telephones (Palm and Farrington 1976; de Sola Pool 1977; Langdale 1978, 1981); post offices (Witthuhn 1968; Alvin 1974; Löytönen 1985); high technology (Rogers 1985); satellite services in sparsely populated rural areas (Morairty 1981; Holmes 1984); and newspapers (Kariel and Rosenvall 1978; Brooker-Gross 1985). Few have looked at the role of the media (Schramm and Atwood 1981; Schramm 1964), including the role of the media in popular culture (Abler 1975b; Thomas 1984; Burgess and Gold 1985). Related studies have examined the role of national and international business transactions and firm behavior, many as they are tied to cities and urban systems (Pred 1973, 1974, 1975; Brooker-Gross 1980; Bakis 1985, 1986; Cloher 1978; Brunn 1983; Kellerman 1984; Marshall 1984; Robinson 1984; Törnqvist 1984; Goddard et al. 1985; Langdale 1985; Danields 1986; Hepworth 1986; Janelle 1986; Bakis 1987; Moss 1987a and 1987b; Langdale 1984c, 1989; Wheeler and Mitchelson 1989a and 1989b). Few have been the studies that have examined spatial dimensions of communications and information in an international political context (Mackay 1958; Minghi 1963; Smith 1980; Brunn 1981; Doucet 1983; Jussawalla and Cheah 1983).

Our purpose in editing a volume of original essays was to identify scholars who have written on the topics related to the geographies of information, communication, telecommunication, postindustrial societies and economies, and the human consequences of shrinking worlds of the next century. Some of those cited above are among the authors. We contacted a number of scholars in North America, Europe, Asia and elsewhere. Most of those we approached were willing to provide essays.

We are aware of gaps in our volume, especially the absence of authors from Latin America, Asia (especially China and Japan), Africa and the Middle East, USSR, and former East Europe and detailed examinations of the geographies of communications and information in these regions. Invitations to prepare and submit chapters were issued in 1986. Had we issued those same invitations in 1990, we would have had a greater international coverage of topics and probably of authors as well. A number of international conferences and special sessions at larger national meetings suggest the growing importance of these topics in regions and countries not covered in our volume. The interested scholar can keep somewhat abreast of the research directions and conferences organized under the International Geographical Union's Commission on Geography of Telecommunications and Communications. The president of the commission is Henry Bakis who regularly edits a newsletter and *Netcom*, annual volumes containing original papers and abstracts of papers at professional meetings of geographers, regional specialists, and others. These multilingual publications are invaluable sources. We have not cited numerous related papers from the *Netcom* volumes in the literature review above.

In a sense our volume represents the 'state of the art' of current research on the geographies of communications and information. That assumption is based on the absence of competing books or monographs on a variety of subjects and the lack of a single major volume that attempts to integrate the political, cultural, social, and economic dimensions of postindustrial service economies. That book has yet to be written. Our own book may have a short shelf-life, in that somewhere someone is preparing that single volume or other editors are assembling a new edited work. We see our book as an initial effort that we hope will stimulate subsequent single or multiple authored works. We have yet to see a series of regional geography books that address the major themes and concepts of postindustrial economies or about the regional, cultural, political and social geographies of the twenty-first century. Even single technologies and innovations (television, telephone, satellite dishes, as examples) and cross cultural influences (European, Japanese, and American) of television, radio, cinema, and mass advertising, have yet to be tackled.

Organization of book and topics covered

This book is organized around four major themes, each of which treats a number of related topics. We consider the topics discussed in the ensuing chapters are those on the cutting edge of those working on the geographies of information and communications. The first section, 'Geography and communications,' includes chapters on the conceptual, theoretical, and cartographic challenges awaiting geographers embarking on these subjects (Gould), on hardware and software technologies and their impacts (Abler), and the configurations of interdependencies at regional and global scales (Janelle). The second section, 'Information economies,' contains specific studies that examine selected services and telecommunications industries (Forsström and Lorentzon; Daniels; Langdale); including banking (Code), capital flows and investments (Hepworth and Pickles). The third section, 'Communications, technologies, and regional development,' addresses development issues related to the dynamic geographies of communications and information in Singapore (Corey), the Pacific Basin (Lewis and Mukaida), Israel (Kellerman), Latin America (Howenstein), and New Zealand (Forer and Parrott). A final set of four chapters, in a section entitled 'Social dimensions of information and communications,' looks at political, social, and cultural ramifications. Specific topics include the electronic cottage and homeworking (Gold and Holcomb) religious broad-casting (Stump), and regional integration efforts in the Commonwealth Caribbean (Lyew-Ayee).

The reader of the chapter titles and chapters is likely to have her/his appetite whetted by the topics covered in our volume. That we expect. Someone may wonder why there are no studies included on propaganda mapping, the regional content of regional and national newspapers, the social and political content of television (whether regulated or not regula-ted by a government), the diffusion of satellite dishes or cable television, cross cultural influences (cusine, entertainment, music) wrought by global satellites, mass advertising for global markets, or the international invest-ment in communications technologies to former Second and Third World countries.

All these and others topics are most legitimate areas for subsequent research and we would hope that the reader will finish reading the essays with the notion that much more awaits our study. We could not agree more with such an assessment. It is our aspiration that the research agenda in the geography of communication and information will include studies done in single regions or even in small towns and villages, among specific social or cultural groups, salient differences between men and women, urban and rural children, youth and elderly; about information firms and multinational corporations, and about the role of the state and non-govern-mental organizations who are so much a part of the information age. We also need studies on the processes operating in technology (high and low) transfer, and innovators and laggards, and on the social and political impediments to technology transfers. As geographers we would be prudent to consider how we might map worlds that are shrinking and those that are not. What about cost and time-cost maps rather than distance maps? How are cores and peripheries to be portrayed?

The geographies of information and communications are changing daily at all scales and it behooves geographers with interests in economic, social, political, and environmental agendas to begin to examine some of those fascinating geographies in our own backyards, our towns and cities, and on global scales. It is our hope that this volume will stimulate much needed research on a host of exciting topics for the coming decade and into the next century.

May 1990   Stanley D. Brunn and Thomas R. Leinbach,
                              University of Kentucky, Lexington, Kentucky

References

Abler, R. F. (1971), 'Distance, intercommunications, and geography,' *Proceedings, Association of American Geographers*, 3:1–4.

——(1974), 'The geography of communications,' in M. E. Eliot Hurst (ed.), *Transportation Geography: Comments and Readings* (New York: McGraw-Hill), pp. 327–46.

——(1975a), 'Effects of space-adjusting technology on the human geography of the future,' in R. Abler et al., *Human Geography in a Shrinking World* (North Scituate, MA: Duxbury Press), pp. 35–56.

——(1975b), 'Monoculture or miniculture? The impact of communications media on culture in space,' in R. Abler et al., *Human Geography in a Shrinking World* (North Scituate, MA: Duxbury Press). pp. 122–31.

Allen, B. (1977), 'Formal and informal information systems and rural change in Papua New Guinea,' *Australian Geographer*, 13:332–38.

Alvin, R. A. (1974), 'Post office locations and the historical geographer: a Montana example,' *The Professional Geographer*, 26:183–86.

Bakis, H. (1980), 'Components for telecommunication geography,' *Annales de Géographie*, 496:657–88.

——(1981), 'Elements for a geography of telecommunications,' *Geoforum*, 4:31–35.

——(1985), 'Telecommunications and the spatial organization of firms,' *Revue Géographique de l'Est*, 25:33–46.

——(1986), 'Telecommunications and the localization of activities within firms (France),' *Materialien zur Raumordnung (Bochum)*, 32:10–17.

——(1987), 'Telecommunications and the global firm,' in F. E. I. Hamilton (ed.), *Industrial Change in Advanced Economies* (London: Croom Helm), pp. 130–60.

Bataillon, C. (1977), 'Mass communications and spatial organization in the countries of the third world,' *Espace geographique*, 6:109–11.

Baum, G. et al. (1985), 'Trans-border data flows in West Germany,' in H. P. Gassmann (ed.), *Trans-border Data Flows* (Amsterdam: North Holland), pp. 83–102.

Bell, D. (1973), *The Coming Post-industrial Society* (New York: Basic Books).

Brooker-Gross, S. (1980), 'Uses of communications technology and urban growth,' in S. D. Brunn and J. O. Wheeler (eds.), *The American Metropolitan System: Present and Future* (New York: Wiley and Halsted Press), pp. 145–60.

——(1985), 'The changing concept of place in the news,' in J. Burgess and J. R. Gold (eds.), *Geography, the Media and Popular Culture* (New York: St. Martin's Press), pp. 63–85.

Burgess, J. and Gold, J. R. (1985), *Geography, the Media and Popular Culture* (New York: St. Martin's Press).

Brunn, S. D. (1981), 'Geopolitics in a shrinking world: a political geography of the twenty-first century,' in A. D. Burnett and P. J. Taylor (eds.), *Political Studies from a Spatial Perspective* (Chichester, U.K.: John Wiley), pp. 157–72.

——(1983), 'Cities of the future,' in S. D. Brunn and J. F. Williams, (eds.), *Cities of the World: World Regional Urban Geography* (New York: Harper and Row), pp. 453–87.

Chen, H. and Kuo, E. (1985), 'Telecommunications and economic development in Singapore,' *Telecommunications Policy*, 9:240–44.

Chu, G. et al. (1985), 'Rural telephone in Indonesia and Thailand: social and economic benefits,' *Telecommunications Policy*, 9:159–69.

Cloher, U. (1978), 'Integration and communications technology in an emerging urban system,' *Economic Geography*, 54:1–16.

Currien N. and Gensollen, M. (1985), 'Reseaux de telecommunications et amenagement de l'espace,' *Revue geographique de l'Est*, 25:47–56.

Daniels, P. (1985), *Service Industries: a Geographical Approach* (London: Methuen).

——(1986), 'Foreign banks and metropolitan development: a comparison of London and New York', *Tijdschrift voor economische en sociale geografie*, 77:269–87.

——(1987), 'Producer-services research: a lengthening agenda,' *Environment and Planning A*, 19:569–71.

Davey, G. J. (1984), 'Telecommunications development in the South Pacific,' in C. C. Kissling (ed.), *Transport and Communications for Pacific Ministates: Issues in Organization and Management* (Suva, Fiji: University of the South Pacific, Institute of Pacific Studies), pp. 15–24.

Doucet, M. (1983), 'Space, sound, culture and politics: radio broadcasting in southern Canada,' *Canadian geographer*, 27: 109–27.

Falk, T. and Abler, R. (1980), 'Intercommunications, distance, and geographical theory,' *Geografiska annaler*, Ser. B, 62:59–67.

Gaspar, J. C. *et al.* (1977), *Telecommunications and Regional Development in Portugal*, (Arbejdsrapport-Aarhus Universitet, Geografisk Institut).

Gillispie, A. and Hepworth, M. (1986), 'Telecommunications and Regional Development in the Information Society,' Newcastle Studies on the Information Economy, (University of Newcastle upon Tyne, United Kingdom).

——and Pye, R. (1977), 'Telecommunications and office location,' *Regional Studies*, 11:19–30.

——and Williams, H. (1988), 'Telecommunications and the reconstruction of regional comparative advantage', *Environment and Planning A*, 20:1311–322.

Goddard, J. B. and Gillispie, A. (1986), 'Advanced telecommunications and regional economic development,' *Geographical Journal*, 152: 383–97.

——*et al.* (1985), 'The impact of new information technology on urban and regional structures in Europe' in A. T. Thwaites and R. P. Oakey (eds.), *The Regional Impact of Technological Change* (New York: St. Martin's Press), pp. 215–42.

Gottman, J. (1983), *The Coming of the Transactional City* (College Park, MD: University of Maryland, Institute of Urban Studies).

Gould, P. R. *et al.* (1984), *The Structure of Television* (New York: Pion).

——and Lyew-Ayee, A. (1985), 'Television in the third world: a high wind on Jamaica,' in J. Burgess and J. R. Gold, (eds.), *Geography and the Media* (New York: St. Martin's Press), pp. 33–62.

Guttman, W. (1986), 'Telecommunications and Sub-Saharan Africa: the continuing crisis,' *Telecommunications Policy*, 10:325–40.

Hägerstrand, T. and Kuklinski, A. (eds.), (1971), *Information Systems for Regional Development*. (Lund: C. W. K. Gleerup, Lund Studies in Geography), No. 37.

Halina, J. W. (1980), 'Communications and the economy: a North American perspective,' *International Social Science Review*, 32: 264–82.

Helleiner, G. and Cruise, H., (1982), 'The political economy of information in a changing international economic order', in M. Jussawalla and D. Lamberton (eds.), *Communications, Economics, and Development* (Elmsford: Pergamon Press), pp. 100–32.

Hepworth, M. (1986), 'The geography of technological change in the information economy,' *Regional Studies*, 20:407–24.

——(1987), 'Information technology as spatial systems,' *Progress in Human Geography*, 11: 157–80.

——(1990), *Geography of the Information Economy* (New York: Guilford Press).

Holmes, J. (1984), 'The domestic satellites and remote area communications,' *Australian Geographical Studies*, 22:122–28.

Innis, D. Q. (1953), 'The geography of radio in Canada,' *Canadian Geographer*, 1:89–97.

Janelle, D. (1968), 'Central place development in a time-space framework,' *The Professional Geographer*, 20:5–10.

——(1973), 'Measuring human extensibility in a shrinking world,' *Journal of Geography*, 72, 5:8–15.

——(1986), 'Metropolitan expansion and the communications-transportation trade-off,' in S. Hanson (ed.), *The Geography of Urban Transportation* (New York: Guilford Press), pp. 357–85.

Johnson, B. (1968), 'Utilizing telegrams for describing contact and spatial inter-actions,' *Geografiska annaler*, 50B: 48–51.

Jussawalla, M. and Cheah, C. (1983), 'Emerging economic constraints on trans-border data flows,' *Telecommunications Policy*, 10:285–96.

Kariel, H. G. and Rosenvall, L. A. (1978), 'Circulation of newspaper news within Canada,' *Canadian Geographer*, 22:85–111.

Karunaratune, N. D. (1984), 'Telecommunications infrastructure and economic development of Pacific Island nations,' in C. C. Kissling (ed.), *Transport and Communication for Pacific Ministates: Issues in Organization and Management* (Suva, Fiji: University of the South Pacific, Institute of Pacific Studies), pp. 15–24.

Katz, R. (1986), 'Explaining information sector growth in developing countries,' *Telecommunications Policy*, 10:209–28.

Kellerman, A. (1984), 'Telecommunications and the geography of metropolitan areas,' *Progress in Human Geography*, 8:222–46.

——(1985), 'The evolution of service economies: a geographical perspective,' *The Professional Geographer*, 37:133–42.

Kim, T. J. (1987), 'Growth and change in the service sector of the United States: a spatial perspective,' *Annals of the Association of American Geographers*, 77:353–72.

Kissling, C. C. (ed.) (1984), *Transport and Communication for Pacific Ministates: Issues in Organization and Management* (Suva, Fiji: University of the South Pacific, Institute for Pacific Studies).

Laboda, J. (1974), 'The diffusion of television in Poland,' *Economic Geography*, 50:70–82.

Lamberton, D. M. and Mandeville, T. D. (1982), 'Substitution of communication for transportation,' *Singapore Journal of Tropical Geography*, 3:162–69.

Langdale, J. V. (1978), 'The growth of long distance telephony in the Bell system,' *Journal of Historical Geography*. 9:145–59.

——(1979), 'The impact of the telegraph on the Buffalo agricultural commodity market, 1846–1848,' *The Professional Geographer*, 31:165–69.

——(1982), 'Competition in the United States long distance telecommunications industry,' *Regional Studies*, 17:393–409.

——(1984a), *Information Services in Australia and Singapore* (Kuala Lumpur and Canberra: ASEAN (Australian Joint Research Project)). Paper No. 16.

——(1984b), 'Computerization in Singapore and Australia,' *The Information Society*, 3:131–53.

——(1985), 'Electronic funds transfer and the internationalisation of the banking and finance industry,' *Geoforum*, 16:1–13.

——(1989), 'The geography of international business telecommunications: the role of leased networks,' *Annals of the Association of American Geographers*, 79:501–22.

——(1990), 'Telecommunications and international banking and finance: Asia-Pacific perspectives,' in *Proceedings, Pacific Telecommunications Council*, pp. 759–62.

Lerner, D. (1963), 'Towards a communications theory of modernization,' in L. W. Pye (ed.), *Communication and Political Control* (Princeton: Princeton University Press), pp. 327–51.

Lerner, D. and Schramm, W. (eds.) (1967), *Communication and Change in the Developing Countries* (Honolulu: East-West Center Press).

Löytönen, M. (1985), 'Spatial development of the post office network in the province of Mikkeli, Finland, 1860–1980,' *Fennia*, 163:1–112.

McAnany, E. (ed.) (1980), *Communications in the Rural Third World*. (New York: Praeger).

Mackay, J. R. (1958), 'The interactance hypothesis and boundaries in Canada,' *Canadian Geographer*, 11:1–8.

Marshall, J. N. (1984), 'Information technology changes corporate office activity,' *GeoJournal*, 9:171–78.

Masuda, R. (1981), *The Information Society as Post-industrial Society* (Washington: World Future Society).

Minghi, J. V. (1963), 'Television preference and nationality in a boundary region,' *Sociological inquiry*, 33:65–79.

Morairty, B. M. (1981), 'Existing and alternative methods of providing health care and education services in sparselands of America,' in R. E. Lonsdale and J. H. Holmes, eds., *Settlement Systems in Sparsely Populated Regions: the United States and Australia* (New York: Pergamon Press), pp. 295–321.

Moss, M. J. (1987), 'Telecommunications, world cities, and urban policy,' *Urban Studies*, 24:435–46.

——(1987b), 'Telecommunications and international finance centres,' in J. F. Brotchie *et al. The Spatial Impact of Technological Change* (London: Croom Helm), pp. 75–88.

Nicol, L. Y. (1983), 'Communications, economic development, and spatial structures: a review of research' (Berkeley, University of California, Institute of Urban and Regional Devleopment), Working Papers No. 404 and 405.

Oberschall, A. (1969), 'Communications, information and aspirations in rural Uganda,' *Journal of Asian and African Studies*, 4:30–50.

Ó hUallacháin, B. (1989), 'Agglomeration of services in American metropolitan areas,' *Growth and Change*, 30, 3:34–49.

Openshaw, S. and Goddard, J. (1987), 'Some implications of the commodification of information and the emerging information economy for applied geographical analysis in the U.K.,' *Environment and Planning A*, 40:1423–439.

Palm, R. and Farrington, J. (1976), 'Telephone use in New Zealand inter-city contacts: the businessman's view,' *Area*, 8:139–42.

Parrott, N. and Forer, P. (1986), 'The information sector in New Zealand,' *New Zealand Geographer*, 42:25–30.

Pred, A. (1973), *Urban Growth and the Circulation of Information: the U.S. System of Cities 1790–1840* (Cambridge, MA: Harvard University Press).

——(1974), *Major Job Providing Organizations and Systems of Cities* (Washington, DC: Association of American Geographers, Commission on College Geography), Resource Paper 27.

——(1975), 'On the spatial structure of organizations and the complexity of metropolitan interdependence' *Papers, Regional Science Association*, 35:115–42.

Ranck, F. (1984), 'Telecommunications in Papua New Guinea,' in C. C. Kissling (ed.), *Transport and Communications for Pacific Ministates: Issues in Organization and Management* (Suva, Fiji: University of the South Pacific, Institute of Pacific Studies), pp. 49–60.

Robinson, F. (1984), 'Regional implications of information technology,' *Cities*, 1:356–61.

Rogers, E. (1969), *Modernization among Peasants: the Impact of Communications* (New York: Holt).

——(1976), *Communications and Development: Critical Perspectives* (Beverly Hills, CA: Sage).

——(1985), *The High Technology of Silicon Valley* (College park, MD: University of Maryland, Institute of Urban Studies) (especially 'The emergence of information societies,' pp. 1–10 and 'Information exchange in continuous technological innovation,' pp. 21–33).

Salomon, I. (1985), 'Telecommunications and travel: substitution or modified mobility,' *Journal of Transport Economics and Policy*, 19: 219–35.

Saunders, R. J. et al. (1983), *Telecommunications and Economic Development* (Baltimore: Johns Hopkins Press).

Semple, K. (1985), 'Quaternary place theory: an introduction,' *Urban Geography*, 6:285–96.

Schramm, W. (1964), *Mass Media and National Development: the Role of Information in Developing Countries* (Stanford: Stanford University Press).

——and Atwood, E. (1981), *Circulation of News in the Third World* (Hong Kong: Chinese University Press).

Smith, A. (1980), *The Geopolitics of Information* (New York: Oxford University Press).

Soja, E. (1968), 'Communications and territorial integration in East Africa: an introduction to transaction flow analysis,' *East Lakes Geographer*, 4:39–51.

Thomas, P. (1984), 'Through a glass darkly: some social and political implications of television and video in the Pacific,' in C. C. Kissling (ed.), *Transport and Communication for Pacific Ministates: Issues in Organization and Management* (Suva, Fiji: University of the South Pacific, Institute of Pacific Studies), pp. 61–76.

Thorngren, B. (1970), 'How do contact systems affect regional development?' *Environment and Planning*, 2:409–27.

Tong, H. (1984), 'Citibank global telecommunications network (GTN)' in R. C. Barquin and G. P. Mead, (eds.), *Towards the Information Society* (Amsterdam: North-Holland), pp. 85–93.

Törnqvist, G. (1970), *Contact Spaces and Regional Development* (Lund: C. W. K. Gleerup, Lund Studies in Geography), No. 35.

——(1973), 'Contact requirements and travel facilities,' in A. Pred and G. Tornqvist (eds.), *Systems of Cities and Information Flows* (Lund: C. W. K. Gleerup, Lund Studies in Geography), No. 38, pp. 85–121.

——(1974), 'Flows of information and the location of economic activity' in M. E. Eliot Hurst (ed.), *Transportation Geography: Comments and Readings* (New York: McGraw-Hill), pp. 346–57.

Warntz, W. (1967), 'Global science and the tyranny of space,' *Papers, Regional Science Association*, 19:7–19.

Wheeler, J. O. and Mitchelson, R. L. (1989a), 'Information flows among major metropolitan areas in the United States,' *Annals of the Association of American Geographers*, 79:523–43.

——(1989b), 'Atlanta's role as an information center: intermetropolitan spatial links,' *The Professional Geographer*, 41:162–72.

Wellenius, B. (1984), 'On the role of telecommunications in development,' *Telecommunications Policy*, 8:59–66.

Witthuhn, B. O. (1968), 'The spatial integration of Uganda as shown by the diffusion of postal agencies, 1900–1965,' *East Lakes Geographer*, 4:5–20.

PART I

Geography and Communications

1 Dynamic structures of geographic space

PETER GOULD

Communitito: an introduction in which we make a connection

If we are prepared to think 'communication' in the title of this book in a properly broad and general sense, we can make a very strong case for placing a 'geography of communication' at the heart of all human geographic inquiry. What does 'communication' really mean? Its old, strictly etymological roots obviously lie in the Latin *communicare*, and from the manner in which great writers like Cicero and Caesar used this word, we know that it originally meant to share, to join and to unite. Something that is shared – everything from a kiss to an international shipment of grain – moves from person to person, or from place to place. What could be more human and geographic than sharing? But sharing something means that people and places (and especially people in places), are united in some way – they are connected together, related in some fashion. No longer are they simply things existing apart and on their own, just floating around in an unstructured way, but things joined together through sending and sharing, and by these connections forming structures.

Later on, as we try to dig a little deeper into connections between things, we shall talk more formally about relations in or between sets, but for the moment let us pursue the meaning of *communicare* at a more intuitive level. The sound of it reverberates through many words of the English language – communicant, communion, community, communism – and reaches back to *munus* and *moendus*, to send or to give as a present, a meaning we still hear in 'remuneration', a giving back (re-) for something already given. But if you give a present, then you share something that you have with another, thoughts in a love letter (personal communication); anxiety in a telephone call – 'How's Dad after surgery?' – (family communication); AIDS (a *communicable* disease); foreign aid (national communication); multinational contract (business communication); seismic disturbance (physical communication – earth's crust or atomic test?); wine shipments from Bordeaux (aesthetic communication); automobile imports (material communication); protest over terrorism (diplomatic communication); a new theory (scholarly communication); this essay (a pedagogic communication) . . . all sharing, all joining, uniting, connecting into structures.

A geography of connections

So much for 'communication': but what about that other part, that
'geography of . . .'? Well, without the 'communication' there can be no
'geography of'. You cannot have a *geography* of anything that is uncon-
nected. No connections, no geography. No connections means mere
checklists without any relations between the items. No relations means
long and boring compilations of unstructured facts – what is the rainfall of
Kathmandu, the longest river in the world, the capital of North Dakota?
At first sight, this connection business seems so obvious that you may
wonder whether it is worth hammering the point so hard. But sometimes
obvious things are so close to us that they escape our attention. If we lose
sight of connections, things literally fall apart – including geography,
which becomes an unraveled mess in our hands. As human geographers,
we have at the forefront of our concern the way connections are made
between peoples and places in all sorts of different spaces, and how these
spaces are being constantly restructured and reshaped by the human
presence. It is the way in which these human spaces are structured,
projected and played out over the surface of the earth that influences the
way things move – letters, telex messages, telephone conversations, flows
of money, raw materials, information, diseases, films, television programs,
manufactured goods, ideas . . . these are only some of the many things that
lie at the heart of human geographic inquiry. And these are also some of
the things that give substance to that particular geographic way of looking
at the world that has come to be known as 'spatial analysis' and 'spatial
organization'.

It is important to emphasize this fundamental concern for human spatial
organization and reflect upon it briefly within the larger context of
geographic thinking today. The adjective 'spatial' may only be a synonym
for 'geographic' (Gould, 1985), but it points to an awareness of structural
dynamics that is at the heart of modern geographic inquiry. Discard this
tradition, this 'way of seeing', and geography becomes like a three-legged
donkey, stumbling when the going gets rough, and collapsing sometimes
under the load. Today there is an expanded awareness that *place* is of the
utmost importance for our understanding, and it is the varied but quite
concrete particularities of specific places that lie behind the geographer's
concern for regional analysis and regional geography. But even *a* place, or *a*
region, only takes on human and geographic meaning *in relation to* other
places and regions, and relations mean, once again, connections over
geographic and all sorts of other *spaces*. No place or region exists
meaningfully in and of itself, disconnected and floating in a void, but always
in relation to, and connected with, others. 'No man is an *Island*, entire of it
self', wrote the poet John Donne, who obviously never forgot his
geography of place and space.

Spatial structures

But sometimes people, even geographers, do forget these basic ideas and need reminding. Anthony Gatrell (1985) writes an essay entitled 'Any space for spatial analysis?' and shows us how the meaning changes, depending upon whether we emphasize the *any* or the *space*. The former (*Any* space . . .) reminds us that we must not limit our analysis and thinking to plain, oldfashioned geographic space but should be prepared to think in all sorts of other spaces that help us understand that marvelous, multitudinous and complex world we call the human condition. The latter emphasis (Any *space* . . .) tells us that there may be less room today in geography for that spatial tradition, that some may even want to chop off that leg of the donkey. Ron Johnston (1986), for example, expresses grave doubts when he discusses the tradition of spatial analysis as a geographic 'fix', but even he has to admit that 'the value of a discipline that highlights the spatial elements is . . . undeniable'. Others within the Marxist tradition also insist that the 'spatial' must be fully injected and incorporated into any social theory that tries to make sense of the interconnected world we all live in today (Harvey, 1982; Massey, 1984a, 1984b). The same concern for the 'spatial' informs the wider project of Derek Gregory and others (Giddens, 1985; Gregory and Urry, 1985; Pred, 1981; Soja, 1985), who also emphasize the importance of infusing social theory with geographic insight, even as geographic inquiry is infused in a reciprocal way with an awareness of the importance of social structures. While sharply criticizing the geometric emphasis of the spatial analytical tradition as too simple, Gregory (1982) makes it clear that to cast off and dismiss geometric and spatial concern is to impoverish geographic and social inquiry as a whole. No matter how the connections, structures and geometry of geographic space are altered, no matter how much geographic space is shrunk by cost, or 'collapsed' by time, it always forms the underlying platform, the backcloth, upon which things of the human world exist and move. We shall have to think very hard about the basic concepts that inform our descriptions of all sorts of spaces in a reasonably general, and therefore somewhat abstract, way but in any move to the abstract level it is always useful to have some specific and quite concrete examples upon which we can hang our thinking.

Thinking about places and spaces geographically

Geographers are sometimes viewed as a bit strange because they think about places and spaces in ways that are rather different from other people. Most people never get the chance to move beyond the traditional map hanging on the schoolroom wall, or the road map spread out on their laps as they drive along trying to find their way to Shy Beaver, Pennsylvania. There is nothing wrong with such maps, particularly if you do not know where you are, or how to get to Shy Beaver, but these (geo)graphic images of simple physical space are not always directly and immediately relevant to the human geographer today. Not that such maps should be disparaged, for they were created at immense human labor and cost over centuries (Wilford, 1981), something we tend to forget these days with our satellite imagery. Moreover, they do represent, in a stylized and compressed form, the actual physical world with its real places that forms the ultimate stage over which the human drama is played. But that human drama today is no longer played on a stage simply given and structured by Nature. The spaces relevant to human and geographic understanding today have been drastically structured and restructured by human beings in an ongoing and reciprocal relationship (Goudie, 1986), and they now strongly influence the people and things that exist within them. What we are talking about, of course, is the geographic impact of technology and the way it alters the time and cost relationships between places, so warping, collapsing and even changing the dimensionality and other properties of the space formed by the connections between the people and the places.

Today, we are so very much 'at home' in our technological world that we forget how geographically dramatic new forms of communication and transportation have been. In 1840 it might have taken a hundred days to communicate from San Francisco to London, or around 8,640,000 seconds. Today a direct-dial telephone makes the same connection in about ten seconds, so in 'telephone space' this part of the world has shrunk by a factor of about 864,000. That is roughly the same as saying that Piccadilly Circus in London is now only a few meters from Union Square in San Francisco, so the two cities are practically overlapping in a strangely warped and folded 'time–distance' surface (Marchand, 1973).

The spatial dynamics of air travel

To look at the world as a geographer means that we have to break out of the habit of thinking exclusively of the old-fashioned map, although we do not always have to discard it entirely. For example, suppose we think what the technology of air travel does to State College, Pennsylvania. If we had a private helicopter, the isochrones (*iso-* equal, *chronos* time) would form concentric circles (Figure 1.1), radiating out from State College like the ripples on a millpond, and we would have no need for any rubber sheet stretching or squeezing. But this apparently simple 'accessibility surface' makes an important point; it is characteristic only of a very rich person who owns, or who can hire, a helicopter, so the 'travel time' space depends significantly on whether you are rich or poor. If we were to map the *cost*, rather than time, of accessibility from State College in terms of the proportion people would have to spend from their annual incomes, then the United States would expand for a poor person and shrink for a rich one.

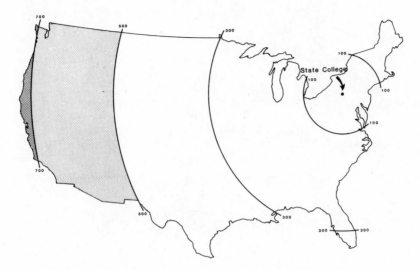

Figure 1.1 The isochronic surface of travel time from State College, Pennsylvania, for the owner of a helicopter

For a rich person, a trip to Seattle is 'nothing'; for an unemployed person trying to feed, cloth and shelter a family, a new job in Seattle may be unreachable. No human space – geographic, time, cost, income, etc. – is a constant, but is experienced in quite different ways depending upon who you are. Many people, for example, might travel on scheduled airlines and rent a car at their destination, drastically altering the isochronic map (Figure 1.2). Now the isochrones reflect the way the scheduled airlines actually connect the commercial airports of the country, in other words the *structure* of air transportation, to produce a strange and pockmarked accessibility surface.

Notice that this apparently static accessibility surface is really highly dynamic and changeable. Not only does it alter somewhat throughout the day (if you were to leave State College after ten o'clock in the morning some places would come closer in time, while others would move farther away), but it changes from season to season as airline schedules change. It can also alter drastically from day to day. Let a cold front or cyclonic disturbance move a snowstorm west to east across the map in January, and a ripple of distortion crosses this communication space, moving some places much farther away, or even disconnecting some parts of the country entirely. As a geographer, you might think of large chunks of the map being successively clipped out by Nature, to be temporarily discarded and finally replaced as the snowploughs clear the roads and runways. Is this not exactly what we mean when we say we have been 'cut off'? Never tell an exhausted, snowbound traveler in the Cheyenne, Wyoming, airport that the physical environment no longer has much impact on the human world.

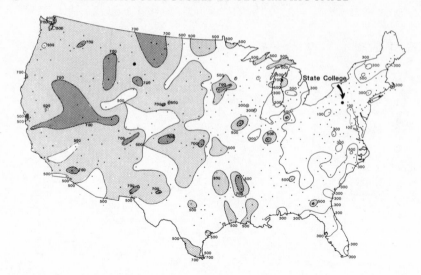

Figure 1.2   The isochronic map from State College, Pennsylvania (April, 1987) reflecting the actual structure of commercial airlines. Departure at 6:50 a.m. to Harrisburg, or 7:05 a.m. to Pittsburgh, is assumed

It would be time-consuming, but not intrinsically difficult, to make such an accessibility map for each of the 496 airports with scheduled commercial flights in the United States, and then add these all together to produce a national air travel time surface showing the relative accessibility of places for face-to-face human communication. A map somewhat analogous to this has been made for Europe (Gould, 1985; Törnqvist, 1978), indicating that Paris is the peak of accessibility, while towns like Oporto in Portugal, or Tbilisi in the Soviet Union, are very much at the periphery of this communication space, with only 3 percent of Paris's 'connectivity'.

But the constantly shifting and pockmarked accessibility surface from State College makes yet another point. It is simply that we cannot 'read it backwards'. In other words, it may take 505 minutes to go from State College to Billings, Montana, but we cannot assume that it takes the same time from Billings to State College (it actually takes 480 minutes). Travel times between places are not symmetrical, not simply because tailwinds help some aircraft while headwinds hinder others, but because schedules depend on meshing thousands of arrivals with thousands of departures every day, and these times are always subject to delays of mechanical breakdowns, bad weather conditions, traffic congestion at airports and unethical commercial behavior – like overbooking, so that people with reservations are thrown off airplanes as if they were just bundles of freight. In a profit-maximizing, technological world, people tend to be treated more and more as things. Most people have had the discomfiture of experiencing an air network as a delay amplification system, when one plane is late and all the connecting flights down the line are missed. So travel in a fluctuating time–space generally means that the travel times between two places are not symmetrical, and all our usual ideas about measuring distance fall apart (Gatrell, 1983).

Spatial stress

This lack of symmetry and distortion of straight-line distances are exactly what we experience in a typical American city with a gridiron of avenues and streets. Our movements in New York, for example, are confined to the square lattice of blocks, so that the distance from the corner of 7th Avenue and 48th Street to the corner of 3rd and 52nd is actually eight 'blocks'. Notice that in this lattice, or Manhattan space, we no longer have a single and shortest path between these two corners. Instead we have forty-five different but equal routes, and it is a nice exercise to draw a picture to try to count them.

Now suppose some streets (perhaps Broadway) cut diagonally through the square lattice. How do we experience and measure distance now? Not as a simple straight line, and not as the awkward zigzag on a square lattice, but something in between. As far as real people in a real New York City are concerned, their experience of distance depends entirely upon how the space of the city is structured by its connecting streets. We are assuming, of course, that we are talking about pedestrians. But why only pedestrians? Because nearly every city and town these days has one-way streets for vehicular traffic, and that means that the distances or times between many pairs of places are *not* symmetrical. This means that automobile drivers are moving around in one sort of space while pedestrians are walking around in another. As far as people are concerned, geographic space is really two human spaces with completely different properties, and no one ever really thinks about it – except geographers (Marchand, 1977). It is impossible to represent the non-metric and asymmetric space experienced by motorists in a one-way system on a sheet of paper in this book because the space is now highly stressed and can only be drawn by severely distorting it – even more than the way a map projection always distorts some property of the actual curved surface of the earth.

Notice that this sort of spatial stress is not the figment of an overexcited geographic imagination, but is actually experienced by the traffic that moves through and on such a highly asymmetric and non-metric space. Motorists find it terribly frustrating and stressful to drive through it, even if they happen to know the system of one-way streets quite well. So stressful can these systems become that people react against them. For example, in Caracas, Venezuela, drivers who see a short one-way street leading to their destination can seldom resist turning into it illegally. Often they get away with it, unless another driver happens to turn into it from the other end. Then the human stress really rises as one driver knows that he is right, and has the law on his side, while the other has lost face in a real machismo sense and has to back his car up ignominiously. In Grenoble, France, the non-metric stress of the one-way streets produced so much human frustration that a local election was lost mainly on the issue of 'rationalizing the system', i.e. reducing the stress of the space in both a mathematical as well as in a thoroughly human sense. In Cambridge, England, the one-way system makes it practically impossible for a visitor or tourist in a car to find the way, and after circling around and around in this deliberately designed non-metric 'maze', motorists finally learn to park their cars and walk. The rumor that they receive a food pellet on leaving the parking lot has not been confirmed. In Lund, Sweden, the town forbids all private vehicular traffic to enter the center, so creating an urban space for motorists rather like a donut. The hole in the town center is experienced as an object that one cannot go through, but has to drive around. Thinking of empty holes as solid objects in a space, and solid objects as empty holes (have you walked through a tree lately?) is all part of good spatial, i.e. geographic, thinking.

Spaces defined by connections

Even though we have focused on certain distance (i.e. metric) properties of spaces, we have also seen that many spaces of great interest to human geographers are not strictly metric at all, because we cannot measure distances between places symmetrically and as people actually experience them. In fact, for many spaces of great geographic importance the very notion of metricity may not be pertinent. What may be much more important is the simple fact of how people and things are connected together. It is the sheer *connectivity* of things that creates many spaces of interest to a geographer, spaces that a mathematician would call topological spaces.[1] For example, many of the things we observe going on in the world of actual geographic space – factories closing down, people losing their jobs, flows of raw materials being cut off, finished products being redirec- ted, people selling their precious homes at a loss and moving elsewhere to find work . . . in brief, all the searing human consequences of restructuring old industrial regions – these things are seldom the result of decisions taken in the local space by the local people of the region. Few regions today have much control over their own economies, and what we are seeing are the impacts of other topological spaces 'mapped onto' the stage of real earth space. If this seems abstract and obscure, ask yourself *who* makes these decisions affecting the lives of thousands of people (Massey, 1985; Scott, 1985), and then ask *in what space* are such decisions formulated with little or no regard for those who have to experience the forces of such wrenching structural change. A map of unemployment may reveal deeply impacted and economically distressed areas, some of which may have been in a chronic state for decades, but it never reveals the multidimensional topological space of decision making that affects the spatial distribution of employment opportunities.

'A multidimensional, topological space of decision making . . .', is this mouthful just jargon? Despite its apparent pretension it is not, and later we will try to formalize this rather abstract and obscure notion by making it well-defined and operational. For the moment, however, let us rely on our common sense and intuition. Decisions to open and close industrial plants, offices, storage facilities, mines, shipyards, factories – almost any commercial employment opportunity – are made by companies, often in consultation with, not to say the encouragement of, the state (Scott, 1982). At the very least, the state legitimatizes a company by giving it legal status, and it regulates a company in varying degrees, taking it when it makes a profit and sometimes bailing it out when its management makes a mess of things. But we have long since passed the era when the economy was dominated by small and independent firms. Today the commercial world is formed from many highly connected structures of companies, governments and financial institutions, and a lot of communication, affecting millions of people, is transmitted over these connections. It is these structures, how these elements are connected together, that form the topological spaces in which decisions affecting whole regions and nations are taken. And notice that these exist at a variety of levels: a single 'company', national or multinational, at, say, the $N + 2$ level, may consist of many smaller companies at the $N + 1$ level, each consisting of many branch plants at the N level . . . and so on down the hierarchy. But such companies are not simple disembodied things: they are controlled, managed and mismanaged by people, who also exist at varying levels in this structural hierarchy (Gould, 1982a). At the higher levels, personnel of companies connect with governments and financial institutions, and people in these public institutions connect with the private companies (Haxey, 1939; Stockman, 1985). Many politicians, for example, also serve on company boards of directors, and high-ranking military personnel often join boards of directors on retirement. After all, it is important to have connections. Why? Because connections make structures, and without structures there is nothing for decisions, and lots of other things, to move on.

What other things? Money, particularly money in the form of capital investments, which may be from private sources (individual investors), financial institutions (banks), organizations (pension plan funds), companies (corporate investments), or the state itself (in partnership or ownership). But whatever the source of the flows of capital, the major goal is almost invariably profit. After all, why invest and not try to get the best return? So what we see is the formation of capital flows as traffic being transmitted on the underlying structure of connections between the commercial and governmental worlds, a structure that acts as a support or backcloth for these flows. It is on these structures that decisions are made and transmitted, and it is the impact of these decisions to move capital from less profitable to more profitable *places* that we see in the ebb and flow of job opportunities, healthy and depressed regions, hope and misery, decency and inhumanity. In 1984, one man, worth $300 million, used 'leveraged buy-outs' to take over and strip the companies under his control (TRB, 1986). Two years later, he was worth $2.5 *billion*. Companies were sold off, factories closed, jobs were lost, but the profits were enormous. After all, this is the way the system is meant to work. Profits are more important than people.

Traffic moving on a backcloth

This specific example of capital existing and being transmitted as traffic on a backcloth structure makes a much more general and important point. It not only represents an example of communication with crucial implications for real people in real places, but it also raises again the fundamental question of communication on all sorts of other space-forming structures,

Figure 1.3   The global response (measured as requests for reprints) to a scholarly communication transmitted over a professional geographic structure

Figure 1.4   The global response (requests for reprints) to a scholarly communication transmitted over a general scientific structure

and how the connectivity of these structures affects the transmission of
things from certain locations to others. We could take scholarly communication as an example, because ideas and their transmission are essential in an
academic world in which publication takes on its fundamental meaning of
'making public'. A scholarly insight has to be shared; it has to be offered as
a 'present', a communication communicated to a community.

There are many structures in the world of scholarship, most of them
actually made up of connections between people sharing certain disci-
plinary interests (Gatrell, 1984). As a result, 'scholarship space' tends to be
quite fragmented, so a communication may generate a different response
depending upon which piece of the structure, that is to say, which part of
the disconnected backcloth, it is offered to for transmission. One communication (Gould, 1981b) was offered to a geographic structure (Figure 1.3),
and produced one sort of response, while another (Gould, 1982b) was
transmitted on a scientific structure (Figure 1.4) and produced another.
Projecting the response of these two structures into geographic space
discloses a rather marked difference between them, particularly when one
considers the nineteen language differences represented by the people in the

scientific structure. Of course, it is always possible that the former communication may have been worthless, and so elicited the proper response of silence, while the latter may have had some value to an international scientific community. On the other hand, it is not impossible that the strength of the intellectual relationships forming the backcloth structures were also important.

This specific example of the way in which responses to transmitted traffic may be conditioned by the degree of connectivity illustrates yet another general and important idea for thinking about communication. A communication is often made in order to elicit a response, so that the traffic transmitted on a structure may well give us considerable insight about the underlying structural connectivity and the way it changes. All forms of management, for example, are based on communication of some sort, and they usually require a response, even if only to confirm that a message has been received. Most management communications actually travel up and down hierarchical structures, formed by connections both within and between the various levels. Official organization charts often show the connections branching like upside-down trees from the higher to the lower levels, but real organizations and their communications seldom work this way. In other words, hierarchical structures 'on paper' seldom reflect the way things actually operate, and sometimes these 'official' descriptions often do more harm than good because they mislead. Take the problem of managing a huge irrigation system, with several dams and many hundreds of miles of irrigation channels. The management engineers at the $N + 2$ level certainly ought to know what the requirements are of the system they are trying to control (Chapman, 1983). But 'knowing' means that someone from a local area in need of water has been able to get a message transmitted up from the $N - 2$ level farmers, through the $N - 1$ level headman, to the N level local irrigation office, and so on up the hierarchy. In many parts of the world this sort of transmission of information for control in real time is impossible, with the result that many irrigation systems produce at only one-third of their capacity. In India, for example, many of the districts that seem to produce most efficiently are those in which there is a traditional 'rest house' (simple overnight living quarters for touring officials), because information can then be gathered directly by irrigation personnel at higher levels in the management hierarchy making their periodic inspection visits. The performance of the system discloses the information transmitted, which is directly related to structural properties.

Similarly, at the international level, the transmission of traffic may well reflect quite subtle changes in the connectivity of a 'multidimensional topological space' formed by structures of governments and companies. After the revolution of 1972 in Portugal, there were quite marked and negative changes in attitudes by a number of trading partners and these had the effect of literally crushing Portugal down (Gould and Straussfogel, 1983). However, after a year, when Portugal was seen to move left, but of course not *too* left (after forty years of Salazar, where else could you move?), the transmission of commercial trading traffic at former levels reflected the reconstructed connections.

Epidemics as traffic on a backcloth

In a series of highly original studies across geography and medicine, Pyle (1986, 1969) has shown how successive waves of cholera epidemics (transmitted traffic) disclosed a restructuring of geographic space by technological developments in the United States, developments in transportation that radically affected the underlying geometrical structure, or what we are calling the backcloth. For example, the successive waves of 1832, 1849 and 1866 occurred during the introduction of the steamboat and railway. As a result of the changing connectivity, early epidemics, that spread by a process of contagious diffusion, controlled by a high friction of distance, became more and more influenced by the gradually strengthening inter-urban connections. For this highly 'restructural' reason, later cholera epidemics spread from city to city in a process of hierarchical diffusion.

The same restructuring impact of transport technology is seen in the case of influenza epidemics, with waves moving through Europe in the late nineteenth century twice as fast as one hundred years before (Pyle and Henderson, 1984). Even political connections, and the international structures they create, can influence the way a traffic of influenza is transmitted across geographic space. In 1580 a wave of influenza moved through North Africa, and then spread northwards through the Iberian, Italian, and Balkan peninsulas (Figure 1.5). It then suddenly seemed to jump to the Netherlands, very far away in geographic space, especially in those days of slow land and sea travel. Historically, of course, the reason is clear: the new 'landing grounds' for the influenza invasion were at that time the *Spanish* Netherlands, and it was the intense connections by sea between imperial Spain and subjugated Holland that formed the relevant and pertinent backcloth structure that controlled the transmission of the disease.

Notice that this fundamental idea of the structure of the space controlling the transmission of things existing on and in it is by no means confined to traditional geographic scales. In the 'microgeography' of a cervical canal the spermatozoa transmit themselves as highly directed traffic in order to communicate with the ovum in the uterus. What they sense is the cervical space structured by an acidity gradient, a gradient that could actually be mapped as a series of pH contour lines. If the cervical space were neutral (pH = 7), it would literally be structureless, implying that the relevant directional backcloth no longer existed for the transmission of the traffic. In this case, the spermatozoa could only do an unstructured, two-dimensional random 'swim', thus greatly reducing the chance of fertilization. This apparently strange example makes the crucial point that geographic think-

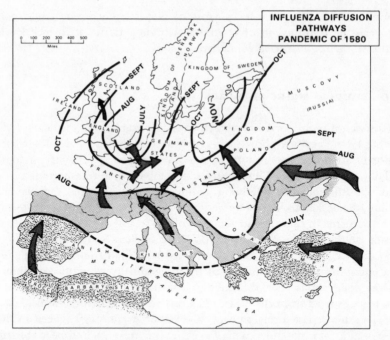

Figure 1.5   The diffusion of influenza through Europe in 1580 strongly shaped by political connections between Spain and the 'Spanish' Netherlands.(By permission of Gerald Pyle)

ing, at its fundamental level of conceptual concern for spatial and structural matters, can allow us to see many things of great variety from a rather different perspective, no matter what the scale or subject. It also raises important questions of spatial search; for example, the next move in a complex, multidimensional space of a chess game (Atkin, 1972; Atkin, Hartston and Witten, 1976), or how to find life rafts or submarines at sea. These examples raise the much more general problem of *spatial diagnostics* of all sorts, questions that are always easier to answer when the structure of the space, the topology of the backcloth, is known. In a 'geography of communication', we always work with fundamental ideas and basic concepts that have powerful 'carry over' values for many other fields of inquiry. This is one of the great strengths of the 'geographic perspective'.

Traffic can change a backcloth

The relation between an underlying supporting structure (backcloth), and the things (traffic) that exist on, and are transmitted over, such structures, is not a deterministic one. The human structures of communication that are of concern to geographers are always allowing and forbidding, but not necessarily requiring, structures (Gould, 1986). Moreover, in the human world these structures are obviously not fixed eternally like the space–time geometry of the physical world but are subject to deliberate or accidental change – perhaps a slow and gradual restructuring, or sharp, revolutionary and cataclysmic breaks (Marchand, 1974). Nevertheless, if all communication, considered as transmitted traffic, requires some sort of structure or backcloth to support it, then we should be aware that sometimes the transmission of traffic may alter the underlying structure in turn.

Fire, for example, is a case in point, although few people think of it spreading as a communication process in these general, conceptual terms. A forest fire is transmitted on a structure that must be, by definition, inflammable, yet it destroys the structure as it is transmitted. Every forest firefighter is aware that you can only stop the transmission of fire traffic by changing the structure of the backcloth by somehow disconnecting parts of the unburnt forest. This is exactly what a fire break means, a break in the structure of the underlying geometry, which in this case exists at a single hierarchical level. In contrast, fire in buildings might exist in topological spaces at different hierarchical levels. A hot lightbulb touches a lampshade that bursts into flames that catches the curtains that fall on the sofa that . . . furnish the house that Jack built. Not quite. Things like lightbulbs, lampshades, curtains and sofas form a *set* of furnishings at the $N - 2$ level that are contained within a floor, a ceiling, doors, windows and walls that make up a set of rooms at the $N - 1$ level that make up the house that Jack built at the N level. And Jack's house may well be one of a set on a street or in a neighborhood at the $N + 1$ level, and sets of neighborhoods or streets make up the town at $N + 2$, and so on. Perhaps cows kicking over lamps in Chicago in 1871 knew all about the transmission of fire traffic on hierarchically structured backcloths made up of connections between sets of things.

This example is not fanciful: fire is greatly feared for good and obvious reasons, and knowing something about how fires spread can be of great importance when it comes to designing better buildings and effective fire control. Fire is *communicated* from one inflammable element of a set to another. The point is that we always have to start with real and well-defined sets of things and think through carefully the actual relations that connect them together. A fire in certain pine forests, for example, spreads differently from one in a forest of deciduous trees, because highly dangerous 'crown fires', bursting into flame at the top of resinous pines, can send swirling eddies of burning gases to set trees alight relatively far away. Firefighters fear these most of all because they can become trapped by the fire jumping behind them (Haines and Smith, 1983; Smith, Haines and Main, 1986). In our structural thinking, the pine trees are no longer connected to those just adjacent to them, as they would be in a deciduous forest, but to more extended neighborhoods of trees around them. The relation structuring the set of trees in the pine forest produces a tighter, more highly connected backcloth that greatly influences the rate of traffic transmission. Similarly, in thinking about fire in buildings, we have to start with actual and concrete descriptions of real houses with real rooms and real furnishings (fireproof curtains and fabrics? firewalls and doors? fireproof insulation? fireproof safes?).

Rabies in a cellular space

While the example of fire as transmitted traffic is essentially one from the physical world, backcloth–traffic relations may greatly influence the world of living things as well. Rabies, for example, is a deadly disease for animals and humans if not treated, and in Europe it seems to be transmitted mainly by foxes. A continuous geographic space, restructured by the behavior of foxes, is transformed into a cellular space (Couclelis, 1985, 1986; Tobler, 1979), a region divided into generally non–overlapping partitions. Foxes, like wolves, mark their territories by urinating around the boundaries, and these signs are usually respected. Since foxes tend to be highly territorial, and respectful of the territories or cells around them, the question arises: how is a disease like rabies communicated across such an apparently partitioned, and so disconnected and fragmented space? The answer is that the behavior of the foxes alters under the influence of rabies, and this altered behavior actually *restructures* the cellular space (Källén, Arcuri and Murray, 1985). Once again, we have an example of the way the transmission of traffic changes the backcloth required for its transmission.

If a fox has rabies, it may change its behavior in two quite distinct ways. The incubation period may be short, in which case the fox or vixen dies quickly, and remains within his or her territory, thus effectively stopping the spread of the disease in its tracks. However, if the incubation period is long, the fox may become spatially disoriented and start moving around as though it were doing a two–dimensional random walk. This disoriented movement may bring it to the boundary of an adjacent fox territory, a boundary normally respected. Now, however, the rabid and disoriented fox moves over the boundary, connects the formerly partitioned cells, and so alters the structure of the space. The neighboring fox meets the challenger, fights, gets bitten, and so carries the rabies virus into its own cell or territory. The transmission may also be aided by young rabid foxes splitting off from the home den and carrying the disease further as they try to establish their own territories. Like a forest fire, the only way to stop the spread is by disconnecting the structure with a 'rabies break'. With sufficient warning, this may sometimes be achieved for limited areas by dropping chicken heads inoculated with rabies vaccine from helicopters in a *cordon sanitaire* around the area of known outbreaks. In this way, foxes eat the chicken heads and form a ring of immunity around the infected region.

Telephone, air travel, cost, symmetric, asymmetric, pedestrian, motorist, decision-making, financial, scholarship, management, trade, cervical, forest, building, fox, human ... *spaces*, a whole spectrum of different communication spaces, connected by relations of different sets of elements, to form structures that influence the existence and transmission of traffic upon them. Some are relatively stable, but slightly oscillating spaces, like the structure of air transportation and telephone space. Others are constantly and dramatically changing, like the communication space of a Third World country dependent on laterite roads. These have the effect of expanding the country in time–space during the wet season and shrinking it in the dry season. With all of these different examples of communication behind us, it is time we thought a little more formally, abstractly and deeply about the things they have in common so that we can sharpen the concepts we bring to bear as 'geographers of communication'.

Fundamental Concepts of Communication

As we worked our way through a variety of examples of communication, certain words appeared more and more frequently in our discussion, like *set, relation, hierarchy, backcloth, traffic, transmission*, and so on. Some of these terms have very precise and fundamental mathematical meanings, which implies that mathematicians have taken the trouble to think hard about such concepts in order to come up with clear and unambiguous definitions. In any science, ambiguity has to be removed if possible so that people place the same meaning upon the terms they use in their common discourse. Otherwise we are just ships sailing past each other in the night. This is not the place to introduce all the details of a large body of methodological work, and in any case a number of basic introductions are available (Atkin, 1974, 1977, 1981; Gould, 1980, 1981a; Johnson, 1981, 1982). Nevertheless, we have reached the point where we must try to collect some of these basic ideas together and look at them carefully. Our concern is not mathematical at all, but rather conceptual.[2] We have to return to an older, more fundamental concern: to describe the things, the people and the relations in our human world faithfully, using ways that are appropriate to our descriptive task, rather than employing forms of mechanical mathematics generated out of a desire to describe the world of physical things two hundred years ago (Gould, 1987).

Defining sets and relations

We have seen that all structures are formed by connections between elements of sets, so our first task is to define these sets. Most people's first reaction is that this is a naively obvious and simple task, but in a concrete empirical study it may well turn out to be the most difficult job of all. It often happens that in the later stages of a structural analysis things seem to 'go wrong'; all the thinking and computing may result in a 'descriptive text' that is simply a mess, one which is either interpreted as being totally banal, or not interpretable at all.[3] When this happens, it is usually a case of going 'back to the drawing board', a return to the starting point of set definition that may be discouraging. In fact, such a return is quite typical of genuine scientific inquiry, because it never knows for sure whether something will be illuminated by probing in a particular direction before it has been tried. Good science is always exploration, a movement from the known to the unknown. In fact, the new starting point is not really the same as before: by previously trying and failing we have actually made some progress because now we know not to start with the same definitions as before. Perhaps most of science consists of 'failures', even though these seldom get reported in the journals.

Sets can be defined in two basic ways. First, we can define them by actually listing the elements one by one – a set of houses, people, farms, countries, directors, institutions, census districts, points . . . whatever the elements happen to be. This is the way we usually do it when we are trying to describe, illuminate and understand some real and concrete situation, rather than undertaking a theoretical speculation couched in conventional mathematics. In contrast, we might define a set by trying to make a more general statement; for example, 'the set of all road links in Toronto'. However, this sort of general statement implies that the things like 'road links' and 'Toronto' have already been well and operationally defined. We might well ask what is, *exactly*, a 'road link'? And where, for the purposes of the study, does 'Toronto' begin and end? Too frequently, these sorts of general definitions lead to ambiguity, paradox, contra-diction, or uncertainty (Couclelis, 1983), things logicians like to play around with, but not terribly useful for scientists, including geographers of communication. The reason is that the general types of definitions tend to imply that they hold across all space and time, and so are eternal truths. In contrast, definitions made by listing things tend to be more useful here and now.

But simply defining sets by themselves is not enough. A set is just a . . . well, just a set, a carefully specified collection of things that have been defined in that way because someone thought it would be useful to do so. But structure, as we now know, comes from connections, and in our terms this means relations on and between sets. Relations are nothing more than very free and unconstrained ways of specifying connections between, or on, elements of sets (Gould, Johnson and Chapman, 1984), so they are just what we need to describe things as they actually are, rather than what a particular mathematics forces them to be. To make these ideas clearer, let us consider a small, hypothetical example and its graphic expressions just to get the feel of things.

The structure of a seminar or department

Suppose we have a set of people (relatively easy to define), and a set of intellectual concerns (perhaps not so easy to define), and you can make up your own story about them – perhaps they define some aspects of a seminar, an academic department, a family, co-workers in a research field, and so on. For the purposes of this illustration, we shall assume all the intellectual concerns are at the same hierarchical level, although this obviously need not be so. Two people may share the same interest in music, and so be connected at the N + 1 level, but then be disconnected at the N level because one likes Bach while the other likes Schoenberg. We assume things like Rock, and other thumping noises, are not music, but rhythmic aberrations without enduring human value (Bloom, 1987). This approach not only makes us aware of the sets and their structuring relations, but points to the possibility that structures at one hierarchical level of generality may be quite different at another. Suppose the intensity of a person's concern for a particular interest can be scaled from 3 (intense concern) to 0 (no concern whatsoever), so that we can represent the connections between our sets as a matrix, say S (see p. 22). And before we continue, notice that this is a totally general representation. We could substitute anything we liked here: either different sets, like farmers who give and receive advice (Gaspar and Gould, 1981), or even the same set, like trees in a forest related by proximity, or a set of women in a village related by friendship, or cities linked by airline connections, or countries linked by diplomatic exchanges, or players in a team game (Gould and Greenawalt, 1982). Notice also how our everyday language often reflects these formal terms of sets and their connecting relations: we break diplomatic relations, we *cut* someone out of our will (no inheritance traffic transmitted on that fragmented backcloth of kinship!), a team 'falls apart' (Gatrell and Gould, 1980), a student fails to 'tie it all together' in an exam, we 'snip out' a gene in genetic engineering, we feel 'disconnected' in another culture and language, and so on. It turns out that sets, relations, connections and structures are all around us, and they are always necessary to transmit some form of communication.

With this description in front of us, how can we visualize the structure of our seminar or department? In fact, we can define a number of structures, depending upon whether we choose very intense interests (evaluated as 3), or less intense ones (evaluated as 2 or 1). When we choose to define the structure on the basis of only the most intense interests, we are actually 'slicing away' all the others. So a particular structure depends upon which slicing value we use. If we decided to choose 3, then only the 3s in the matrix S define a connection, and these form geometrical figures called simplices. For example, Person A is defined by intellectual concerns a and c, and so becomes a one-dimensional simplex (Figure 1.6). Notice that no other person shares these concerns at this level of intensity, so one-dimensional A is not connected to anyone else. But no connections means

INTELLECTUAL CONCERNS

PEOPLE	a	b	c	d	e	f	g	h	i	j	k	l	m	n	o	p	q	r	s	t	u	v	w	x	y	z
A	3	2	3	1																						1
B	3	1			2																					
C						1	2	1																		
D							2	2																		
E			3						2	1	3	3		2												
F									1																	
G									2	2	1	3	3													
H																		2	1							
I																	2	1	3							
J																	2	1	3							
K									2	3							2	1	3							
L									2	3							2	1	3							
M				3	2																2	3	1	2	3	

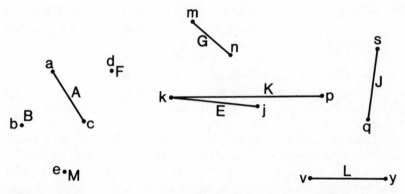

Figure 1.6   The structure of a seminar defined by intense (3) intellectual interests

no possibility of communication: Person A can only talk to him or herself (is that perhaps what thinking is?), but cannot discuss either or both of these concerns with anyone else. In fact, if we define our seminar or department with a slicing value of 3, we have a very fragmented structure. Persons C, D and H have been 'sliced out' altogether and have disappeared, and the only connections are between the one-dimensional people-simplices K and E (through their intense concern for interest k), and Persons I and J (through their concern for s). Everyone else is talking-thinking to themselves. Is this good or bad in a seminar or department?

Given this well-defined backcloth, we might live on, or be supported by, this structure? What traffic could this geometrical representation carry? Perhaps K and E, and I and J, could collaborate on some papers of mutual interest, but these pieces of traffic would be highly specialized, and actually constrained to the zero-dimensional vertex of shared concern. Or perhaps these two pairs of professors might team-teach a course, with Professors E and K bringing in material from their other concerns with j and p respectively, while Professor I can only pinch-hit perhaps for Professor J on topics. A traffic of students living on this structure, undergraduates or postgraduate students, might feel a sense of intellectual fragmentation if they wanted to integrate different interests into their own research for a thesis or a dissertation. But notice that 'to integrate' means to connect! Do doctoral students forming a dissertation committee actually create some sort of connective tissue in a department, strengthening the structure in some way? Of course, students with highly specialized interests themselves might like to have experts in these topics, but other problems can arise from specialization. Do students living on such low-dimensional pieces of a structure turn into carbon copies of the low-dimensional professors they work with? In a seminar, do such people just stick to their own interests and never broaden their horizons? And in a department, what happens to the student-traffic existing on a part of the faculty structure when a faculty member moves to another university? It is not uncommon for graduate students to move too, otherwise they feel they have had the backcloth-rug pulled out from under them. All sorts of stresses are felt when structures change or disappear.

Naturally, we do not have to define the backcloth of the seminar or department by choosing only the most intense concerns. If we change our slicing value to 2 or more, then a structure with greater connection appears (Figure 1.7). People E, G and K can now communicate reasonably complex (i.e. higher-dimensional), ideas through their common concerns which now form the two-dimensional face < j, k, n > between these three-dimensional people-simplices. In a department, a connecting face like this might well attract students who wanted to find such a piece of structure for their own intellectual 'home'. Again, we could define our seminar or department with even slight interests (1 or more) in which case everyone connects into a single component, except Person L (Figure 1.8). Is she or he in the right seminar or department? And does this *question* imply something about the way we expect a seminar or department to be? Is such an expectation justified, or even desirable?

It is worth noting that sometimes people are bothered by the apparent arbitrariness of choosing a particular slicing value to define a backcloth structure. They seem to need the security of some mystical theory to tell them what structure is the 'best' one. Such displays of intellectual insecurity are due to the pseudoscientific 'potty training' that many receive today, for it is quite clear that no theory (whatever we might mean by this word in a human context) is going to inform us what structures are relevant in any specific, and often previously unexplored, empirical domain of concern (Gould, 1988). As we saw before, science, real science, is exploring the

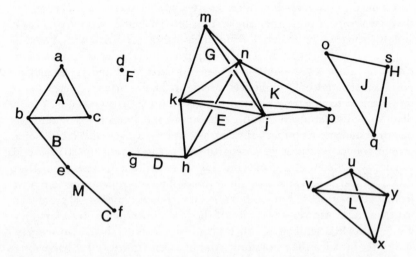

Figure 1.7   The structure of a seminar defined by both intense (3) and moderate (2) interests

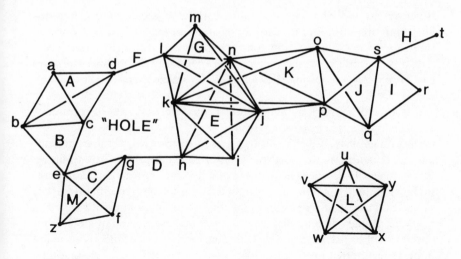

Figure 1.8 The structure of a seminar defined by intense (3), moderate (2), and weak (1), interests

unknown, trying to find interpretable structures, and often failing many times before an interesting (i.e. interpretable), 'story' can be told. Interpretation – interpretation shared by others – is the heart of all scientific, indeed all human, inquiry. We define and explore structures until we find those that 'make sense', and then we try to convince others of a truth we have found by the old, and honorable, art of rhetoric.

What are the structures of communication?

Although this example is very simple, it does make a number of important points, and it raises some important questions about conducting research on communication. First, notice that while traffic, in the form of discussions, research articles, joint research, students, etc., can live on certain parts of this backcloth, it cannot be *transmitted* unless a connection of concern is shared. Communication, thought about as transmitted traffic, requires very specific pieces of structure to be in place. Simplices in a backcloth that share one-, two-, three- or more-dimensional faces are said to form *stars* connected by *hubs* made up of shared vertices (Johnson, 1983). Obviously, if these connecting hubs were to be removed from the structure, they would leave great gaping holes, and these would be felt as obstructions or 'topological objects' in the communication space.

There is nothing mysterious about these holes or voids in structures: we have met them before as Chicago was closed down by a snowstorm, as a *cordon sanitaire* was formed by chicken heads, as the pedestrian center of Lund, as fireproof safes in a burning building, and so on – all holes in structures around which any transmitted traffic of people, rabies, automobiles and fire must move. This is what it means to have a 'common language' to describe real and concrete structures of human interest. We suddenly see different cases of communication as special instances of more general processes requiring structural ideas to make them operational and well defined.

It is a useful exercise to think about common experiences of communication in these explicit structural terms. What are the gaps and holes in the fabric of a seminar? What about that person–simplex who 'knows his stuff but can't communicate'! Is Professor Simplex incapable of transmitting intellectual traffic because he defines himself at too high a value of interest, so slicing himself out of the seminar, and failing to connect with his student simplices? Poor Professor Simplex, floating there all by himself, lonely as a cloud in a high-dimensional communication space. And poor student simplices, who are trying to get to the same level (slicing value). How many students realize that professors are often so intellectually insecure that they dare not drop their slicing values? Perhaps good teachers are those who remember their own struggles at an earlier time to increase their understanding. Perhaps so much teaching in mathematics is ineffective because those now teaching mathematics had a particular gift for it when they were young, and so are incapable of understanding what the difficulties are? Even a mathematical curriculum is a structure, perhaps one formed by a relation between courses and particular elements of subject matter, all hierarchically arranged at their proper levels. Is the structure of a backcloth describing such a curriculum in the form of a long chain, a chain down which students (a chain gang?) plod as transmitted traffic to emerge at the other end exhausted and with little enthusiasm left – what Herbert Simon (1969) has called 'teaching by recapitulation of the field'? Or has a department taken the rare step of wiping the blackboard clean, and then really thinking about the structure of a modern mathematics curriculum (Open University, 1971)?

As you read through this book, ask yourself; what is being communicated? What are the structures required for the transmission of this communication traffic? What are the sets that form these backcloths? What are the connecting relations? Is everything at the same hierarchical level? Who or what are the stars and hubs? What and where are the 'topological objects'? How, concretely, and in the specific empirical context of the problem, might these 'holes' be filled? And as you ask these sorts of questions, realize that by thinking about many apparently different aspects of communication, you, the reader, are connecting the chapters of this book with a methodological and conceptual relation, so creating a structure along which you are the transmitted traffic.

And in the process, look out for those holes!

Notes

I would like to acknowledge the help of Kate Petersen and all the people of Professional Travel, State College, Pennsylvania, for their generous help in allowing me access to their computer scheduling system and national timetables to compile the isochronic map of Figure 1.2.

1   Which means that they conform to a set of precisely stated axioms. We do not need to worry about these here, but if you are interested in exploring further, see Munkres, 1975, p. 76, or any sound elementary text on topology.

2   The effective description of structures in the human sciences reaches back to the early topological work of Kurt Lewin, 1936, and then appears again in the graph theoretic tradition of the mathematician Berge, 1958, the geographer Kansky, 1963, and the architect Alexander, 1966. Many of the structural descriptions we shall use here are isomorphic with such graph theoretic ideas (Johnson, 1981), although the representations are derived from the more general mathematical tradition of algebraic topology. The purely mathematical extensions are of little concern here: what is important is the body of ideas, concepts and terms that seem to be useful when it comes to describing, in a well-defined and operational sense, quite real and concrete structures in the human, biological and physical worlds.

3   The apparently inappropriate phrase 'descriptive text' is used here with good reason. Jurgen Habermas, 1971, has noted three perspectives that can be taken in human inquiry: the *technical*, the *hermeneutic* and the *emancipatory*. When we create a map, an equation, a remotely sensed image, a diagram, or some sort of computer output, we are really adopting the technical perspective in order to create a 'text', i.e. something to be interpreted. But the things we create out of that technical perspective just lie there mute and without meaning unless and until someone takes the hermeneutic perspective and gives them meaning in an act of interpretation. In any research, one tries to create texts or descriptions that are meaning-full, i.e. that are illuminating because they can be interpreted and given meaning. Although the word *hermeneutic* originally arose out of the tradition of interpreting theological texts, it is now used in a much more general sense precisely because anything that is humanly meaningful requires an act of interpretation. Sometimes descriptions derived from the technical perspective may be interpreted with such insight that they let us see ourselves and our societies in a new, perhaps deeper, way. Thus we achieve an emancipatory perspective, one that allows us to see ourselves more truthfully and in a new light. Obviously, these three perspectives are not distinct and separate, but may often inform one another in an actual piece of human research and inquiry (Gould, 1982c).

References

Alexander, C. (1966), *Notes on a Synthesis of Form* (Cambridge, Mass.: Harvard University Press).

Atkin, R. (1972), 'Multidimensional structure in the game of chess', *International Journal Man–Machine Studies*, 4, pp. 341–62.

——(1974), *Mathematical Structure in Human Affairs* (London: Heinemann).

——(1977), *Combinatorial Connectivities in Social Systems* (Basel: Brikhauser).

——(1981), *Multidimensional Man* (Harmondsworth: Penguin).

Atkin, R., Hartston, W., and Witten, I. (1976), 'Fred CHAMP, position chess analyst', *International Journal of Man–Machine Studies*, 8, pp. 517–29.

Berge, C. (1958), *Théorie des graphes et ses applications* (Paris: Dunod).

Bloom, A. (1987), *The Closing of the American Mind: How Higher Education Has Failed Democracy and Impoverished the Souls of Today's Students* (New York: Simon & Schuster).

Chapman, G. (1983), 'Underperformance in Indian irrigation systems: the problems of diagnosis and prescription', *Geoforum*, 8.

Couclelis, H. (1983), 'On some problems in defining sets for q-analysis', *Environment and Planning B*, 10, pp. 423–38.

——(1985), 'Cellular worlds: a framework for modelling micro-macro dynamics', *Environment and Planning A*, 17, pp. 585–96.

——(1986), 'Artificial intelligence in geography: conjectures on the shape of things to come', *The Professional Geographer*, 38, pp. 1–11.

Gaspar, G., and Gould, P. (1981), 'The Cova da Beira: an applied structural analysis of agriculture and communication', in Pred, A. (ed.), *Space and Time in Geography* (Lund: Gleerup), pp. 183–214.

Gatrell, A. (1983), *Distance and Space: A Geographical Perspective* (London: Oxford University Press).

——(1984), 'The geometry of a research speciality: spatial diffusion modelling', *Annals of the Association of American Geographers*, 74, pp. 437–53.

——(1985), 'Any space for spatial analysis?', in Johnston, R. (ed.), *The Future of Geography* (London: Methuen), pp. 190–208.

Gatrell, A., and Gould, P. (1980), 'A structural analysis of a game: the Liverpool v Manchester United Cup final', *Social Networks*, 2, pp. 247–67.

Giddens, A. (1985), 'Time, space and regionalization', in Gregory and Urry, op. cit., pp. 265–95.

Goudie, A. (1986), *The Human Impact on the Natural Environment* (Oxford: Basil Blackwell).

Gould, P. (1980), 'Q-analysis, or a language of structure: an introduction for social scientists, geographers and planners', *International Journal of Man-Machine Studies*, 12, pp. 169–99.

——(1981a), 'A structural language of relations', in Craig, R., and Labovitz, M. (eds), *Future Trends in Geomathematics* (London: Pion), pp. 281–312.

——(1981b), 'Letting the data speak for themselves', *Annals of the Association of American Geographers*, 71, pp. 166–76.

——(1982a), 'Electrical power failure: reflections on compatible descriptions of human and physical systems', *Environment and Planning B*, 8, pp. 405–17.

——(1982b), 'The tyranny of taxonomy', *The Sciences*, 22, pp. 7–9.

——(1982c), 'Is it necessary to choose: some technical, hermeneutic and emancipatory perspectives on inquiry', in Gould, P., and Olsson, G. (eds), *A Search for Common Ground* (London: Pion), pp. 71–104.

——(1985), *The Geographer at Work* (London: Routledge & Kegan Paul, 1985).

——(1986), 'Allowing, forbidding, but not requiring: a mathematic for the human world', in Casti, J., and Karlqvist, A. (eds), *Complexity, Language and Life: Mathematical Approaches* (Berlin: Springer Verlag), pp. 1–20.

——(1987), 'A critique of dissipative systems in the human realm', *European Journal of Operational Research*, 30, pp. 211–21.

——(1988), 'What does chaos mean for theory in the human sciences?' in Golledge, R., and Gould, P. (eds), *A Ground for Common Search* (Santa Barbara: University of California, Department of Geography), pp. 11–30.

Gould, P., Johnson, J., and Chapman, G. (1984), *The Structure of Television* (London: Pion).

Gould, P., and Greenawalt, N. (1982), 'Some methodological perspectives on the analysis of team games', *Journal of Sport Psychology*, 4, pp. 283–304.

Gould, P., and Straussfogel, D. (1983), 'Revolution and structural disconnection: a note on Portugal's international trade', *Economia*, 7, pp. 435–53.

Gregory, D. (1982), 'Solid geometry: notes on the recovery of spatial structure', in Gould, P., and Olsson, G. (eds), *A Search For Common Ground* (London: Pion), pp. 187–219.

Gregory, D., and Urry, J. (eds) (1985), *Social Relations and Spatial Structures* (London: Macmillan).

Habermas, J. (1971), *Knowledge and Human Interests* (Boston: Beacon Press).

Haines, D., and Smith, M. (1983), 'Wind tunnel generation of horizontal roll vortices over a differentially heated surface', *Nature*, 306, pp. 351–2.

Harvey, D. (1982), *The Limits to Capital* (Oxford: Basil Blackwell).

Haxey, S. (1939), *Tory MP* (London: Victor Gollancz).

Johnson, J. (1981), 'Some structures and notation of Q-analysis', *Environment and Planning B*, 8, pp. 73–86.

——(1982), 'Q-transmission in simplicial complexes', *International Journal of Man-Machine Studies*, 16, pp. 351–7.

—— (1983), 'Q-analysis: a theory of stars', *Environment and Planning B*, 10, pp. 456–69.

Johnston, R. (1986), *On Human Geography* (Oxford: Basil Blackwell).

Kansky, K. (1963), *Structure of Transportation Networks* (Chicago: University of Chicago, Department of Geography).

Källén, A., Arcuri, P., and Murray, J. (1985), 'A simple model for the spread and control of rabies', *Journal of Theoretical Biology*, 116, pp. 377–83.

Lewin, K. (1963), *Principles of Topological Psychology* (New York: McGraw-Hill).

Marchand, B. (1973), 'Deformation of a transportation surface', *Annals of the Association of American Geographers*, 63, pp. 507–21.

——(1974), 'Quantitative Geography: Revolution or New Tool?' *Geoforum*, 17, pp. 15–18.

——(1977), 'Planning pedestrian flows around a subway station: a French case study of the time-distance decay function', *Geographical Analysis*, 9, pp. 42–50.

Massey, D. (1989a), *Geography Matters!* (Cambridge: Cambridge University Press).

——(1984b), *Spatial Divisions of Labour: Social Structures and the Geography of Production* (London: Macmillan).

——(1985), 'Geography and class', in Coates, D., Johnston, G., and Bush, R. (eds), *A Socialist Anatomy of Britain* (Cambridge: Polity Press), pp. 76–96.

Munkers, J. (1975), *Topology: A First Course* (Englewood Cliffs: Prentice-Hall).

Open University (1971), *Mathematical Foundation Course* (Milton Keynes: The Open University Press).

Pred, A. (1981), 'Social reproduction and the time-geography of everyday life', *Geografiska Annaler*, 63B, pp. 5–22.

Pyle, G. (1969), 'Diffusion of cholera in the United States', *Geographical Analysis*, 1, pp. 59–75.

——(1986), *The Diffusion of Influenza* (Totowa, NJ: Rowman & Littlefield).

Pyle, G., and Henderson, K. (1984), 'Influenza diffusion in European history', *Ecology of Disease*, 2, pp. 173–84.

Scott, A. (1982), 'The meaning and social origins of discourse on the spatial foundations of society', in Gould, P., and Olsson, G. (eds), *A Search for Common Ground* (London: Pion), pp. 141–56.

Scott, J. (1985), 'The British upper class', in Coates, D., Johnston, G., and Bush, R. (eds), *A Socialist Anatomy of Britain* (Cambridge: Polity Press), pp. 29–54.

Simon, H. (1969), *Sciences of the Artificial* (Cambridge, Mass.: MIT Press).

Smith, M., Haines, D., and Main, W. (1986), 'Some characteristics of longitudinal vortices produced by line-source heating in a low-speed wind tunnel', *International Journal of Heat Mass Transfer*, 29, pp. 59–68.

Soja, E. (1985), 'The spatiality of social life: towards a transformative retheorization', in Gregory, D., and Urry, J. (eds), *Social Relations and Spatial Structures* (London: Macmillan), pp. 90–127.

Stockman, F. (1985), *Networks of Corporate Power* (Cambridge: Polity Press).

Tobler, W. (1979), 'Cellular geography', in Gale, S., and Olsson, G. (eds), *Philosophy in Geography* (Dordrecht: D. Reidel), pp. 379–86.

TRB (1986), 'Boesky's disease', *New Republic*, 3747, p. 4.

Törnqvist, G. (1978), *Oresundsforbindelser* (Stockholm: Statens Offentliga Utred-ningar).

Wilford, J. (1981), *The Mapmakers* (New York: Alfred Knopf).

2 Hardware, software, and brainware: mapping and understanding telecommunications technologies

RONALD F. ABLER

Geographers will be in a better position to provide insights into the nature of communications technologies and the ways societies use them if more geographers address telecommunications topics and if they have clear notions of what to study and how to do so. This chapter suggests topics for mapping and analysis, stresses the role of external regulation as a determinant of the geography of telecommunications industries and services, and argues that telecommunications technologies interact with societies in ways that are not congruent with traditional cause-and-effect conceptual frameworks.

Where are the communications geographers?

Research on geographical aspects of telecommunications (the word is used herein to denote all two-way communication over distance) is scarce. Despite the fundamentally geographical natures of telecommunications and the industries that provide such services, and despite extensive publicity and concern in many nations over deregulation of telecommunications industries, the topic remains underdeveloped in geography. Nor have large numbers of other social scientists expressed great interest in the topic. Given their importance in both advanced and developing economies, there exists a serious lack of detailed research on telecommunications in general and on geographical aspects of telecommunications industries in particular.

The absence of research is matched by a corresponding lack of instruction that might interest more students in telecommunications topics. In the United States, the courses on communications recently established at the University of Kentucky by Professors Brunn and Leinbach and at Syracuse University by Professor Monmonier and my course on the Geography of Communications Systems at Penn State are the only regularly taught offerings on the subject. More university courses and seminars are urgently needed to produce graduates who will address the many unanswered questions about how telecommunications networks function, how they interact with economies and societies and how geographers can think most incisively about them.

Mapping Hardware

Geographical analysis of interconnecting systems must be based on knowledge of the physical networks that make telecommunications possible. To be sure, the specific location of pieces of telecommunications hardware is increasingly irrelevant to consumers of telecommunications services. Yet network structure continues to constrain industry operations even when such constraints are not apparent to consumers. More to the point, technological differences among networks are a major basis for competition within telecommunications industries.

Transmission media

Telecommunications rely upon transportation media, wires, coaxial cables, microwave channels, earth satellites, and fiber-optic cables to move information from one place to another. Each medium has its own advantages and disadvantages. The transportation media used by mail and express services are slow, but they are accessible to small users, they can still deliver large volumes of information cheaply and they are critical when actual documents are required. Wires are simple and sturdy, but can transmit little information. Coaxial cables can carry large bandwidths and are reliable; they are also expensive. Microwave carries large bandwidths less expensively, but microwave signals are easily intercepted and linkages can be unreliable under certain atmospheric conditions. Satellites offer cheap transmission at the price of certain technical problems. Fiber-optic links are expensive, but offer superb signal quality and seem to offer almost unlimited transmission capacity, to the degree that some analysts believe fiber-optic links will eventually replace satellites.

Some transmission media are more flexible than others. Mail services can carry any kind of hard copy information, including high quality pictures and maps. Coaxial cable, microwave and fiber optics can carry any kind of electronic signal, including television pictures. Other media are limited. Wires carry voice signals well, data transmitted at low speeds satisfactorily and most other kinds of information with great difficulty or not at all. While it is technically feasible to transmit television pictures or data at high speed on a pair of wires (the transmission medium that connects to most residences), it is extremely expensive to do so, and cheaper and better alternatives exist.

Despite great technical progress, transmission technology still determines what kinds of services can be offered at what costs at what places. Therefore, knowing which telecommunications media link which places will remain a concern of telecommunications geographers. Regions of a nation that possess little more than perfunctory transportation or wire *plant* (to use the industry term) will be handicapped in obtaining services that require high-speed or broadband transmission capacity (facsimile or data connections, for example). Should the population densities of such regions be low enough to discourage investment in superior transmission technologies, their handicaps – in terms of availability or cost – may be longstanding or permanent. Regions possessing high densities of high capacity links such as microwave and fiber-optic media on the other hand, will enjoy access to a great variety of services at low cost.

Throughout the history of telecommunications, the highest capacity transmission media have connected the largest communications nodes first and then diffused down the urban hierarchy to interconnect smaller places as their costs could be justified by smaller volumes of traffic. That process cannot and will not change as long as there is any relation between distance and transmission cost. In broad outline then, large cities and their immediate environs will continue to enjoy advantages with respect to communications services. Rural areas and small towns will continue to be disadvantaged by higher communications costs, the unavailability of certain communications services, or both. That being the case, knowing what is possible and probable for specific places and regions will require descriptions and analyses of the kinds of transmission links that interconnect, are available within, or traverse all places and regions of interest.

Detailed maps of unit area ratios among total transmission media of various kinds would provide an overview of the development of the telecommunications complex at any given time and enable geographers to identify broad regions of communications advantage and handicap. At regional and local scales, network maps will continue to provide useful insights into what is possible at what kinds of costs for particular locations.

Nodes

Mapping and analyzing nodes of various kinds documents the degree of access people and places have to the communications *potentials* networks offer, especially at local levels. In 1972, for example, Professor Jean-Claude Thomas, then of Catholic University, produced a series of manuscript maps showing the locations of telex terminals in Washington, DC. The maps clearly documented the clustering of information activities in the then emerging office district in the vicinity of K Street and Connecticut Avenue Northwest. More recent work of this kind has used the locations of modem devices (Forsström and Lorentzon, 1987) to identify information-intensive areas of Gothenburg, Sweden.

Smart buildings and *teleports* reflect the locations of information-based activities (transactions, to use Jean Gottmann's felicitous term) at the same time that they attract such activities. Smart buildings are furnished by their builders with the latest in intra- and inter-office telecommunications facilities, from express courier pickup boxes to satellite dishes. Teleports write the concept of smart buildings even larger by providing bulk access to microwave, fiber-optic and satellite transmission channels. Because smart buildings and teleports enable the firms and offices they serve to bypass local telecommunications carriers and thereby realize significant savings on long-distance communications costs, they could become critical *hosting locations* (to use another of Gottmann's terms) for communications-intensive activities of a commercial and administrative nature. Monitoring the construction and fortunes of such facilities should be a major component of the geographical analysis of telecommunications. The locations and densities of satellite sending (uplink) stations are similar indices of intense telecommunications activities that have received little attention.

At the opposite extreme from substantial facilities such as smart buildings and teleports lie mobile communications. Mobile telephone service based on cellular radio technology is expanding rapidly in the United States and elsewhere. It is likely that cellular services will soon diffuse down the metropolitan and urban hierarchy, and that other telecommunications services will be available in addition to voice telephone connections. Indeed, proposals are already before US regulatory bodies to extend cellular access outward from cities along interstate and other major highways. It is conceivable that cellular service could provide communications in sparsely settled regions at lower cost than hardwired technologies. Thus a variety of intensely geographical issues of access, cost and use imply that the evolution of mobile communications services in general, and cellular mobile telephone services in particular, should be given high priority on research agendas.

As important as they are, the locations of some kinds of telecommunications nodes are difficult to document. When licenses or permits are required, records are readily available. There would be little difficulty in mapping the growing number of satellite uplink stations, for example. Other nodes are refractory. Modems and satellite receiving stations can be used without permits or registration in some nations, and are therefore virtually impossible to locate even if they are regularly used at a single location rather than carried about. Similarly, records of mobile telephones may be unavailable or proprietary, thus barring their use by interested researchers. Hard data on who is using what kinds of terminal equipment when and where has historically been harder to obtain in nations with private telecommunications industries than in those where telecommunications services are provided by government agencies. We can probably expect data on access nodes to become more scarce as deregulation of telecommunications services becomes more widespread internationally.

Flows

Ideally, flow analysis should be a standard component of the geography of telecommunications. Telecommunications links provide a potential for communication. Nodes provide access to that potential. Flows of messages among places reflect the degree to which that potential is realized, and fluctuations reveal how communications patterns among places change from time to time. In practice, such studies are conspicuously absent, especially in North America. In twenty-five years of research on American telecommunications, I have been able to obtain but two flow matrices, one for letter flows and one for telephone calls, and both at a gross level of aggregation. For the most part, telecommunications flow data are considered to be proprietary by those who hold them, and they are rarely released for research purposes. Letter flow data in the United States contain information on time elapsed from origin to destination, a sensitive matter. A propensity for private telephone companies in North America to keep flow data confidential has been reinforced over the last two decades by first the prospect and then the reality of competition. In a competitive environment, accurate and detailed information about communications demand at and among places is valuable to actual or potential competitors. Keeping such data confidential is a rational policy. For the foreseeable future, it is unlikely that North American researchers will be able to match the kinds of flow studies that can be done in Europe (Charlier, 1987; Grimmeau and Sortia, 1987) where government telecommunications monopolies take a less protective attitude toward flow data. To the degree that deregulation and competition are introduced in Europe and other areas with government telecommunications monopolies, we can expect reduced access to flow data in those regions.

Despite genuine difficulties in obtaining reliable data, difficulties which will likely grow worse rather than better, it is important that geographers and others interested in the past, the present status and the future of telecommunications continue to pursue local, regional, national and international research on the geography of links, nodes and flows. Learning the geography of hardware is prerequisite to knowing where access nodes are, why they are where they are, and where and when people have access to networks. Learning the geography of flows fosters an understanding of the degree to which the communications potential provided by hardware is realized, and thereby makes it possible to identify localities and regions where communications potential remains unrealized owing to the absence of hardware.

The evolving geography of communications hardware results from economic and technological forces that constrain and drive the telecommunications industries. Changes in technology and in the economic geography of technology are constantly redrawing maps of telecommunications networks and use at local, regional, national and global scales. As important as hardware is, and as much as geographers are drawn to it because it is tangible and therefore easy to map, we should recognize that the technical and economic forces expressed in the geography of telecommunications hardware are in fact a secondary and decreasingly important explanation of the maps of telecommunications. A greater force, even less noted by geographers than hardware, continues to draw and redraw telecommunications maps.

Regulatory and policy software

The term *hardware* is an appropriate characterization of the physical channels by which telecommunications among places are effected. The global telephone network has historically been the world's largest machine. It is now well on its way to becoming the world's largest computer. If the net is hardware, what of *software*? What is it, and how should it be incorporated into the geography of telecommunications?

In one sense telecommunications software consists of the routing routines that direct messages from one place to another. In the past, that kind of software was largely indistinguishable from telecommunications hardware – it was physically wired into networks. The technological history of telecommunications in the last fifty years is the story of the segregation of routing from hardware. Increasingly, flows are directed and regulated by computers that are functionally and physically separate from passive transmission networks. But *regulation*, a different and higher order kind of software, fundamentally determines the shape and operations of telecommunications systems.

Throughout history, telecommunications industries and services have never *diffused*; they have always been deliberately extended by strong centralized authorities. Throughout history, telecommunications media have been under direct or indirect government control. The United States is, in theory, deregulating its telecommunications industries to eliminate interlocational and interregional subsidies. Deregulation of telecommunications follows in the wake of deregulation of other movement technologies such as the bus, truck and air transportation industries. In fact, as opposed to theory, the telecommunications industries, like their transportation counterparts, remain administered industries, and like their counterparts, they will likely remain so. In practice, in the United States and elsewhere, the question is not regulation versus deregulation; it is a question of *how much* regulation.

Origins of regulatory and policy software in the United States

The telecommunications software – policies and practices that led to deliberate, subsidized telecommunications for many areas of the nation –

were sometimes explicit. Examples are the extension of postal services to all parts of the country, the reluctance of the federal government to close small post offices and the enlargement of the REA's purview to telephone services. Sometimes policies that led to the provision of telephone services to marginal areas were implicit and appear to be unintended artifacts of pricing policies. That appearance is deceiving.

The American telephone network, for example, expanded rapidly after 1890. Its extension to all parts of the nation was an outgrowth of the background and vision of Theodore Vail, who led the American Telephone and Telegraph Company (AT&T) during two critical periods in its history.

Prior to his first term at the helm of AT&T (1878–87), Vail was the Superintendent of Railway Mail operations for the United States Post Office. In that capacity, Vail directed the application of the new technology of long-distance, high-speed rail operations to telecommunications. In that capacity, he also acquired an understanding of the value of an integrated national network, the goal he pursued throughout his second term (1907–19) as head of AT&T.

Vail's critical role in establishing a mechanism and philosophy for providing basic telecommunications services in peripheral regions is only one example of a general importance of software in the history of communications technologies. In fact, *no* telecommunications technology has been established without software in the form of strong central planning, direction and control. That principle applies as much to the Microwave Corporation International (MCI) fiber-optic network as it did to the Persian postal system and AT&T. Interconnecting networks are always carefully and rationally planned by some central authority or strong figure such as Theodore Vail or Bill McGowan (MCI's longtime leader).

One reason why central direction is needed is that a *utility-penetration paradox* must be overcome before an interconnecting network is of value to potential users (Falk and Abler, 1985). The nature of the paradox is highlighted by the unlikely example of a person who had the only telephone in the world. The device would have no value because it would not enable its owner to communicate with anyone else. Alternatively, a communication device or medium that permits contact with everyone in the world has immense value because so many people can be reached.

Central direction of the establishment and expansion of interconnecting networks is required to overcome the utility-penetration paradox. A threshold must be surpassed before a telecommunications technology can achieve self-sustaining growth. A backbone network must be provided that connects enough people at enough places to convince potential users that purchasing the service is worthwhile in terms of the number of people and places the medium connects. The neonatal mortality among a number of highly touted innovations such as Picturephone and various teletext services demonstrates the consequences of failing to anticipate and overcome the utility-penetration threshold with adequate central direction. Left to diffuse on their own, such services have soon failed.

Today as in the past, if we wish to know who is currently doing the most to draw and redraw the telecommunications maps of nations and the world, we need look no further than the civil administrations, regulatory bodies and the courts that have historically decided what such maps will portray. The map of Local Access and Transport Areas (LATA) in the United States (Figure 2.1), for example, was drawn neither by technological and economic forces nor by communications geographers. It was produced as part of the 8 January 1982 Modified Final Judgement handed down by a Federal District Court in Washington, DC.

The LATA map defines the basic geography of telephone and many ancillary services in the United States. The 187 LATAs delimit areas of service responsibility within which the seven Regional Bell Operating Companies (RBOCs) are authorized to route and transport telephone traffic. Thus each of the LATA regions is a local monopoly territory for the operating company that serves it. The RBOCs cannot carry traffic between LATAs; that function is reserved to long distance carriers, including the vestigial American Telephone and Telegraph Company (AT&T).

Despite deregulation and continued efforts in that direction in the United States, the fact remains that regulation exerts more control over what services are available to what classes of customers at what places at what costs than any other single variable. In other words, the software that dominates telecommunications hardware is regulation more than anything else. It follows that a geography of regulation and deregulation – and of the ideology and geographic perceptions that underlie regulation and policy – is a serious *lacuna* in the geography of telecommunications: regulation is a topic that geographers interested in telecommunications have scarcely noticed, let alone addressed with serious research.

Scholarly brainware

If telecommunications facilities are hardware and regulation and policy are software, they would dominate the ways scholars think about the relationships between telecommunications and societies as relevant *brainware*.

Our brainware is faulty. Scholars and planners have persistently misperceived the ways telecommunications and societies affect each other and they have persistently neglected the importance of policy and regulation in telecommunications because they have persistently based research questions and implementation programs on an untenable assumption: the notion that telecommunications technology *causes* things to happen – socially, economically and geographically – underlies most thinking about telecommunications. That assumption will not withstand scrutiny. The way that assumption needs to be reworked – the way telecommunications

NATIONAL LATA MAP™

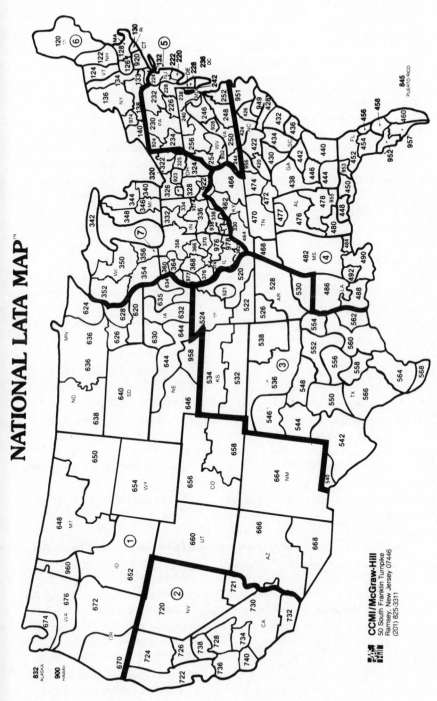

Figure 2.1 US Local Access Transportation Areas (LATAs), 1984.

CCMI/McGraw-Hill
50 South Franklin Turnpike
Ramsey, New Jersey 07446
(201) 825-3311

brainware needs to be rewired – can be illustrated by reference to a topic of considerable current interest: the role of telecommunications in regional development.

Telecommunications and Regional Development

A number of individuals, national governments and international organizations are enthused about using telecommunications technology to promote regional economic and social development (Estabrooks and La-Marche, 1987; Giaoutzi and Nijkamp, 1988). Their thinking, in common with that of most people, is predicated on the assumption that technology can cause people to do things. Based on the experience of the United States, efforts to use telecommunications to promote development are likely to be disappointing.

Throughout US history, telecommunications media have been used to promote regional development. Postal services were extended westward and southward with the expanding settlement frontier. The explicit reason for doing so was to foster commercial activities and national unity. The ability to communicate at a distance was viewed as a critical aspect of local, regional and national affairs. Throughout its national history, the United States government has maintained tens of thousands of uneconomic post offices to meet local commercial and social needs.

Telephone service has also been used to promote regional development. Sometimes such policies were consequences of the nature of telephone networks and the long-distance pricing principles those networks fostered. Sometimes the policies were explicit goals set by the federal government.

Until the advent of coaxial cable and microwave transmission after World War II, telephone lines were accessible for local connection at any point along their routes. Thus bulk circuit capacity built between large towns and cities offered access to sparsely populated regions lying between the larger places. Any place within reasonable distance of the new high-capacity links was afforded easy connection. This provision of access in less-favored regions offers opportunities for studying the effects of introducing advanced technologies in isolated regions.

From 1910 onward, long-distance rate-averaging was another force that made telephone service available to rural and peripheral regions. Long-distance tariffs were based on distance alone, without respect to the location of the call's origin or destination. A 1,000 kilometer connection between Chicago and Pittsburgh, for example, cost a caller the same as a 1,000 kilometer call between two small, isolated places. The *network* cost of making a connection between the two small places could be as much as fifty times the cost of connecting Chicago and Pittsburgh. Subsidizing long-distance service in rural areas with large surpluses earned on links connecting major cities made telephone services available in outlying regions at prices well below actual costs.

The ways long-distance revenues were apportioned between terminal and line-haul costs further subsidized rural telephone companies. Separations overpaid local telephone companies; they were another mechanism that provided telephones and long-distance services at prices below true costs. The net effects of rate-averaging and separations principles have been to make telephone services available in many outlying places where they would not have existed if customers had paid the full costs of building and using telephone networks.

The private-sector subsidies described above were augmented by deliberate use after 1945 of Rural Electrification Agency (REA) programs to provide telephone service in marginal areas. Some 4.5 million telephone subscribers were provided with new or upgraded service by companies that received subsidized loans from the REA (Pierce and Jequier, 1983, pp. 27–8).

The United States enjoys effective, efficient and inexpensive telecommunications. Postal services are universally available within the United States, despite high costs of providing service in isolated regions. Telephone services – and therefore many of the ancillary services that can be provided via telephone circuits – are available in most parts of the country. The percentage of households with telephones approaches 100 percent in many parts of the nation. Even in Alaska, 72 percent of households have access to a telephone.

Regional consequences of telecommunications subsidies

What have the massive, long-term subsidies of telecommunications services in peripheral regions of the United States yielded in terms of development? In all likelihood, a great deal that is socially important. In terms of documentable economic progress, very little. We have neither the theory nor the methods to calculate past and current subsidies, let alone the degree to which those subsidies have yielded benefits commensurate with or exceeding their costs.

It is clear that the subsidies are important to people and their political representatives. There were 77,000 post offices in the United States in 1900. We have since closed some 44,000 of them. Every one of those 44,000 post office closings was bitterly contested by the people the office served. Accordingly, we have had extensive experience with the concerns and issues that arise when post offices are closed in rural areas.

I am confident we shall get more insights into the importance of telecommunications in isolated regions during the remainder of this century. We shall have to cope with the erosion of telecommunications services that will follow deregulation and the consequent abandonment of rate-averaging. Service in marginal regions must inevitably degrade with the adoption of marginal-cost pricing.

Overall, in terms of the ability of marginal regions to compete with metropolitan centers, few demonstrable returns to historic subsidies are evident. Whether measured by sophisticated telecommunications terminal facilities or the locations of information- and communications-based industries and occupations, metropolitan centers continue to win out and rural regions continue to lose out in the United States and elsewhere (Abler and Falk, 1985).

The implicit and explicit policies of subsidizing telecommunications in outlying regions have not prevented the nation's great metropolitan regions from widening that gap between them and disadvantaged regions. Neither will those policies reverse the current trend toward greater disparities between metropolitan centers and rural peripheries in the future. All indicators suggest that information and network societies are metropolitan societies, and that they will continue to be so. Believing otherwise, unless such beliefs are based on solid evidence that contradicts historical patterns, is evidence of faulty brainware.

Related facets of faulty brainware

The hazards of failing to think carefully about how telecommunications and societies affect each other are evident in three largely overlooked aspects of the relationships between telecommunications technology and economic development. First, telecommunications services are traditionally viewed as wholly positive forces, but on regional scales, telecommunications services will threaten certain interests within peripheral regions. Second, we should be wary of uncritically applying the telecommunications experience of advanced nations such as the United States to problems in other parts of the world. Finally, there is no technological imperative. Telecommunications are necessary conditions of regional progress. They have not been nor will they ever be sufficient conditions of development.

THREATS VERSUS PROMISES

One of the hardest geographic lessons to learn is that interconnecting networks run in both directions. Extending and maintaining telecommunications services to peripheral regions will threaten the economies of those areas (Gillespie, 1987). Rural Free Delivery and Parcel Post were the deaths of the crossroads store in the rural United States: better and more varied goods could be obtained at lower cost from mailorder houses in distant cities. More recently, electronic banking has been the death of locally owned banks throughout the United States.

Connecting peripheral areas to metropolitan centers does indeed provide outlying regions with access to the major hot spots of the space economy. But those connections simultaneously increase the reach of entrepreneurs located at the metropolitan hot spots, providing access to markets that were formerly inaccessible (Moss, 1986). Corporate strategies for employing telecommunications (*orgware*) are now become important weapons in the territorial competition in which firms engage (Giaoutzi and Nijkamp, 1988, p. xiii). Business establishments in outlying regions are unlikely to employ telecommunications as effectively as metropolitan competitors. In such a contest, it is unlikely that regional merchants and industrialists will win. Extending advanced telecommunications technologies to outlying areas will leave no places for the information- and communication-poor to hide. In the absence of deliberate policies to the contrary (and perhaps in spite of them) telecommunications will inevitably reinforce the advantages of higher-order metropolitan centers.

As a corollary, it should be obvious that geographic disparities in telecommunications facilities and use will persist. Fiber-optic transmission and teleports are the latest developments in telecommunications transmission and switching technology. As has historically been true, these new high-capacity links connect the most important places, which are also the sites where smart buildings and teleports are under construction. These new technologies will diffuse down the combined metropolitan and network hierarchies as their costs drop and as greater demand justifies their use to connect and serve smaller places. But disparities will persist as new transmission and switching capabilities are invented and applied first at the top of global and national network hierarchies. As long as differences in population density persist, differences in the cost of telecommunications infrastructure and use will persist. Despite their undeniable power, telecommunications technologies cannot repeal the laws of economic geography, as enthusiasts often imply they can.

LESSONS VERSUS MODELS

The cases of the United States and other advanced nations are instructive, but they should be viewed more critically than they have been. When the overburden is stripped away, it seems that many countries are convinced that they have to pattern their telecommunications policies on those that are currently evolving in the United States.

The telecommunications complex that exists in the United States evolved from the peculiar national experience of occupying a continent at the same time that telecommunications media were developing. It evolved as it did because of system-specific software – the distinctive personalities of some of the major actors. It evolved as it did because of historic accidents such as the failure of the United States Congress in 1845 to see any value in its ownership of the patent on Samuel Morse's telegraph, which it obtained by financing Morse's experimental line between Baltimore and Washington. Deeming the innovation little more than a toy, Congress allowed ownership of it to revert to Morse, thus establishing the precedent of private ownership for electronic means of telecommunications.

That the US telecommunications complex and its evolution are unique does not mean that they do not hold valuable lessons for other times and other places. Indeed there are fundamental principles of hardware geometry that transcend national experience and different software schemes. But other nations and other regions face problems different from those in advanced nations such as the US that may well call for different software than that developed for US applications.

In that context, it is important also to recognize that some attractive and desirable software features are mutually incompatible. Software designed to foster deregulation at the same time that it guarantees regional economic and social equity and integrity will not work. Competition will always override regional equity.

To use another analogy, guaranteeing regional integrity in the face of telecommunications competition is like trying to make water run uphill. If the water is confined to a container, it is possible to make it go uphill. It is hard work, but it can be done. If the water is not confined, it will surely run downhill, no matter how much we wish it to go uphill.

Policy makers were able to make water go uphill in the United States for a long time. They could do so because the telecommunications water was contained in policies, practices and regulations that confined it. The United States has now decided to let that water run free. It has deregulated telecommunications and in the process has eliminated the subsidies that paid for telecommunications services in marginal regions. Under those circumstances, water is certain to run downhill; economic activities will be drawn from marginal regions toward metropolitan centers.

If other nations wish to move their water uphill, they must take care not to let it run free. I contend that there is virtually no middle ground. Liberalization or deregulation on the one hand, and regional equity in telecommunications infrastructure and access on the other, are incompatible and antithetical. To reiterate, the laws of economic geography cannot be repealed.

TELECOMMUNICATIONS AS PANACEA

What if somebody built a teleport and nobody called? Telecommunications facilities and services are *enabling* technologies (Kaye, 1987, p. 43). Telecommunications permit individuals, firms and agencies to perform certain tasks. Telecommunication capacity does not guarantee that those tasks will be accomplished. The literature on telecommunications and economic development could give an impressionable person the notion that a teleport is the simple solution to development problems in any region.

We know better than that. We know that telecommunications and development at regional and national scales are closely correlated. We are still not sure which is cause and which is effect (Pierce and Jequier, 1983, pp. 14–17). It is most likely that each is both: telecommunications evolution on the one hand, and economic and social development on the other, proceed in concert with each other. And together they proceed in concert with other economic, social and political developments in a cumulative, mutually causal manner.

We know, therefore, that if somebody built a teleport in Lame Deer, Montana, nobody would call. We also know that nobody possessed of common sense would build such a facility in such a location. The pattern of existing and planned teleports in the United States makes it clear that such facilities are feasible and necessary only in the largest, most information-intensive metropolitan locations.

Software policies designed to achieve and maintain regional tele-communications equity will be defensible if they aim to provide *basic* telecommunications capacity in outlying regions. They will be less defensible if they envision sophisticated teleports in every village and town across a nation or around the globe.

The need for teleports in locations other than the world's largest conur-bations is questionable. Telecommunications – particularly those that make use of satellite capacity – are especially footloose technologies. In the United States, there exist literally dozens of entrepreneurs who can provide bulk transmission capacity via portable satellite antennas at any location in the country on twenty-four hours notice.

Regional development policies must also recognize the primacy of existing regional infrastructure and knowledge bases in the development process. Schemes for providing advanced media must incorporate funds to insure that those media can actually be used. A teleport, for example, requires complicated and compatible computer architecture in the systems it serves in order to work effectively (Enslow, 1987). Therefore policies and plans that look beyond regional equity in basic telecommunications should be based on detailed inventories of existing capabilities and realistic assess-ments of opportunities within affected regions (Gillespie, Andrew *et al.* 1984, p. 178). Policies and plans that assume that advanced capabilities will stimulate their own demand will be open to telling criticism.

Coevolutionary brainware

A mature, articulated geography of telecommunications would enable planners to make intelligent use of scarce resources. It would enable scholars to speak with authority on where, how, and why societies employ those technologies to move information from place to place. That mature geography would embrace the study of telecommunications hardware and regulatory software; absent either, it would be seriously incomplete. More importantly it would have its causal brainware in order. If it does not, it will continue to misconstrue the history of telecommunications, to mis-diagnose the present, and to misprescribe for the future.

Books and articles on telecommunications are larded with arguments about the 'impacts of X communications technology on A, B and C'.

Analysts rarely write about the impacts of A, B and C on communications technology X. The latter is as prerequisite to understanding communications technologies and their use as the former. Some social and economic conditions are clearly associated with some telecommunications hardware and its use. But even given the most extensive and detailed data possible, we still lack the theory and the wisdom to know which variables belong on the ordinates and which on the abscissas of the relationships that undeniably exist. It is most likely that most variables belong on both axes.

Foremost among the prerequisites to a mature geography of telecommunications is the need to abandon the assumption that communications technologies cause cultural, economic, political and social behavior and events and that behavior and events are consequences of the absence, presence or use of telecommunications hardware. Neither historical evidence, current events, nor prospects for the future support that assumption.

Monocausal models of biological evolution have given way to conceptual frameworks that recognize that species often coevolve in ways which make it impossible to understand changes in one species without considering simultaneous change in other species. We will think and teach more effectively if telecommunications and society are viewed as a coevolving complex of processes: telecommunications, societies, economies, governments, human aspirations and human values shape each other, in much the way complexes of living organisms evolved in interaction with each other, each being cause as well as effect of the other's change.

Note

This chapter is based on presentations made at the meeting of the International Geographical Union Study Group on the Geography of Telecommunication and Communication, Seville, 28–30 September 1986; the Conference on Telecommunications and Business, Urban, Regional, National and International Development sponsored by the Canadian Institute for Research on Regional Development and the Canadian Department of Communications, Ottawa, 10–12 November 1986; and the OECD Seminar on Information and Telecommunications Technology for Regional Development, Athens, 7–9 December 1986. Parts of this chapter were previously published in Abler, 1987.

References

Abler, Ronald F. (1987), 'The geography of telecommunications in the United States: local and regional research problems', Le bulletin de l'IDATE, 26, pp. 120–5.
Abler, Ronald F., and Falk, Thomas (1981), 'Public information services and the changing role of distance in human affairs', Economic Geography, 57, pp. 10–22.
——(1985), 'Intercommunications technologies: regional variations in postal service use in Sweden, 1870–1975', Geografiska Annaler 67B, pp. 99–106.

Beniger, James R. (1986), *The Control Revolution: Technological and Economic Origins of the Information Society* (Cambridge, Mass.: Harvard University Press).

Bolling, George H. (1983), *AT&T: Aftermath of Antitrust* (Washington: National Defense University).

Borsos, Charles (1985), 'World network development: the critical points', *Telephony*, 22 October 1985, pp. 43–54.

Charlier, Jacques (1987), 'Les flux téléphoniques interzonaux belges en 1982: une approche multivariée', *Le bulletin de l'IDATE*, 26, pp. 126–30.

Coll, Steve (1986), *The Deal of the Century: The Breakup of AT&T* (New York: Atheneum).

Cowhey, Peter (1987), 'International trade and telecommunications', in Estabrooks and Lamarche, op. cit., pp. 163–78.

Dobell, Rodney (1987), 'Putting informatics to work for economic and social development', in Estabrooks and Lamarche, op. cit., pp. 74–82.

Downs, Anthony (1987), 'Future impacts of telecommunications upon the location of economic activities', in Noothoven van Goor and Lefcoe, op. cit., pp. 151–63.

Enslow, Jr, Philip H. (1987), 'Computer communications utilizing teleport services', in Noothoven van Goor and Lefcoe, op. cit., pp. 53–63.

Estabrooks, Maurice F., and Lamarche, Rodolphe H. (eds) (1987), *Telecommunications: A Strategic Perspective on Regional, Economic and Business Development*, (Moncton, N.B.: Canadian Institute for Research on Regional Development).

Falk, Thomas, and Abler, Ronald F. (1980), 'Intercommunications, distance, and geographical theory', *Geografiska Annaler*, 62B, pp. 59–67.

——(1985), 'Intercommunications technologies: the development of postal services in Sweden', *Geografiska Annaler*, 67B, 1, pp. 21–8.

Fischer, Claude S. (1985), 'Technology's retreat: the decline of rural telephony 1920–1940', paper presented at the Annual Meeting of the American Sociological Association, Washington, DC, 26 August.

Forsström, Ake, and Lorentzon, Sten (1987), 'Data communication and settlement structure – the use of modems within the Gothenburg telecommunications region', *Le bulletin de l'IDATE*, 26, pp. 139–51.

Giaoutzi, Maria, and Nijkamp, Peter (eds) (1988), *Informatics and Regional Development* (Brookfield, Vt: Avebury).

Gillespie, Andrew (1985), 'Telecommunications and the development of the less favoured regions of Europe', *Le bulletin de l'IDATE*, 21, pp. 471–7.

——(1987), 'Telecommunications and the development of Europe's less favoured regions', *Geoforum*, 18, pp. 229–36.

Gillespie, Andrew, and Hepworth, Mark (1987), 'Telecommunications and regional development in the network economy', in Estabrooks and Lamarche, op., cit., pp. 107–28.

Gillespie, Andrew et al. (1984), *The Effects of New Information Technology on the Less-Favoured Regions of the Community* (Brussels: EEC).

Goddard, James B., and Gillespie, Andrew E. (1986), 'Advanced telecommunications and regional economic development', *The Geographical Journal*, 152, pp. 383–97.

Grimmeau, J. P., Sortia, J. R., and Colard, A. (1987), 'Analyse régionale des flux téléphoniques de la Belgique vers l'étranger', *Le bulletin de l'IDATE*, 26, pp. 131–8.

Hanneman, Gerhard J. (1987), 'Telecommunications, teleports, and the new urban infrastructure', in Estabrooks and Lamarche, op. cit., pp. 143–50.

Hepworth, Mark E. (1986), 'The geography of economic opportunity in the information society', *The Information Society*, 4, pp. 205–20.

Irwin, Manley R. (1987), 'The fusion of telecommunications and corporate strategy, in Estabrooks and Lamarche, op. cit., pp. 181–93.

Kaye, A. Roger (1987), 'Current trends in telecommunications technology, systems, and services', in Estabrooks and Lamarche, op. cit., pp. 31–44.

Lamarche, Rodolphe H. (1987), 'Telecommunications and regional development: a new concept', in Estabrooks and Lamarche, op. cit., pp. 85–105.

Lannon, Larry (1987), 'Flexibility and social values', *Telephony*, 26, November 1987.

McPhail, Thomas (1987), 'A history of telecommunications', in Estabrooks and Lamarche, op. cit., pp. 15–30.

Moss, Mitchell (1986), 'Telecommunications and the future of cities', *Land Development Studies*, 3, pp. 33–44.

——(1987), 'Urban development in a global economy', in Estabrooks and Lamarche, op. cit., pp. 33–44.

Noothoven van Goor, J. M., and Lefcoe, G. (eds) (1987), *Teleports in the Information Age* (Amsterdam: North Holland Publishing Company). Proceedings of Teleport '86 World Teleport Association Second General Assembly and Congress, 21–3 May 1986, Amsterdam, The Netherlands.

Paquet, Gilles (1987), 'The new telecommunications: a socio-cultural perspective', in Estabrooks and Lamarche, op. cit., pp. 45–68.

Pierce, William, and Jequier, Nicolas (1983), *Telecommunications for Development* (Geneva: International Telecommunication Union).

Saunders, Robert J., Warford, Jeremy J., and Wellenius, Bjorn (1983), *Telecommunications and Economic Development* (Baltimore, Ala.: Johns Hopkins University Press).

Singer, Benjamin D. (1981), *Social Functions of the Telephone* (Palo Alto: R&E Research Associates, Inc).

Temin, Peter with Galambos, Louis (1987), *The Fall of the Bell System: A Study in Prices and Politics* (New York: Cambridge University Press).

Williamson, John (1987), 'Can the dream ever become reality for the less developed countries?', *Telephony*, 23 November.

Winston, Brian (1986), *Misunderstanding Media* (Cambridge, Mass.: Harvard University Press).

3 Global Interdependence and Its Consequences

DONALD G. JANELLE

Interdependence implies that the parts of a system rely on one another and that events occurring at particular times and places may have impacts elsewhere at the same or in future times. This chapter considers recent trends toward global interdependence and their contrasting consequences for peoples of the developed and developing worlds. It describes some of the principal forces contributing to global interdependence, documents the increasing levels of global transactions in information, and considers the implications of global interdependence for those who share most fully in such trends as well as for those who remain in relative isolation of global interactions.

Contributors to global interdependence

Space adjusting technologies

The need to interact with other people and places is a function of organizational forms and the diversity of resource requirements. The transition from subsistence to commercial forms of organization entailed progressively increasing levels of both labor and regional specialization, where motivations toward productive efficiency were combined with various technological innovations to yield increases in the scale of productive units. This process of an industrial revolution has been accompanied by the search for and the development of more resources and larger markets, spread over a greater proportion of the earth's surface. The collation of events that originated and sustained this transformation is complex and it is difficult to specify all of their underlying contributors. Nonetheless, any interpretation must recognize the prominent roles that transportation and communication technologies have played.

Advances in transportation and communication that reduce the significance of distance are called *space-adjusting technologies*. This notion of a shrinking world can be measured by reference to the concept of *time–space convergence* (Janelle, 1968). It measures the rates at which places move closer together or further away in travel or communication time. Figure 3.1 provides an example of this convergence between Boston and New York

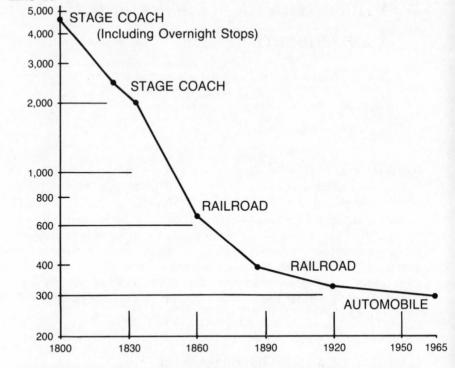

Figure 3.1   Time–space convergence between New York and Boston, 1800–1965

during the period from 1800 to 1965. Whereas 4,700 minutes were required to cover the 210 miles in 1800 by stagecoach, in 1965 automobiles could make the journey in 300 minutes. In a functional sense, they converged at the average rate of twenty-seven minutes per year. Convergence is a direct result of the input of new technologies – more direct roads, faster vehicles, routes bypassing intervening places, and so forth. The histories of mail delivery, international air travel, and telephone connections provide examples of how various space-adjusting technologies have led to the convergence of places.

National mail service

Mail deliveries from New York City to San Francisco took twenty-four days in 1858, but were reduced to twelve days with the introduction of the Pony Express system in 1860. Railroads decreased this to about 106 hours in 1900 and, by 1924, regular airmail service reduced the time to 72 hours. According to Pennsylvania State University geographer Ronald Abler (1975), this progression of new technologies resulted in a reduction of mail delivery time from nearly six hundred hours in 1858 to twelve hours (or less) for most deliveries today. And recently, electronic messages, sent via computers from places of residence or business to regional centers, are delivered within six hours or less, while direct computer linkages by telephone allow for nearly instant connections.

Global air travel

International air travel has also seen reductions in travel times between major cities. Travel from Paris illustrates this shrinkage of distances. The fastest aircraft in 1950, the DC7, traveled roughly 350 miles per hour. Thus, most cities in southern and northern Europe could be reached from Paris in two hours, while flights to eastern North America took seven hours. Due to refueling stops, it took nearly twenty-four hours flying time to reach cities halfway around the world. The Boeing 747 jumbo jets, introduced in the late 1960s, reduced these times considerably. Flying at 640 miles per hour, northern South America, central Africa, southern Asia, and central North America could be reached within six hours from Paris. The supersonic Concorde, which made its debut in January 1976, reduced such long-distance flights almost by half. Travelling at roughly 1,350 miles per hour, the Paris–Dakar–Rio de Janeiro flight takes only seven hours, compared to thirteen hours at subsonic speed. At Concorde speeds, New York, Moscow, Caracas, Lagos and New Delhi are reachable from Paris within four hours, and Sydney and Buenos Aires within seven hours. By the early twenty-first century hypersonic aircraft (HST), burning liquid hydrogen, may shrink spaces even more. These clean-burning, noiseless, delta-winged craft may be able to fly at 4,000 miles per hour (Mach 6), and could make the Paris–Tokyo trip in only 2.6 hours. Paris to New York would require only two hours – 800 times faster than the sixty-seven-day transatlantic voyage of the Mayflower in 1620!

Telephonic convergence

The concept of time–space convergence has been used to demonstrate the effect of technological advances in telephone service (Abler, 1975). Coast-to-coast calls, possible in 1915, did not become common until the 1920s. Placing a call from New York to San Francisco, a distance of 3,000 miles (4,600 kilometers), took fourteen minutes in 1920. New network structures reduced the time to 2.1 minutes in 1930, and to 1.5 minutes in 1940. This was decreased to a little over one minute with direct dialing in 1960 (Fig. 3.2a). Today, touch-tone phones establish contact in less than thirty seconds between almost any locations in North America. Toronto, Cleveland, Miami, Denver and San Diego can be reached from Boston or Vancouver in virtually the same amount of time. It is not uncommon to discover that business or friends several thousand miles away can be reached in less time than someone across the street or in an office in the same building.

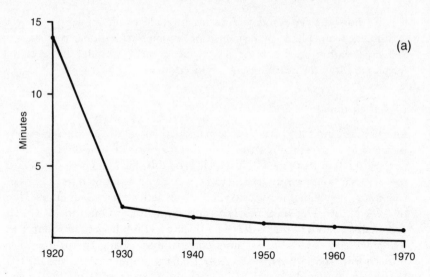

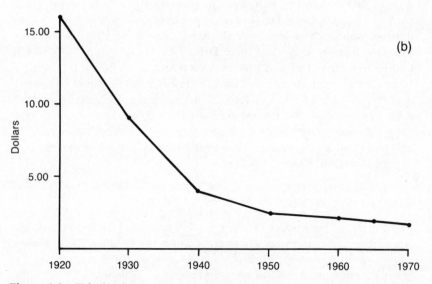

Figure 3.2   Telephonic convergence between New York and San Francisco: (a)
time–space convergence; (b) cost–space convergence
Source: Abler, 1975, pp. 39 and 43

Cost–space convergence

Related to the concept of time–space convergence is *cost–space convergence* –
the reduction over time in the costs of interaction over space. This is
illustrated in Figure 3.2b. A telephone connection from New York to San
Francisco in 1920 cost in excess of $15.00. In 1930 it was $8.00 and by 1940

it was only $4.80. Aided by engineering innovations, the costs declined to $2.50 in 1960 and to $1.35 in 1970. With direct dialing and night rates the costs have been reduced further. In 1987 an unassisted station-to-station call placed at night and during weekend hours cost about 25 cents for the first minute and about 16 cents for each additional minute. During daytime hours the cost was about 60 cents for the first minute and 38 cents for each additional minute, with variations due to the pricing policies of different carriers. The geographic consequence of these cost savings are illustrated by the decreasing significance of distance in the cost of calls from Missouri to other parts of the United States (Fig. 3.3).

An idealized telephonic cost–distance relationship would have a uniform rate regardless of the time engaged in communicating or the distance between the parties. Such a flat-rate service exists in North America. It is called Wide Area Telephone Service (WATS). For a flat monthly fee subscribers (generally businessmen and government offices that make heavy use of phone service) can call any location an unlimited number of times. Thus the long-term trend in the cost curve in phone service is for flatter or more uniform rates over long distances. This cost curve would be similar to postal rates in many countries that have had complete cost convergence since 1863 – that is, equal rates for mail delivery regardless of the distance.

A more general comparison of changes in the costs for transmitting messages is given by Pool (1983), and is illustrated in Figure 3.4. His 1960–77 evaluation of seventeen different communication media shows far more rapid growth in use, and declines in cost, for the electronic media than for print media, and for mass media than for point-to-point media. Without these dramatic changes, it is doubtful that levels of national and international transactions would be as great as they are today. However, the success of these technological facilitators of communication is interdependent with the increasing abilities of people to transcend linguistic barriers to understanding and political barriers to access.

The erosion of political and linguistic barriers

The agents who contribute to greater global interdependence are those decision-making and resource-allocating organizations whose activities transcend national and other cultural boundaries. Through their establishment of functional bonds between resource and market regions, and through their use of advanced space-adjusting technologies, they have accelerated the diffusion of practices and value systems that are compatible with industrial and post-industrial philosophies. These trends were initiated by the global extension of Europeanized modes of organization and viewpoints in the period of colonization and early industrial development. Subsequently, the agents of widespread standardization have been the globally pervasive cultures (usually powerful nation states, such as the United Kingdom in the nineteenth century or the United States in the

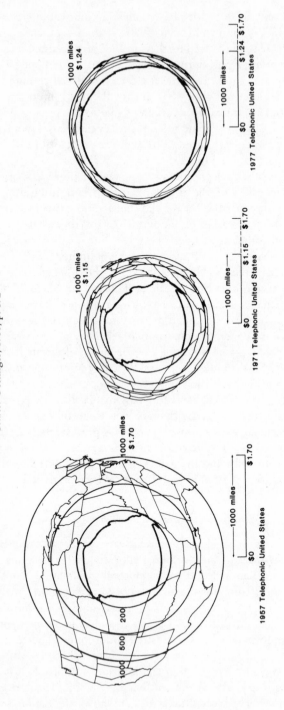

Figure 3.3 Telephonic cost distances from Missouri, 1957 to 1977
Source: Oettinger, 1980, p. 192

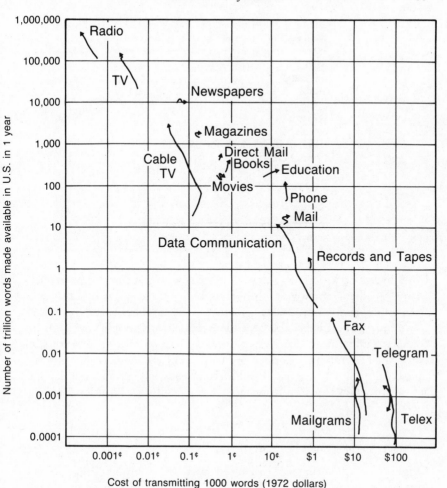

Figure 3.4 Trends in the volume and costs of communication media, 1960 to 1977
Source: Pool, 1983, p. 610

twentieth century), international organizations (particularly the United Nations), and multinational corporations. Since World War II, multinational corporations have been the most significant economic agents in facilitating transactions across national frontiers.

Multinational firms

International Business Machines (IBM), with its headquarters in a suburb of New York City, is a prime example of a firm with international linkages. It has subsidiaries in 137 countries. But, aside from being multinational, such firms may be multilingual, employing thousands of people

Table 3.1 Principal languages of the world – 1987 (in millions of speakers).

Mandarin (China)	788
English	420
Hindi	300
Spanish	296
Russian	285
Arabic	177
Bengali (Bangladesh; India)	171
Portuguese	164
Malay–Indonesian	128
Japanese	122
German	118
French	114

Source: Culbert, 1987.

from diverse backgrounds. They adopt firm-wide standards of procedure. Many of their employees, being highly mobile, develop associates throughout the world, and maintain ties by telephone and computer communication. It is possible that these residentially mobile and globally connected individuals identify with the corporation more than with spatially based communities and nation states. They are suggestive of possible global cultures that transcend loyalties to nations and regions, and that cut across groupings based on language, religion and other traditional indicators of culture.

International languages

Language is the principal structured way of communicating among members of a group. On the world scale, 151 languages are spoken by at least 1 million people, ranging from the 788 million who speak Mandarin Chinese and 420 million who speak English to approximately 1 million who speak languages such as Aymura (in highland Bolivia and Peru), Estonian and Tiv (in east Central Nigeria). Twelve languages are spoken by more than 100 million people (Table 3.1) and, in general, these tend to be the principal languages of the most highly populated countries. However, many countries, for example India, the Philippines, Indonesia and the Soviet Union, show considerable linguistic diversity.

The 'babble of languages' complicates the vision of an interdependent world; however, communication among those of diverse cultural–linguistic backgrounds has been facilitated by the use of only a few languages in international dialogue. The United Nations recognizes only six official languages for its discussions and publications (Arabic, Chinese (Mandarin), English, French, Russian and Spanish).

Some have suggested that one international language should be adopted.

A frequently cited example is Esperanto. It was developed in 1887 by the Polish philosopher Ludwig Zamenhof and is based on roots derived from Germanic, Slavic and Romance languages. Esperanto's estimated 8 million or more users claim that it is easy to learn. However, it seems unlikely that a universally accepted single language will be developed. Rather, the languages now used by the United Nations will probably continue as the basis for communication at international levels.

In terms of their global dispersal, the Germanic and Romance languages are most widely distributed. Indeed, English and French may be regarded as *international languages*. Aside from the prime status of English in North America, the United Kingdom and in Commonwealth states, it is a strong second language in most of Europe, Asia and Latin America. French is a primary or second language in much of West and North Africa, in the Middle East, Southeast Asia and parts of Canada, particularly Quebec.

The significance of English is illustrated by the degree to which books written originally in English are translated into other languages. Books in English accounted for 42 percent of the 44,000 translations (worldwide) in 1981. This compares with 13.5 percent from Russian, 11.4 percent from French, and 9 percent from German. Regional variations in this pattern (shown for selected countries in Table 3.2) reflect political alliances (for

Table 3.2   Number of books translated from major languages, 1981.

Country	Total	English	French	German	Russian
Egypt	142	116	5	7	7
Ethiopia	13	5	—	—	5
Nigeria	10	10	—	—	—
Canada	364	277	25	7	1
USA	1,086	10	257	287	91
Mexico (1979)	269	243	12	6	—
Panama (1980)	9	7	—	—	—
Brazil	844	565	106	56	—
Bangladesh	27	16	4	—	1
India	577	199	8	8	16
Iran (1980)	7	5	1	—	—
Japan	2,754	2,001	241	246	100
Indonesia (1980)	372	297	16	11	1
Saudi Arabia (1979)	2	2	—	—	—
USSR	7,171	624	171	319	4,035
Belgium	1,071	598	147	203	9
Hungary	419	44	26	39	50
Italy	1,871	917	436	241	50
Poland	591	104	50	72	107
United Kingdom	1,035	19	276	278	84
West Germany	4,904	3,028	667	209	119
Australia	39	15	1	5	2

Source: UNESCO, 1986, Table 7.16, VII, 120–4.

example, Eastern European translations from Russian), the international orientation of cosmopolitan countries, such as the Netherlands and West Germany, and the isolation of low-income developing countries, such as Bangladesh and Peru.

The emergence of languages that allow for communication in most parts of the world (even the development of standardized computer languages) have combined with space-adjusting technologies to permit new forms of organization (for example, multinational firms) and to form the framework for greater global interdependence.

International transactions

It is possible to document several trends toward increasing levels of international exchange. Trade in raw materials and manufactured goods is an obvious example; but the global exchange of information (films, mail, telephone traffic and mass media) reflect most critically the changing pattern of global consciousness and organizational structure.

Information flows

The increasing importance of information exchanges among nations is shown by the global networks of mail and telephone traffic, and by international trade in films and television programming.

Mail

The volume of mail sent between places is one measure of their interdependence. Messages are sent from and to individuals, businessmen, industries, hobby groups, professional organizations and governments. Information on the prices of commodities, investment holdings, diplomatic maneuvers and forthcoming meetings are all relayed by mail. Postal services also are used for advertising, for shipping goods and sending money. A map of all domestic and international mail movements would look like a gigantic set of cobwebs, thickest in and between Europe and North America, and, except for coastal populations and major urban nodes, much finer in Africa, Asia and Latin America.

The United States was the world leader (113 billion) in the number of domestic items mailed in 1983. Combined with about 3 billion foreign mailings, this amounted to nearly 320 million pieces sent daily, or more than 400 items sent per person per year (against a world average of 11.1). The heaviest mailings are from federal welfare offices, major wholesalers and retailers and major billing companies. An extensive domestic system (39,445 post offices in 1983) connects isolated farmsteads and the smallest hamlets with the system of larger urban places.

Nations with the highest literacy rates tend to handle the largest volumes of mail domestically as well as internationally. Thus, the Japanese mailed more than 16 billion, West Germany and France, nearly 15 billion each, and the United Kingdom, 12.5 billion. India, with 11 billion items, sent less than twice as much mail domestically as did Canada (6.5 billion), even though it has 30 times more people. Developing countries sent comparatively few items domestically compared to European nations or the United States. Some developing countries received more correspondence than they sent: for instance, Bolivia received five times more, and Indonesia and Zaire received more than twice what they mailed out. Mineral and timber activities in these countries, financed by outside capital, may explain the larger volume of incoming mail from international headquarters, banks and global markets. In general, most nations received more foreign letters than they sent; but tourist-oriented countries (e.g. Spain and Mexico) are exceptions.

Telephone traffic

The pattern of international telephone traffic bears some resemblance to the global mail networks, but differs in three major respects. First, since mailings occur at points where letters and packages are picked up and postmarked, there are far fewer collection and reception nodes than is the case for the more widely dispersed telephone installations. Second, phone service is dependent on a power-generating source and on the abilities of individuals to pay for phone installation and service. Therefore there are likely to be few phone calls placed and received in developing countries, especially in rural areas. Third, an advantage of phone service is that a low literacy level does not prevent contact with friends, family or business. For this reason the usage of phone service in developing countries is increasing more rapidly than mail service.

Some perspectives on international phone traffic can be obtained by looking at the number of calls originating and terminating in the United States. With nearly 170 million telephones in 1983, it has about 38 percent of the world's total. Ninety-five percent of all her households are equipped with at least one telephone and the average household makes 120 calls per month, almost half within a two-mile radius. But, an increasing number are long-distance (more than 36 billion in 1982) and international calls. Most of the 237 million international calls in 1983 were to Mexico, Canada, major European trading countries and with tourist-oriented countries in the Caribbean. Calls of a military nature might explain the large number to Europe and East Asia. For most countries, the proportion of incoming and outgoing calls was roughly equal. Exceptions were Cuba, Egypt, Hong Kong and Argentina, which received more than they placed; Kuwait, Venezuela, the USSR and France placed more than they received.

A major development in international telephone traffic has been the increased volume facilitated by low-cost satellite networks. In general, all areas of the world are sharing in these gains in interpersonal communications. However, governments, businesses and international agencies account for most of the increases.

Mass media and instant news

In contrast to the most selective participation in private telephone communications, developments of mass-media technology have exposed a large proportion of humanity to nearly instant awareness of events occurring throughout much of the world. Personally owned television sets and newspaper subscriptions serve this purpose in most developed countries, while wall posters and transistorized radios are media found in many developing countries. The inexpensive acquisition and widespread availability of transistor radios by millions of people throughout the world has placed them in touch with global events.

Two major concerns about the international dissemination of information by mass media are its extent and impact. A very small number of countries are responsible for transmitting most of the information received worldwide. On a daily basis the United Press International (UPI) and Associated Press (AP) in the United States, TASS in the USSR, Reuters News Service in Great Britain, and Agence France-Presse (AFP) in France, collect, process and disseminate millions of pieces of information to large and small countries. The operations of such organizations are typified by those of AFP.

The French newspaper organization has an information-gathering network of global dimension, with 171 full-time journalists and 1,200 stringers in news production centers in 167 countries and territories, and customers in 152. More than 12,000 newspapers and sixty-nine national news agencies make use of AFP's services.

Information is channeled by mail, radio, telephone, telegraph, or satellite to the central office in Paris where it is then sorted, selected, rewritten and translated into French, Spanish, English and German. Information is then transmitted to customers around the world. Each day roughly 1 million words are sent by cable teleprint, 900,000 by radio teleprinter and 975,000 by satellite; and about fifty photos are distributed by telephoto transmission. Similar global networks exist for other world press agencies and international newspapers, including the *New York Times* and *Christian Science Monitor*.

The leading exporter in television programming is the United States. It produces three times more television programming than that of France, West Germany and the United Kingdom combined. Sales of US television programs to other countries are channeled primarily through multinational corporations, mostly to Latin America, West Europe (notably the United

Kingdom), Canada, Australia and Japan. Most of the United Kingdom's exports go to English-speaking Commonwealth countries around the world (especially Australia and Canada) and to the United States, whereas French television programs are sold primarily to the Middle East, French-speaking Africa and Quebec. West Germany's television programs are distributed about equally to countries in Latin America and Asia. The ratio of programming exports from the developed to the developing countries is about 100:1.

To counteract the global mass-media dominance by a handful of wealthy countries, a number of nations have developed regional television programming. Nordvision serves the Scandinavian countries. An Asian Broadcasting Union has been developed. A French-sponsored network provides programming for France and Canada (Quebec especially), and a similar Spanish network links Spain and Latin America. Political leaders in the developing world have expressed interest in establishing regional news agencies and a 'Third World News Bureau' to reduce perceived Western bias and distortion. This idea comprises part of what is termed the need for a *New World Information and Communication Order* (NWICO). The issues arising from these concerns relate to the protection of a society's cultural development from alien influences, a topic that will be considered later.

The difficulties of maintaining sovereignty over one's culture in an increasingly interdependent world are made even more evident by the rapid escalation of new information and communication technologies; technologies developed and controlled by multinational firms headquartered mostly in North America and Western Europe. These technologies represent integrated networks (cables, satellites and computers) for transmitting written and verbal messages, symbols, pictures and coded data.

New communication technologies

Nearly instantaneous transcontinental and intercontinental relays of information became possible in limited quantities in the late 1800s through the invention and rapid diffusion of cable technology. The present global pattern of oceanic cables reflects the structure of economic linkages among world regions. The most dominant connections are between Western Europe and North America. In contrast, no major cables link Africa with South America, or Australia with South Asia. The high construction costs and limited capacities of cables made them susceptible to competition by satellite systems in the 1970s. Depending on the number of earth stations constructed, the ocean terrain and cable length, cable costs at that time were about seven to ten times that of satellites. However, by the early 1980s, the cost-effectiveness and capacities for transmission over fiber-optic cables had improved greatly. Unlike the first transatlantic telephone cable, which could handle only fifty-six simultaneous conversations, new transatlantic and transpacific glass-fiber cables will allow up to 40,000 simultaneous conversations, with transmissions of 246 million bits per second.

Distortion-free and cost-effective transfers of digital signals make fiber-optic cable transmission ideal for data, facsimile and video. Therefore these systems are expected to expand rapidly in the service of household, business and government needs, at local, national and international levels. A recent example is the New York City Teleport. It is a clustered communication facility that serves local businesses through a satellite earth station, a glass-fiber cable network, and electronic switching equipment.

Although cables and satellites compete for international and transcontinental communication markets, they are becoming increasingly interdependent. The users of satellite transmissions require wired linkages to the earth stations. The principal advantage of satellite systems is that their linkages are not restricted by geographical barriers; they can be economically set up in isolated areas, and they allow many developing countries to bypass the expense of establishing land-based wired networks to connect distant locations. Since satellite networks can be rapidly reconfigured to meet changes in demand or to serve temporary needs, and since they can service mobile units, such as ships at sea, as well as fixed sites, their uses have expanded rapidly.

The most extensive service of international satellite communications is provided by INTELSAT (International Telecommunications Satellite Organization). It started in 1964 as an agreement among eleven countries to establish a global communication system. Today, more than 100 countries have invested in INTELSAT and share access to its satellites through more than 400 earth stations. A similar international service is provided by the INTERSPUTNIK Satellite System for countries allied with the Soviet Union.

Since the launching of the Early Bird Satellite (INTELSAT I) in 1965, the operating costs of successively more advanced generations of satellites have been reduced and their life expectancies have been extended. Whereas the first transatlantic prime-time live color-television program cost $22,350 for one hour, INTELSAT IV-A (launched in 1975) delivered the same service for $5,100 and its lifetime in orbit (seven years) was nearly five times as great. Similarly, the annual charge for a telephone circuit dropped from $64,000 in 1965 to $9,360 in 1980. Furthermore, the volume of information carried by more recent systems is much greater. The INTELSAT I system, with only one satellite and 240 two-way circuits, allowed connections only between two stations at a time. In contrast, the INTELSAT V system (since 1979) has 12,000 circuits and can accommodate multiple access. The more recent INTELSAT VI series of satellites allow for 40,000 telephone channels. Through such advancements, any place that is properly equipped can transmit and receive signals from INTELSAT satellites.

Other improvements that have expanded the use and flexibility of satellite communication systems include the development of smaller-diameter, earth-station reflectors that can be adapted for local uses, the construction of mobile reflectors that can be used in areas of hilly terrain and on large ships. The simultaneous transmission of several messages over the same frequencies has extended greatly the capacities of these systems. Understandably, this powerful technology is providing a basis for establishing domestic communication systems for many countries, some of which could not otherwise afford the more expensive land-based networks.

In recent years many countries have established domestic satellite systems by leasing transmission capacity from INTELSAT. Examples include Argentina, Chile, Denmark, Sudan and Zaire. Others have their own satellite systems – the Soviet Union, Brazil, Australia, Canada (the first country to establish a domestic satellite service in 1972) and Indonesia. At regional levels, many European nations have joined the EUTELSAT system, several Asian countries have obtained service through Indonesia's PALAPA system, and the ARABSAT Satellite System serves several Arab states. Systems to serve Latin America and Africa are in various stages of development.

The principal domestic satellite systems in the United States are Western Union's WESTAR, RCA's SATCOM and the COMSTAR satellites provided by Comsat General Corporation. WESTAR was the country's first private domestic service for telephone and television transmission in 1972, and played a leading role in servicing offshore oil rigs. SATCOM has been important in pay TV transmission and in improving communications to Alaska. COMSAT is used mostly for long-distance telecommunications.

Increasingly, special purpose and privately owned satellites are being launched. NASA's series of Applications Technology Satellites (ATS) have been used to study microwave transmission and to conduct experiments in direct broadcasting of educational programs by satellite to isolated rural areas in Alaska and Appalachia and in several developing countries. Another example is the Satellite Business Systems (SBS), operational since 1981. It was the first system to allow small-dish access to digital transmissions of telephone and telex messages, facsimiles, data and teleconferences. Headquartered in McLean, Virginia, SBS serves businesses throughout the United States. But some business firms have launched their own satellites for internal communications, others have entered the race for the direct broadcasting market, and even religious groups have launched their own satellites (televangelism).

Another major trend in the geography of communications is the increasing use of computers to link and interface various telecommunication systems – visual, audio and verbal. Two examples of computer-based networks, established since 1970, are the National Crime Information Center (NCIC) and EURONET. NCIC, associated with the Federal Bureau of Investigation headquartered in Washington, DC, has a large computer-based file providing information on fingerprints, criminal arrest records, stolen vehicles and missing and wanted persons. The network links major cities and permits almost instant access via on-line data terminals in local law enforcement offices.

EURONET is an on-line information network designed to promote the rapid exchange of scientific, technical, legal and socioeconomic information among public and private users in the twelve European Community member countries. This international telecommunications network uses *packet-switching technology* – that is, telephones linked via computers to teletype terminals directly or via other packet-mode interfaces. Among the databases available via EURONET are computing software for education from a computing center in Manchester, medical library data from Copenhagen and London, physics and astronomy abstracts from Karlsruhe and data on chemical and nuclear research from Paris. Many of the databases available are abstracts in fine arts, geology, medicine, law, ecology, energy, engineering and literature. In order to handle more than 2 million on-line queries per year, EURONET has designed a system that permits easy interactive retrieval in a simple command language that draws on multilingual thesauri and automatic translating systems.

The development of computer-integrated communication systems has the potential for influencing life at all levels of human organization. Aside from international and national level information systems, their uses are being extended to the household and individual levels. Technological advances in hardware miniaturization and growing efficiencies in low-cost production have facilitated greater accessibility to such communication capabilities. A wide variety of information is being made accessible by telephone, television and computer through interactive systems that link households, businesses, multinational firms, governments, voluntary organizations and international agencies into an integrated network. It is expected that it will be years before the world's dominantly analog-based telecommunication networks can be replaced with the required digital systems. Nonetheless, in only two decades, satellite, communication and computer technologies have contributed to a rapidly increasing consciousness of global interdependence.

Possible futures in telecommunications and information processing are listed in Table 3.3. Some of these have potentials for evil as well as good purposes. Therefore the monitoring of computer and satellite futures will merit the attention of government, consumer groups and individuals in both the developed and developing world. It could be argued that the futures of television, telephone, computer and satellites, and especially their integration, are as important to understanding a society's economic and political futures as are breakthroughs in agriculture and industry. The spatial dimensions of change in telecommunications and information handling are significant in both the futures of developed and developing countries. However, even in a world of increasing opportunities for electronic exchange, there are needs for direct person-to-person contact.

Table 3.3 Advances in telecommunications and information processing.

extensive national and international interactive systems
greater use of human–machine and machine–machine systems
growth of electronic data services
a cashless and credit-card oriented society
international phone directories and direct-dialing anywhere
global toll-free information numbers
shopping, banking and voting by phone
microsized computers, televisions, radios and telephones (cigar-size)
nonhookup phone contact (to person, not location)
teleconferencing, telemedicine and telerecreation
computer communications
electronic mail service and electronic journalism
global multiple television channel selection
'seeing eye' satellites and remotely piloted vehicles
automatic translation devices on television

There seems little doubt that increasing levels of trade and tourism, exchanges of cultural events, and sporting competitions have all expanded the consciousness of global interdependence. An important example is the growing significance of international conferences.

World conferences

A distinct feature of international cooperation and decision making that has resulted from a consciousness of global interdependence is the growing number of nations participating in world conferences. These began in earnest with assemblies of the United Nations and affiliated organizations, such as the World Health Organization and the Food and Agricultural Organization. These conferences have discussed political problems and pressing social concerns. In the last two decades, there have been world conferences convened on the environment, population, food, housing, oceans, women, settlement, trade and water. Nearly 135 of the possible 160 nations that could participate were represented at most conferences. Among the participants were government leaders, planners, scholars and heads of non-governmental organizations.

The sites for these world conferences reveal both their global nature and the growing recognition of developing countries and their roles in resolving international problems. Earlier meetings on the state of world affairs usually took place in European or North American capitals. New York, London, Paris, Rome and Geneva were recognized as world capitals and as the most logical centers for conferences. But, since newly independent nations now comprise a sizable proportion of membership in the United Nations and in other international organizations, decisions to hold meetings in major cities of the developing world attest to their importance in world discussions. Recent conferences sponsored by the United Nations have been held in Mexico City, Caracas and Nairobi. These are now recognized as world capitals along with Beijing, Brasilia, Lagos, New Delhi, Singapore and Cairo.

When all nations, or at least those agreeing to participate, convene at a single site to discuss a specific problem, the site might be considered a *world city*. It is a 'world' in that the delegates meet at one location. Participants from nations of varying size, economic potential, military strength, social development and cultural complexity converge at that one city, be it New York for the annual United Nations General Assembly or Caracas for a Law of the Sea Conference. Representatives from 130 to 150 nations sit in the same room, listen to supporting and opposing views through simultaneous translation, and strive to resolve problems they face, individually and collectively. Linguistic, cultural, religious, economic and political barriers that may be strong outside the 'city' may be diminished or become nonexistent in this setting. Officials and planners from neighboring nations and from countries that are remote from each other may discuss similar or different perspectives of the problems that brought them together. These miniworld conferences by their very nature have not always resulted in agreement on resolutions, but they do promote a consciousness of global interdependence.

Summary: evidence of global interdependence

The evidence of increasing global interdependence has emphasized the importance of exchanges among world regions. Specific consideration was given to information flows, mass media, new technologies in communications and conferences. Of course, a more exhaustive treatment would consider, also, world trade, tourism, technological assistance, cultural exchanges, sport competitions, and so forth. All of these play important roles in defining the patterns of world development. But, at this juncture, it is useful to reflect briefly on a few of the explicit and possible consequences of such interdependence.

Consequences of a shrinking world

The opportunities that space-adjusting technologies have provided relate, in part, to a significant restructuring of our thoughts about the proportion of the earth that we regard to be within our spheres of interest and action. However, this expanded access to the world has also posed serious difficulties for coping with 'overexposure' to vast amounts of information and to diverse human cultures and life styles.

Expanding spheres of interest

Advances in communication, through satellites primarily, have brought individuals, businesses and governments in cities throughout the world into more direct contact. When all places can be reached from all other places in as little as five minutes (telephonically), we will have witnessed the emergence of what Marshall McLuhan (1964) described as a *global village*. Physical distance would cease to be a barrier to contact. In some circles – namely international business and government – this global or 'wired village' is already a reality. It has altered the interpretation of what may be meant by the term 'community'.

Communities historically were groups of individuals living in close physical proximity who shared similar social values, religious beliefs and cultural heritages. Commonly shared social spaces (territories) were an important part of the cohesiveness that held groups together. With high levels of mobility, life-cycle changes, easy long-distance travel, instant phone contact to anywhere and extensive television coverage about national and global events, communities are forming that possess less and less spatial cohesion. Groups of artists may define their community by the contact they maintain through annual conferences, publications and telephone. Their sense of community derives from professional (and possibly personal) shared values, allegiances and commitments to pursuing a particular craft or design. Professional communities of research scientists and writers are additional examples of *noncontiguous communities*; those who contribute to research at international levels are examples of *global communities*. Although extensive travels, communication links throughout the world and changes in one's business address may contribute to expanded geographical spheres of interest, such personal contact is no longer as essential in establishing consciousness of the world beyond one's local region.

It could be argued that there is, today, less need to travel to learn about other environments, culture and major problems. Often, accurate and crisp accounts may be gained about major international issues by reading international newspapers and magazines and by watching special lectures and documentaries. Even within one's own society, public opinion polls, editorial analysis, columns by regional journalists and the positions advocated by consumer groups provide interpretations to current issues. In the information-oriented societies of the post-industrial era, magazines from anywhere can be delivered to one's address, information on business opportunities and legal advice is available by telephone, and even college credits can be earned via television programming without leaving the home. In essence, it is possible for the world, with its vast problems and great diversity, to collapse or coalesce upon each and every individual who shares in its technologies of communication.

Overexposure to the world

Each day millions of pieces of information are carried to individuals, organizations, governments and offices through the mails, by radio and television, and via newspapers and magazines. The volume and diversity originating in and received by the more than 160 countries during a day or week is more than one individual can absorb. Often a situation of *information overload* develops where individuals, firms, agencies and governments are overwhelmed by its sheer quantity and are unable to make effective decisions with any sense of confidence. A recent example was the 'Black Monday' stock-market crash on 19 October 1987. Such problems may stem from either too much of the wrong kinds of information or too little time to evaluate carefully relevant information. Much information that reaches a destination may only be a fraction of total truth about the event or the change in a situation; yet, on that basis alone, often key decisions are made. Instant reaction is often required in response to changes in government leadership or military skirmishes, to executive kidnapping, or to oil price increases. During crises there is often little time to filter out the sources of the problems, its ramifications or even the reliability of the informant, individual reporter, or the office that relayed the report. The result is, frequently, that errors are made in diplomacy, stock investments, and so forth because too little of the appropriate information was provided or too little time was available for suitable analysis to take place. Time, often an essential element in making wise decisions, is not always available in sufficient amounts for governments, corporations and individuals.

For people who are experienced in receiving, assembling and digesting large amounts of information, an overload may not exist. Also, those who choose to ignore the increasing volume of information may not be concerned about their inability to grasp what is happening in the world around them. However, information reports on some events may be too difficult to ignore, as for example reports about political conflict, environmental disasters and commodity price hikes. Because of the great number of such events, reports on some issues may occupy only a few seconds of the evening news. The fact that some actions (e.g. kidnapping, violent acts and major policy positions) are reported but not analyzed often tends to weigh the events incorrectly or out of context.

Individuals who have had lengthy formal educations and who are regular observers of television documentaries and readers of global newspapers may not be likely to make snap judgments on critical issues that have been reported in piecemeal fashion. But, for those who are less well versed with events, information overload may lead to confusion. Thus, traditional rural areas of the developing world have had recent and substantial increases in information levels. In some instances their proliferation of communication contacts with other 'worlds' has ushered in more changes and ideas more quickly than could be easily absorbed.

The reaction by individuals and groups experiencing overload incude (1) becoming comfortable with what is occurring, (2) refusing to comprehend what is going on because 'things' are so muddled and complicated, and (3) resigning oneself from 'the world' voluntarily and permitting others to study and resolve world problems. Frustration, boredom and general cynicism about individual leaders and politics also represent ways of coping with information overload. In addition, these reactions characterize the responses to what Alvin Toffler (1970) called *Future Shock*.

Future shock is defined as a premature arrival of the future or as a situation where too much happens too fast for the person to be able to sort out effectively the changes occurring. The shock is a direct result of an evolving globally interdependent system, of rapid scientific and techno-logical changes and of time–space convergence.

The very rapidity of such change is seen by Toffler to encourage the ad hoc making of decisions, changes in life styles, new social mores, shifting national political allegiances, and increasing scales of international trans-actions – all taking place so fast and so frequently that individuals cannot comprehend what is happening.

The problems of information overload and future shock must be bal-anced against the many advantages that can be realized through greater interconnections of people and places. However, although an increasing proportion of the human race is sharing in both the problems and potentials of a shrinking world, most of the world's population remains in relative isolation.

Underexposure to the world

Many people do not share in the potentials of modern space-adjusting technologies. Some have deliberately and consciously chosen to maintain a slower pace and to focus inward on their own group needs. Examples include such religious communities as the Old Order Mennonite and Amish populations in North America. Even nations have attempted to selectively disengage from certain forms of global interdependence. For instance, Albania has isolated herself from the market economies and from many of the centrally planned socialist countries. However, apart from deliberate choice, many people in the rural areas of the developing world can travel no faster than what their own legs will allow. Many regions are not served by modern transportation and communication systems. Thus, while progress in transportation and communication is occurring in many parts of the world, it is not a uniform process. That is, time–space convergence is not universally applicable to everyone. *Nonshrinking societies* are characteristic especially of rural parts in the developing world.

Figure 3.5 attempts to illustrate the contrasting convergence processes within urban systems of the developed and developing worlds. The con-vergence of settlements of different sizes are shown for three hypothetical

Developed World   Developing World

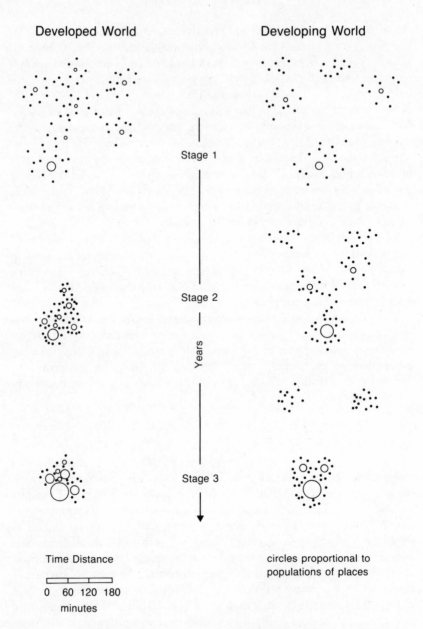

Stage 1

Stage 2

Years

Stage 3

Time Distance   circles proportional to
                                   populations of places
⌐──┬──┬──┐
0   60 120 180
   minutes

Figure 3.5 Sequential patterns of time–distance relationships among urban
centers in the developed and developing worlds

periods of time. For the developed society, characteristic of Japan, Western Europe and other highly industrial regions, the gradual convergence of places may be attributed to the introduction of faster transport systems. By stage three almost all places can be reached from any other place within one hour of travel time. This compares to five-hour journeys between some of the places in stage one.

The settlement system of the developing society depicted in Figure 3.5 has one major population center, two smaller ones and several dispersed rural places. Convergence between the urban centers has occurred between the stages, but not as rapidly as in the developed economy. The difference may be due to a lack of financial resources for modern transport technologies (new aircraft, new automobiles and faster trains) and for a diversified transport system. For this reason only those rapidly growing major cities show pronounced patterns of convergence with other places. In contrast, some rural places are as far apart in time–distance in stage three as in the first stage.

Rural residents, who are often even unaware of the transportation advances that serve large urban centers, may be as remote from the world today as their ancestors were centuries ago. Parts of West Africa, the Middle East and Southeast Asia are among those areas where sharp contrasts exist between accessibility advantages of rural and urban dwellers. Often those in capital cities discover they can travel faster and easier to cities several hundred miles away than rural residents who may be linked with the 'outside world' via the same trade routes used two or three centuries ago. For example, it is easier to travel between Lagos, Accra and Abidjan in Nigeria, Ghana and the Ivory Coast than to smaller cities in the northern parts of each country. Cairo, Benghazi and Algiers have easier and faster access to Europe by air than to small cities and villages in central and southern Egypt, Libya and Algeria. However, there are many important examples of where traditional ground-based communication systems are being introduced and where they are being supplemented with new technologies.

Examples of the expansion in important ground-based information systems include the extension of postal services, the construction of new clinics and libraries and more widespread use of mobile health, education and entertainment facilities. For many people, these systems offer their first regular access to such basic services. The Indian government has an extensive program of mobile film units visiting small cities and towns as well as rural villages. Mobile health and bookmobile units in a number of African countries visit on a regular or semiregular basis areas that cannot afford clinics, physicians and libraries. Often these mobile units are the major source of information about remedies for local health problems or about the outside world. Where human and financial resources permit, governments, often with the aid of missionary groups, have established more permanent systems.

Some developing countries, such as Guinea, have embarked on programs to provide radio, telephone and television service to small cities and towns. In 1978 Guinea contracted a Japanese firm to construct a 710 mile (1,135 kilometer) thirty-six station radio communications network linking the major cities and towns, and providing 960 telephone channels and a television channel. This network forms part of a pan-African telecommunication network (PANAFTEL) that is being developed to provide improved communications between the cities and countries in West and Central Africa. Since the mid-seventies, the PANAFTEL network expanded from 230 radio circuits to more than 15,000 circuits by 1986 (Butler, 1986). Presently nine PANAFTEL countries have domestic satellite networks with 120 earth stations, and forty-eight of fifty countries operate earth stations for communication among countries of the region. In addition, forty-two of these countries had international automatic telephone exchanges by 1986. Similar network expansions are underway elsewhere in Africa, Central America and the Middle East.

Many governments view improved telecommunications and rapid dissemination of printed materials as fundamental requirements in their efforts to increase literacy and to enhance the development potential of their nations. But in spite of their efforts, significant gaps characterize the information available to peoples of the developed and developing worlds.

A cross-section of the information sources available to developed and developing countries illustrates important differences in such key indicators as daily newspaper circulation, proportions of populations with radio or television receivers and annual visits to movie theaters (Table 3.4). In some countries there is only one daily newspaper (e.g., Benin, Barbados and Fiji). A number of poor developing countries (including Swaziland, St Vincent and Tonga) have fewer than ten cinemas. Of course, for most countries, these information sources are found in urban centers more than in rural areas, and this discrepancy contributes to regional disparities within countries that are often as great as those between nations. In the developing world, they may even contribute to the flood of migrants from rural to urban centers.

The impact of providing equal information and services to rural areas may help to stem the tide of rural migrants to cities of the developing world. Prospective migrants may recognize there is less reason to 'go' to the national capital or to secondary urban centers to obtain what they want if satellites provide education, health and agriculture information in one's home village. The village's 'dish', that brings in information from national or international sources via satellite, is a further indication that places do not have to be 'wired' to receive vital information and services. In this regard developing countries may be able to bypass the necessity and expense of establishing wired networks (telegraph, telephone and television cables). Thus, with greater ease than with transportation, the new technologies in communications offer significant hope that current disparities in access to

Table 3.4 Information resources for selected countries.

	Postal service[1] (persons per office) 1983	Telephone[1] (persons per receiver) 1983	Daily news-papers[2] (circulation per 1,000 population) 1984	Television[2] (receivers per 1,000 population) 1983	Film[2] (annual visits per capita) 1983
Egypt	6,696	85.6^e	78	44	0.9
Ethiopia	89,029	408.1	1	1.2	—^{f}
Nigeria	28,860	129.0	6	5	0.1^d
Canada	3,025	1.8	220	481	—^{f}
USA	5,743	1.5	268	790	5.1
Mexico	5,087^a	12.8	120	108	2.8
Panama	10,671	10.3	78^b	122	—^{f}
Bolivia	11,572^a	29.7	50	64	5.7^b
Brazil	17,455	13.9	46	127	—^{f}
Afghanistan	36,447^a	516.1^d	5	3	—^{f}
India	4,815	228.5	20	4	6.6
Iran	11,608	19.5	22	55	—^{f}
Japan	5,113	1.9	562	563	1.3^d
Philippines	24,303^a	79.1	22	26	—^{f}
Saudi Arabia	16,048	8.2	12	264	—^{f}
USSR	2,951^a	10.2	422	308	14.9
Belgium	5,309	2.5	223	303	2.1^e
Hungary	3,322	8.0	254	371	6.5
Italy	3,993	2.6	82	385^c	2.9
United Kingdom	2,556	2.0	414	479	1.1
Australia	2,938	1.9	337	429	—

Sources: [1] *Britannica World Data*, 1986, pp. 898–903; [2] UNESCO, 1986, Table 7.19, VII 138–40; Table 10.4, X, 22–6; Table 9.3, IX, 12.17.

Notes: ^{a} 1978, ^{b} 1979, ^{c} 1980, ^{d} 1981, ^{e} 1982, ^{f} not available.

vital information, both within and between nations, may be reduced. However, it is difficult to know at this stage if such a development would contribute to a more long-term balance in the world's distributions of people and resources.

Of course, isolation from the global village may occur for reasons other than a lack of access to transportation facilities. As noted previously, communication technologies now have the potential for reaching even the most remote and the poorest of regions. However, some societies have barred certain kinds of information from the air waves, sometimes filtering out reports that reflect poorly on their society, that contradict official policies, or that are judged to be detrimental to the cultural development of their people. This type of imposed isolation through the manipulation of information may be seen as beneficial if it allows for a more gradual transition in the exposure to ways of life that are very alien to an isolated group. But, on the other hand, it could contribute to rather biased and uninformed judgments that may be detrimental to a community, whether it be local or global. These situations relate to important issues concerning the degrees of control that local regions should have over their own destinies, the appropriate geographic scales of economic and political organization, and the degree to which regions should share in the global resource base. In part, these issues relate to the ways in which the principal decision-making organizations have adapted their structures and functions to the emergent patterns of global interdependence.

Organizational response to a shrinking world

The consequences of interdependence cannot be understood apart from an awareness of how it is related to patterns of economic and political organization. Increasing levels of interdependence have facilitated the development of new ways for organizing human and material resources. In general, these changes have tended towards the increasing size, complexity and degree of specialization among organizational units, and towards an increase in the geographical range of their operations. These responses have corresponded to the general declining significance of distance as a barrier to human interaction and they have provided opportunities for fundamental changes in organizational patterns and principles. Examples include extensions of market areas, exploitation of high-quality resources at more distant locations, drawing upon larger labor pools and engaging in more specialized divisions of labor. These kinds of changes have been inherent to an industrialization process that has emphasized the importance of bigness and the economies of large-scale operations. These transitions may be seen, in part, as a search for optimal forms of organization in terms of the size and number of operational units and of the linkages among them. However, other viewpoints disagree with the desirableness of such trends. Bigness and global dominance in the hands of a few nations and multinational firms are not regarded as optimal by those nations that are seeking to establish markets for their new industries or that lack resources to meet their own needs. In addition, some see elements of vulnerability built into present-day, large-scale and global patterns of organization.

The late British economist, E. F. Schumacher (1974), proposed an alternative to complete reliance on bigness. His *Small Is Beautiful* emphasized the need to incorporate and to nurture a 'sense of community' at more local levels. Regional diversity of enterprises, as opposed to large-scale regional specialization, would permit more diverse forms of production and services, greater degrees of self-sufficiency and less reliance on a globally based economic system. Hence the interdependence of enterprises within the region would help to form stronger bonds of internal structure. Such a pattern is possibly more congruent with the needs of many new nations than is the present system, which subjects them to intense competition from First World countries and their corporate establishments. Another argument invoked by supporters of this view is that large-scale systems are more vulnerable to the effects of human-made and natural disasters and are less capable of withstanding and surviving major changes in their operational environments. Such changes might result from political unrest, terrorist activities, global warfare, large-scale crop disease, or from a collapse in the global economy. If vulnerability to such changes depends on the level of complexity in the system of organization, then the global functional interdependence among the large-scale and specialized regions means that the entire system may be threatened by the disruption of activities in any single region or in the transportation and communication links among them.

Aside from vulnerability to unforeseen calamities, current forms of economic organization have contributed to polarization in the degrees of regional development, both within nations and at the global scale. In part, this is related to the dominant control of world resources in the hands of a few nations and their multinational firms, and to the concentration of this control within urban core regions. These few areas have been able to consolidate most of the potentials of global interdependence to their own advantage. It is not surprising, therefore, that many new and resource-poor nations view any trends toward global interdependence as threats to their own cultural development and, given their lack of access to modern technologies, as constraints to achieving an equitable distribution of global resources.

Media imperialism and cultural sovereignty

Amidst the onslaught of new technologies and economies, many regional and localized cultures throughout the world face not only problems of future shock but problems of maintaining their cultural identity. This problem is of particular significance for rural-based cultures and for societies who have yet fully to embrace the Industrial Revolution. The achievements of widespread radio and television communication systems have meant exposure to outside values, and alternative ways of life. Many national leaders view 'global culture as Western culture', and the leaders of many Western countries view 'global culture as American culture'. Some see communication (particularly mass-media) technologies as threats to their own sense of identity. From this fear stems demands for a *New World Information and Communication Order* (NWICO), one that provides satisfactory coverage of events in all countries, that is not biased in the service of major Western nations, and that is sensitive to the desires of leaders in developing countries to exercise some control over both the systems and the contents of communication media.

According to the 1980 report of the International Commission for the Study of Communication Problems (MacBride, 1980) the existing information-communication order is controlled by news agencies and film producers of the West, operated mostly in the interest of Western customers, such as multinational firms and powerful nation states, and expresses mostly the values and attitudes of post-industrial and urbanized Western cultures. In essence, the current global information-communication order is seen as a principal diffusionary agent that serves the market and political needs of the Western world. It is viewed as a form of media or cultural imperialism that threatens the cultural integrity of many groups and nations. Thus it is in this context that Western dominance over television programming, film making and distribution must be evaluated and that recent advances in satellite broadcasting technologies must be considered. In addition, it is useful to focus on the control over the content of these media and on the censorship of their dispersal.

The impress of globally pervasive cultures

Only a few Western nations dominate the production and dissemination of information and entertainment media for world audiences. Many developing countries, eager to develop a daily television programming schedule, find it convenient to incorporate the foreign programs. Even if subtitles and dubbing are needed, they are cheaper to use than producing their own. It is not surprising, therefore, that more then 60 percent of the television hours in such countries as Nigeria, New Zealand, Iceland, Malaysia and Guatemala are imported. Most of this programming comes from the United States, the United Kingdom and from a few other West European nations. The scale of mass-media exports from these countries has reached such proportions that questions are raised about what impact such American-produced programs as *Peyton Place*, *Dr Kildare*, and *Legend of Jesse James* will have on those watching them, particularly if the scenes portrayed and messages presented are incongruent with their own experiences and expectations. *Sesame Street*, which began in the United States in the late 1960s, is seen in more than 80 countries and in many languages. But is it as relevant to the needs of children in poorer developing countries and for rural children as it is for the children in urban America?

Series available for international distribution have included *Peyton Place* with 514 episodes (enough for almost ten years of half-hour weekly shows) and *The Virginian*, with 373 hours. *Rin Tin Tin*, *I Love Lucy*, *What's My Line*, *Perry Mason*, *Kojak*, *Six Million Dollar Man* and *Dallas* are other major exports with dubbed dialogues. Concerns of alarm and outrage have been expressed in some countries by critics who question the effects they will have on personal and societal development. For example, how do recent migrants to the cities of West Africa or South Asia identify with the often all-white, male-dominated, violence-prone, cowboy image so often depicted in American-produced movies and television series? From watching many all-white, pre-1960 television programs and movies, it may not be difficult for African viewers to assume that blacks are almost non-existent in the United States. Furthermore, what attitudes to British Commonwealth members in the Caribbean, Southeast Asia and East Africa develop about servants and upper classes from watching the British-produced adult drama *Upstairs-Downstairs* that is shown in forty-five countries? And how do people in the nearly four dozen importing countries react to showings of *Monty Python's Flying Circus*, or *The Forsythe Saga*?

Gatekeepers

A useful construct with which to examine questions concerning the information flows of mass media is the *gatekeeper model*. According to this model there are editors, film or television producers, and ministries who are responsible for making decisions about what information, garnered by reporters and film crews, should reach the public. Depending on the values, tastes and biases of individuals who are in these information decision-making capacities, and on the degree of openness tolerated by a society, the amount and kinds of information that reach the masses may be relatively complete and accurate or censored. The gatekeepers are in powerful positions to determine what information will reach readers and viewers. In closed societies the gatekeeper may discourage coverage of embarrassing national and international issues or block competitive groups from airing their views. Heavy *information filters* have existed in a number of Middle East countries as well as in the Soviet Union, East Europe and China. However, even in the more open societies of West Europe, Australia, New Zealand, Japan, Canada and the United States, gatekeepers are influenced by subscribers, advertisers, critics and media ratings.

Satellite broadcasting

The issue of control over media dissemination has taken on heightened interest with the increasingly widespread use of satellite communication systems and with their improved technical capacities. Of particular concern is the development of direct television broadcasting from satellite to household receivers.

Broadcasting services are now available through community receivers in many parts of the world and even direct household reception is a reality in some regions. Since such systems have the potential to bypass state (gatekeeper) controls over the content of incoming communications, their political implications have evoked heated debates before international forums. Nor surprisingly, the arguments for and against the establishment of international controls over direct broadcasting reflect a polarization between countries of the developing world and the most technologically advanced nations. What are their arguments?

Direct broadcasting via satellite offers considerable economic benefits to the United States. It owns the largest share of the INTELSAT system, dominates the markets for communication hardware and software services, is the largest user of international telecommunications and has significant control over the global dissemination of news releases and films. Direct broadcasting would permit penetration of new and expanded global markets; it would encourage the obliteration of language barriers and promote the standardization of global aspirations for its products. It can be argued that it needs a truly global communication capability in order to sustain the growth of its enormous productive capacities.

In keeping with its competitive ethic, the United States advocates the affirmation of a free-market approach to communication development and, in keeping with its democratic ideals, for the unimpeded free flow of information. However, the nations of the developing world, and even some developed nations, see things differently. Their position stems from a sense of insecurity with a technology that is beyond their control and from a sense of impotence in being unable to exploit its advantages on their own terms.

An efficient domestic communication mode for direct broadcasting is of more immediate relevance to a developing country than is a fully integrated global system. It could be used to promote national cohesion and to expedite national objectives for improving education, health, food production and population control. Satellite systems offer a quicker, more economical and simpler means of achieving these communication capabilities than do terrestrially based networks – particularly for large countries such as Brazil and for fragmented nations such as Indonesia. But the structural reality of a communication system begs for messages to transmit and programs to broadcast. This is where broadcasting and commercial agents from the United States and from a few other countries are eager to help out. This is the kind of help that developing nations need but fear. Will Western-produced news releases and films promote attitudes and opinions contrary to, and incompatible with, their own cultural values and national policies? Will reliance on other countries impede the development of indigenous skills for educational and entertainment programming? Will the lure of Western commercialism undermine their local consumer industries and entice the movement of scarce funds abroad? Will they become unwilling receptors of propaganda warfare between the superpowers and victims of internal interference by other nations? The essential issue is one of uncertainty as to whose ideas and ideals will be promoted to which audiences and for what purposes.

The reconciliation of two seemingly incompatible principles – the free flow of information and respect for the cultural sovereignty of others – will not be easy. The issue is analogous to free trade versus protectionism; but in this case the transfers of mental images rather than goods are in dispute. The established nations did not face this threat in their formative years. Yet nations less than three decades old, many of them facing severe social and economic handicaps, are beset with a technological presence which, while promising to be an effective weapon against illiteracy and poverty, threatens their sense of control over their own cultural development.

The communication media, now buttressed through powerful satellite systems, provide a basis for people throughout the world to be exposed to similar ways of thinking and acting. In combination with the acculturative and assimilative roles of urbanization and the mixing of peoples through migration, the provision of global forums, and the increasing concentration of economic power in the hands of multinational firms, the thrust of cultural convergence must be reckoned with as a powerful force. Equally powerful gatekeeper roles would be required if those concerned with protecting their present cultural systems are to succeed in bucking this pervasive influence.

Conclusion

This chapter has outlined some of the consequences that stem from the rapid introduction of space-adjusting technologies. Transportation and communication innovations have promoted time–space convergence and have allowed opportunities for fundamental changes in the ways that people perceive, organize and use both human and physical resources. A principal outcome has been the increase in economic and political interdependence of people, firms, agencies and nations at regional and global levels. This has brought with it possibilities for more diverse and richer lifestyles; however, the impacts have not been distributed equally among the world's people and places. National core regions have profited more than peripheral areas; global core regions in the developed world have garnered benefits at the expense of former colonies and newly developing nations.

Levels of global transactions have expanded greatly. Postal services, telephone links, film distribution, mass media and news services have all contributed to a growing sense of consciousness of global issues. However, these same advances have reinforced older patterns of dependence of small and resource-poor nations on larger and wealthier countries. These same technologies have also facilitated more intense competitive struggles in the quest for resources, territory and political power. The ideologies of the North are widely dispersed in developing countries that find themselves caught in a power struggle and beset with images of a world that shows little resemblance to the realities of life in their countries.

On the optimistic side, the technological and institutional changes of recent decades may yield a convergence of opportunities between the developed and developing worlds. They could help to free the innovative potentials of peoples in all parts of the world to solve their basic problems and to share more fully in the benefits of life. The diversity of human cultures offers important opportunities to share experiences based on differences in values, technologies and customs. Although such diversity has generated difficulties in our human relationships, it also constitutes a valuable global resource that can contribute to improving the quality of human life and the chances for survival. Examples follow.

Throughout history, advances in agriculture, literature and music, transport technology, weaponry, printing, medicine, chemistry, and many other fields have come from different cultures. The concentration of early agricultural advances in the Middle East and industrial technologies in Europe and North America are cases in point. However, these advances have also been assisted by the influences of innovations from other parts of the world. And with more rapid and easier communications on a global scale, coupled with higher literacy rates in developing countries, the innovations that will be significant in humankind's future advancement will come increasingly from a more geographically dispersed set of innovation centers.

Advances in space and communication technologies will likely come primarily from the developed nations that can now afford the funding of such projects. But, also, the developing countries can be expected to increase their innovative contributions to solving world problems. Already they have contributed to the production of new crop hybrids and livestock breeds for tropical areas and to devising small-scale and energy-efficient farm and industry technology. Medical advances will likely come from all societies, not only from those with heavy investment in the chemistry of the mind and body (rich countries). In some non-Western societies, the psychological and social facets of physical and mental health have been considered and treated as inseparable for centuries. In this latter group are cultures where shamanism, witchcraft and voodooism have contributed remedies unusual to Europeanized cultures. Yet, they may be found to have some use in medical advances for societies throughout the world.

The wider dispersal of innovation origins among and by peoples of different backgrounds helps to expand the human capabilities for coping with many current and future problems. In this sense, the diversity of cultures allows for a wider variety of approaches to solving major problems. The global diffusion of space-adjusting technologies may help to speed the realization of the intrinsic value associated with having a world based on different (but convergent) paths of cultural development.

Although there is considerable basis for optimism in the use of new technologies for sharing ideas on a global scale, the world today remains split in a most fundamental way between the 'haves' and the 'have nots'. This gap above all others represents possibly the most basic division of humankind. More than language, religion or nationality, it reflects two very different ways of life – a 'culture of plenty' and a 'culture of poverty'. The increasing global interdependence of a shrinking world makes such gaps more apparent and more and more intolerable. New technologies provide means for uniting the human family in a commitment to eradicating this gap. Yet they also provide a basis whereby the rich societies can enhance their dominance over the poor. The nature of this gap and the prospects for its destruction deserve close analysis.

References

Abler, Ronald (1975), 'Effects of space-adjusting technologies on the human geography of the future', in Ronald Abler, Donald Janelle, Allen Philbrick, and John Sommer (eds), *Human Geography in a Shrinking World* (North Scituate, Mass.: Duxbury Press), pp. 35–56.

Bleazard, G. B. (1985), *Introducing Satellite Communications* (Manchester: The National Computer Centre).

Britannica World Data (1986), 'Comparative national statistics' (Chicago: Encyclopedia Britannica), pp. 898–903.

Butler, R. E. (1986), 'Telecommunications for development: the ITU contribution', *Development* (quarterly of the Canadian International Development Agency), summer–spring, pp. 48–50.

Culbert, Sidney S. (1987), 'The principal languages of the World', in *The World Almanac and Book of Facts 1987* (New York: Pharos Books), p. 216.

Gerbner, George and Marsha Siefert (eds) (1984), *World Communications: A Handbook* (New York: Longman).

Janelle, Donald (1968), 'Central place development in a time–space framework', *The Professional Geographer*, 20, pp. 5–10.

MacBride, Sean (1980), *Many Voices, One World, Report by the International Commission for the Study of Communication Problems* (New York: UNESCO).

McLuhan, Marshall (1964), *Understanding Media* (New York: McGraw-Hill).

Oettinger, Anthony G. (1980), 'Information resources: knowledge and power in the 21st century', *Science*, 209, pp. 191–8.

Pelton, John N. (1981), *Global Talk: The Marriage of the Computer, World Communications and Man* (Brighton, UK: Harvester Press).

Pool, Ithiel de Sola (1983), 'Tracking the flow of information', *Science*, 221, pp. 609–13.

Schumacher, E. F. (1974), *Small is Beautiful* (London: Sphere Books).

Smith, Anthony (1980), *The Geopolitics of Information: How Western Culture Dominates the World* (New York: Oxford University Press).

Toffler, Alvin (1970), *Future Shock* (New York: Random House).

UNESCO (1986), *Statistical Yearbook* (Paris: UNESCO).

4 Global development of communication: a frame for the pattern of localization in a small industrialized country

ÅKE FORSSTRÖM & STEN LORENTZON

Introduction

The purpose of this chapter is to depict some relations between international premisses and the development of the location pattern in a small industrialized country. These premisses are supplemented by the restrictions inherent in the national pattern and trends. The case is Sweden, but there are a number of countries in the same position, especially in Europe.

Applications of new information technologies such as private and public networks of datacommunication, teleconferencing and mobile telephone are being introduced. The expanded use of these teleservices is based on digitized telecommunications networks with highly increased capacity. At the same time conventional technology in various fields of transportation is being renewed. Increased passenger capacity and shorter time distances are changing the transport situation in general and in Europe in particular.

These technology-based changes are occurring in Europe at the same time as its internal trade is growing. The goal of the European Community is to have a completely free market of goods and services, capital and labor by the year 1992. The transnational corporations are growing by buying companies in Europe and in North America.

This paper aims to analyse significant innovations and developments in the fields of communication and transport at the global level, in Europe and in Sweden. These innovations are examined in the light of the theory of city-system growth.

The present and foreseeable changes in various communication systems will influence accessibility in the European city-system. Some important issues are:

● the accessibility level is increasing all over the continent,

- there is a growing importance of the central part of the European city-system and
- a focus is developing in the center of this system.

It is suggested that this development is a major factor in the corporative behavior of location. Power, influence and decision making are more concentrated in the core regions in Europe. From the point of view of a peripheral industrialized country these events seem to be both inevitable and avoidable and call for a new national strategy in economic affairs. Commitment will be needed if employment opportunities and economic wealth are to be preserved. One approach is to observe developments in the field of information technology. These developments shape other possibilities of managing existing activities of transportation and organizations in general as an alternative to investments in new transportation systems.

Theoretical base

The traditional view of city-systems, the central place theory, admits only hierarchical and symmetrical relationships between cities of different orders. Thus large-city interdependencies and those between subsystems are impossible, which is contradictory to both the development of communication technology and real world trade and transport experiences. The processes of city-system growth are regarded as following existing channels of communication and trade, consisting of constantly repeated economic linkages of a feedback character.

In an advanced economy it is likely that multilocational organizations control a sizeable share of the economy and that headquarters are situated in large cities, which are tied up in complex organizational structures. New investments anywhere are very likely to start multiplier effects via intra- or interorganizational interdependencies in some large metropolitan complexes. Most often these effects appear as expansion of high-level administrative units. These units are located in metropolitan areas which offer face-to-face contacts, business service availability and high accessibility.

These linkages stem from enlarged high-level administrative units which increase the demand for certain services, especially transport and communication services. When demand exceeds supply the capacity of these services is raised. The circulation of 'specialized information' concerning locational decisions is increased both by the enlarged units and by the increased capacity. As a result the attractiveness of new administrative units is raised (Pred, 1977).

The theory emphasizes three advantages concerning the circulation of specialized information in metropolitan city-complexes. They offer better *intra-* and *interorganizational contact, business-service availability* and *inter-metropolitan accessibility* than ordinary cities. Another significant feature is that these advantages tend to be strengthened in each large metropolitan complex (Hägerstrand, 1970; Törnqvist, 1970).

The concept of 'specialized information' involves all matters which influence explicit and implicit decisions concerning city-system growth and development. It covers every decision to establish or expand a job-providing facility on the one hand, or to buy, sell or otherwise spend money or allocate capital. The function of implicit locational decisions is twofold. The aggregated result of these decisions influences city-growth in a direct way and indirectly by contributing to the metropolitan-biased circulation and availability of specialized information.

A city-system has, according to this view of biased availability, certain properties regulating its possibilities to achieve this kind of information. It is suggested that each system's internal dependencies, interactions and its degree of openness or closure are such properties (Pred, 1977).

In this paper the development of increased *intermetropolitan accessibility* in Europe and information technology implications for better organizational contacts is examined. The city-systems of a peripheral country are described. Their internal dependencies and degree of openness are studied.

New global communication and transport development

High-speed train facilities and projects

In the field of public transportation in Europe a lot of money is spent on R&D. Intra-city transport systems are being developed in France and Germany. They include various proposals in a wide distance ranging from 100 or 200 meters up to tens of kilometers. High-speed moving sidewalks are useful at short distances. Horizontal funiculars of two vehicles are at their best for intermediate distances. Cable propulsed vehicles moving slowly through stations are possible to use up to two kilometers (Bieber and Coindet, 1987). Light railway systems are being revitalized. The networks are being extended and the speed moderately raised. The aim is to increase accessibility within cities (Kuhn, 1987).

Also intercity transport demand is considered. Different technical solutions concerning high-speed trains have been and are being developed in various nations. A new way of thinking is emerging in the headquarters of European Railways. Regional transport services are increasingly being separated from intercity transport services. These intercity railway systems are being modernized and are gradually becoming the core of the national railway corporation business operations and activities (Giannopolos, 1986).

The first super-high-speed railway – the Japanese Shinkansen – has been operating since 1964 (Ino, 1986). Seventeen years later, in 1981, TGV (*Train à grande vitesse*) started its operations on a new line between Paris and Lyon. The distance of 424 km is covered by TGV-trains in two hours with a traveling speed around 270 km per hour. The traffic on this line grew by

150 percent up to 1984. At the same time the air traffic from Paris to Lyon dropped around 45 percent. The additional traffic on the TGV-trains as a whole was 2.2 billion passenger-miles and 6 million passengers in 1984. The growth of TGV is partly due to a passenger switch from airlines and roads, and partly to a greater mobility. Financial scenarios show that all loans for this TGV project will be paid back by the beginning of the 1990s.

A European high-speed-train project is also being developed. In 1983, the Belgian, French and German ministers of transport appointed a group to study the possibility of serving the corridor Paris–Brussels–Aachen–Cologne with a new railway system. After a first report in 1984 the group was enlarged and Dutch representatives were included. The preliminary results show that a rail-wheel technology is preferred to a magnetic levitation system for reasons of cost and rate of return (Walrave and de Tessières, 1987).

One output of this work is a report of traffic forecasts concerning the cities named. These forecasts include international traffic (beside the four involved countries) and thus reassesses the present traffic situation in this central region of Europe. It is important to note that the study presumes connections by rail from the center of one city to the center of another. The implication of this is that the high-speed possibilities of the new railway technology are not counterbalanced by time-consuming trips from terminal to city-center but are fully utilized in an increased overall accessibility in this *particular* region with around 20–25 million inhabitants (Morellet, 1986).

This proposition concerns the center of a complex spatial structure with ramifications in both nearby and remote areas. In the whole European context a possible trans-European high-speed railway system interconnecting 26 cities from Rome to Glasgow on a north–south axis and Paris to Hamburg on an east–west axis is considered. It would contain 6,000 kilometres of railway lines and affect 60 million people. A few parts of this system are being realized. A railway tunnel under the English channel has been decided upon. The German inter-city railway system is being provided with high-speed trains on two lines (Mannheim–Stuttgart; Hannover–Würzburg).

Scandinavia is a remote part of Europe. In order to diminish the consequences of this geographic position the Scandinavian Link project has been launched. First, it suggests a double-tracked high-speed railway system which would connect Oslo–Gothenburg–Copenhagen with the German intercity railway system in Hamburg. Second, it is intended to fill in remaining low-standard gaps in the highways from Oslo, the capital of Norway, passing Gothenburg in Sweden and through Copenhagen in Denmark to the European motorway system in northern Germany. The plan includes the building of two bridges, both with a length of about 15 km, which will join the Scandinavian peninsula to continental Europe (Felande länkar, 1984).

In the United States, inter-city travel is provided mainly by airlines. The most efficient railway is the Metroliner service on the line New York–Philadelphia–Baltimore–Washington. A distance of 360 km is covered in less than three hours. The average speed is more than 120 km/h (*Amtrak Train Timetables*, 1986).

Authorities in Japan have led the way in operating high-speed trains. The first part of the Tokaido Shinkansen opened in 1964. The train is operating on the line Tokyo–Nagoya–Kobe–Fukuoka. The first three conurbations include the most populated areas of Japan within a distance of around 500 kilometers. A second line Tokyo–Niigata with a distance of 270 km was completed in 1986.

A complementary line to the northern areas of the Japanese main island has been built. It is called the Tohoku Shinkansen. The first section (500 kilometers) opened in 1982 between Tokyo and Morioka. Its operating speed is 240 km/h at the moment, while the average speed may be around 200 km/h. During this period the traveling time decreased from 6 hours 20 minutes by conventional train to 2 hours 46 minutes by Shinkansen. The number of passengers by airplane decreased by 60 percent in one and a half years. In 1985 the air service on this route was closed. Cities further north experienced only a slight decrease in the number of air passengers. The net increase of railway passengers, taking the loss of conventional lines into account, is estimated to be 14 percent in 1985. These changes are due to the fact that the terminus of the line has a city center position in Tokyo (Ino, 1986).

Increased telecommunicative capacity in Europe, over the Atlantic and over the Pacific

ACHIEVEMENTS OF INFORMATION TECHNOLOGY (IT)

Two major advances underlie the conception of IT. On the one hand we have had a phenomenal increase in computer and telecommunicative capacity within a few years and on the other the merging of these technologies into one united technology, IT. This merger involves the use of digitized information technology, which means the possibility of high-capacity storage and handling and also high-speed transferring of data, text, picture and voice. It also means better quality telecommunications (Forester, 1985).

However, the realization of these advantages between European countries presupposes a number of adjustments to technical standards. According to a common overview of the technical subsystems involved in telecommunications, the ISO-model, there are seven levels in the whole telesystem from the physical telenet at the bottom up to a specific teleservice. In some cases there are agreements on standards for teleservices, but further negotiations are needed.

In this paper the telecommunicative innovations are of primary concern. Digitized information can be transmitted by ordinary cables and coaxial cables as well as microwaves and optic fibres, which have the highest potential capacity. The technical development leads to a gradually lower transmission cost per bit, mainly on large teletraffic relations (Martin-Löf, 1987).

Intra-European projects

In Europe telecommunications are run by a public administration, together with a united administrative organization for post, telegraph and telephone (PTT). Often the income from telecommunications has subsidized the postal activities. This kind of relationship is being wound up. Besides existing data-bases on a cooperative basis and the Datex system (a data communication network in the Nordic countries, Germany and Austria) two tasks can be mentioned. Firstly, there is work in progress on an intra-European high-capacity network for digitized telecommunication proposed by the EC. It is called RACE (Research & Development in Advance Communication Technologies for Europe) and is modelled as a cooperative project between European electronic cooperatives. Secondly, the same authority has also been working for a common system for mobile telephones.

Atlantic and Pacific telecommunications

The transatlantic link has the most intensive and profitable teletraffic in the world. The premises of this profitable traffic are changing. For years the rate of return has been decided by negotiations. A new development with three phases can be described.

The first phase lasted until the beginning of the 1970s. During this period the dispositions of the Atlantic traffic were decided by a few PTTs in Europe. There was no competition and profits were high.

Engagement in the negotiations by American federal authorities via FCC (Federal Commission for Communications) started the second period. FCC demanded an owner split by 50/50 between European and American interested parties. North American PTTs were guaranteed half of the influence over the Atlantic teletraffic. During this period from 1975 to 1985 telecommunications were served by TAT (transatlantic cables) numbers 6, 7, 8 and the satellites of INTELSAT.

The third phase is characterized by competition and deregulation. It started in 1984 with the split-up of AT&T (American Telegraph and Telephone Company) into one federal company including the trunk lines and seven local companies. The idea was to make the new AT&T competitive with IBM, Japanese and other corporations. This initiative is an outcome of the application of American antitrust legislation. The role of FCC is to continue to supervise ordinary teletraffic. A state of regulated competition has arisen. In the United Kingdom, British Telecom has been privatized and a private tele-company, Mercury, has been formed. These companies have a duopoly on the building of networks, but they are forced to supply telenet capacity to teleservice corporations. In both countries services are divided into two groups; basic and enhanced (value added) services.

PTTs in Canada, France, Spain and the United Kingdom and AT&T have made arrangements for a transatlantic fiberoptic cable system (TAT 9), which will be in operation in 1991. In the United States five private-owned satellite systems and three private cable systems are planned with a capacity of 100,000 and 136,000 telephone connections respectively. Together with INTELSAT's 55,000 and the 22,000 lines of TAT 8 and 9 it adds up to 313,000 lines in 1995. This year the estimated demand is 67,000 lines. The overcapacity will thus be around five times greater than present needs (Martin-Löf, 1987; Näslund, 1986).

An optic fiber system, the third transpacific cable system (TCP 3), is planned from Japan to Hawaii and to California. Its network structure and capacity is decided by twenty-two telecommunication corporations. It will have a capacity of 3,780 telephone lines (Eskola, 1985). In the same way as in the USA and the United Kingdom, one group of telecommunication companies in Japan owns and supplies net services and rented lines. Another group offers teleservices via these lines. Thus in three important countries the monopoly of telecommunication has been exchanged for a free market.

What consequences will the increased capacity over the North Atlantic bring about? If the planned new satellite and optic fiber establishments are realized, they will threaten the economy of INTELSAT. The subsidies from the North Atlantic teletraffic will cease as competition decreases profits. The charges on other links regional and local, will increase. On the other hand, the cost and the charges of long-distance lines in general and particularly over the North Atlantic will diminish. These events regarding the cable systems act as incentives for the localization and expansion of high-level administrative units in the core regions of Europe. The satellite systems may have a decentralizing influence. It will depend on the cost relation between satellite and cable communication.

In Sweden the organizational premises of a cost-adjusted tariff structure are favorable. The traditional separation between ministries and authorities has prevented detail control. The Post Office has always been an independent authority and is thus not subsidized. The Swedish Telecommunications Administration is run on businesslike principles. A 3 percent real rate of return on invested capital is a financial goal. There is no legal network monopoly. Today the monopoly is restricted to exchanges and high-speed modems. The monthly cost of having and using a telephone is very low in an international perspective.

An increasing demand for various teleservices presupposes technical advancements. Digital AXE-exchanges are successively being introduced in the national network. The trunk lines were digitized in 1987. Digital communications are offered on 65 kbit/s all over the country. ISDN (Integrated service digital network) was introduced in 1989.

Table 4.1 Passenger-kilometers performed (millions) on scheduled air services, 1985. Ranking of the 10 leading countries. (Countries and groups whose airlines performed more than 100 million total tonne-km in 1985; most 1985 data are rounded estimates, thus the ranking may change when final data become available.)

| Country or Group of Countries | Rank number in 1985 | Passenger-kilometers performed (millions) Total operations (International and domestic) | | | | | |
		Estimated 1985	Actual 1984	Rank number in 1985	Estimated 1985	Actual 1984
United States	1	527,900	479,579	1	112,040	102,240
USSR	2	186,876	183,273	13	11,742	11,354
Japan	3	64,700	61,401	3	31,600	29,220
United Kingdom	4	63,230	56,874	2	59,800	53,685
France	5	39,500	38,712	4	28,260	27,532
Canada	6	35,400	34,122	9	16,900	15,825
Australia	7	28,185	26,126	8	17,305	16,154
Federal Republic of Germany	8	24,431	24,274	5	22,058	22,009
Singapore	9	21,800	20,325	6	21,800	20,325
The Netherlands	10	18,750	17,446	7	18,730	17,411

Source: ICAO Bulletin, June 1986.

IMPACT

What about the IT impact on localization, especially the impact of telematics? One can learn from experience that nodal networks have spatial implications. Railways created towns, freeways and intersections attract certain kinds of manufacturing and service activities. Telecommunication networks have nodes, but they are automatized. These networks lack nodal effects in the usual sense. Instead, every subscriber is a micronode. The telecommunication development means that:

- the accessibility of the whole system is being increased,
- the dependence on distance within and between organizations is diminishing and,
- the steering effect is mainly indirect.

The direct effects are of minor importance. Some routine functions, such as day-to-day commissions to subcontractors or orders from customers, are changed from postal to telephone and from telephone to data communication. The major indirect effect consists of greater organizational liberty of action created by a variety of high-capacity telecommunicative supplies. The consequences are increased division of labor, increased specialization and a reinforced spatial redistribution.

Accordingly, the decision-making process is becoming more and more complex. The more specialized the information, the greater the need for face-to-face contacts. In the long run the administrative structure of various organizations will be rearranged (Törnqvist, 1986).

Air traffic

International perspective

The total number of passengers in the world was 891 million in 1985. This means an increase of 55 percent from 1976, even if a decrease was registered in 1980 (ICAO Bulletin). The United States has a very big volume of passenger-kilometers performed, far ahead of the next-ranked country, the USSR. As can be seen further in Table 4.1 there are big differences between the total and international operations. In international operations the USSR ranks as number 13 compared with the second rank in total operations.

Thus the regional distribution of air traffic shows the very strong position of North America. Internationally, however, both Europe and Asia and the Pacific rank higher, which is illustrated by the figures in Table 4.2. As also shown in Table 4.2, the shares of the North American and European airlines have fallen remarkably during the last ten years. An opposite trend is observed in Asia and the Pacific, where airlines have increased their shares. In 1985 passengers accounted for about 74 percent of the load, freight for nearly 24 percent and mail for the rest (ICAO Bulletin).

Table 4.2   Regional distribution of scheduled services, 1976 and 1985. (Percentage of total tonne-km performed by airlines registered in each region).

Region	International		Domestic	
	1985	1976	1985	1976
North America*	19.9	21.2	58.0	57.3
Europe	37.5	42.9	27.6	31.2
Asia and Pacific	26.7	19.6	8.3	6.5
Latin America and Caribbean	5.7	6.4	4.1	3.5
Middle East	5.9	5.5	1.0	0.6
Africa	4.3	4.4	1.0	0.9
ICAO World	100	100	100	100

*Canada and United States only
Source: ICAO Bulletin, June 1986.

A city-system is a national or regional set of cities with internal interdependencies. The system is characterized by these interdependencies, interactions and its degree of openness or closure. They determine the growth possibilities of the system as a subset of the global city-system. Any significant change in the economic activities of an urban unit will modify the activities, income, employment, or population of the other urban units of the system. Four major categories of city-systems are identified.

(1) Low internal interaction, or interdependence, and low openness. The first type is exemplified by towns dependent on the exploitation of a surrounding region. (2) Low internal interaction, or interdependence, and high openness. This category is found among ports in colonial situations. (3) High internal interaction, or interdependence, and high openness. Urban units in advanced economies possess strong economic linkages. At the same time several cities are bound to extraregional cities by economic flows. The high-openness type is found in Western Europe, Japan and Australia. (4) High internal interaction, or interdependence, and low openness. The difference in comparison with category (3) is that the member cities of this system have relatively weak relations to urban units outside the borders. The city-systems of the United States and USSR have weak relationships outside their respective systems, and are examples of this low openness type.

In the Swedish national city-system the three biggest cities – Stockholm, Gothenburg and Malmö – are dominating smaller metropolitan centers and urban places. At the same time they interact with their own system by way of sales and purchases. They are thus examples of city-systems with high interdependence or interaction. It remains to get an idea of the degree of interdependence and openness. Three indicators are used: company control of workplaces, the relation between residence and workplace, and migration.

The control of employment in private companies is exercised mainly from these cities. In 1985 the metropolitan control comprised 53 percent of private employment in Sweden. Companies in the county of Stockholm

Table 4.3 International city pairs with highest scheduled passenger traffic – 1985. (On flight origin and destination.)

	City pairs	Passengers (000's)		Both directions
1	London–New York	1,156	1,178	2,334
2	London–Paris	1,124	1,186	2,310
3	Kuala Lumpur–Singapore	684	681	1,365
4	Hong Kong–Tokyo	658	648	1,306
5	Amsterdam–London	593	594	1,187
6	Hong Kong–Taipei	586	584	1,170
7	Honolulu–Tokyo	549	505	1,054
8	Taipei–Tokyo	483	538	1,021
9	Bangkok–Hong Kong	490	459	947
10	Seoul–Tokyo	471	472	943
11	Dublin–London	450	456	906
12	New York–Paris	431	462	893
13	Frankfurt–London	441	427	868
14	Jakarta–Singapore	416	429	845
15	Algiers–Paris	408	420	828
16	New York–Toronto	417	409	826
17	Hong Kong–Singapore	412	390	802
18	New York–Rome	370	407	777
19	Bangkok–Singapore	361	396	757
20	Frankfurt–New York	373	378	751
21	Los Angeles–Tokyo	369	369	738
22	Cairo–Jeddah	357	336	693
23	Copenhagen–Oslo	341	330	671
24	Miami–Nassau	330	338	668
25	Chicago–Toronto	338	329	667

Source: ICAO. *Statistical Yearbook 1986*. Doc. 9180/12.

control 36 percent of the total private employment to be compared with a share of employment by workplace of 20 percent. Further studies indicate that the control functions are concentrated in Stockholm (see next section). Gothenburg has its own decision territory. Companies in Malmöhus county have external locations only to a minor degree. The system properties of these counties are expressed in Table 4.6. All three have a high interdependence, while the openness is high in Stockholm and moderate in Gothenburg and Malmö.

Internationally some connections are used very intensively; especially the routes over the Atlantic and within Europe but also certain links in Asia and the Pacific. See Table 4.3.

The ranking of the airports by international passengers confirms the regional distribution mentioned above. London, New York, Frankfurt, Paris, Hong Kong and Tokyo have some of the biggest airports in the world. See Table 4.4.

Swedish perspective

Internationally, the Swedish position doesn't seem to be very strong. As can be seen in Table 4.5, the number of destinations from Stockholm are very few in comparison with the most intensive air traffic cities of Europe. The total number of international destinations are more than 200 from Paris and London respectively, compared to Stockholm with 32. There are more than fifty non-European airlines in Frankfurt, London and Paris but only four in Stockholm. In Scandinavia, Copenhagen has a much stronger position than Stockholm.

However, Stockholm has a very strong national position. As can also be seen from Table 4.5, Paris is the only city with more national connections than Stockholm. It reflects the unicentric structure of the Swedish industries supported by a long centralistic tradition in the government field (Ahnström, 1973).

The dominance of Stockholm as the major hub of Sweden is illustrated in Figure 4.1. It means for example, that a flight from the next biggest city of Sweden – Göteborg (Gothenburg) – to places north of Stockholm goes via stopover in Stockholm. This concentration of the air traffic has been especially pronounced during the 1980s.

Properties of the Swedish city-system

In order to be able to evaluate the continuing changes of primarily global and European communication facilities, the earlier mentioned cumulative

Table 4.4 Airports having world's highest commercial traffic volume 1986[1]. Ranking by International Passengers Embarked Plus Disembarked Commercial Air Transport.

	Number in 000's	Percent change from 1985	Rank order 1986	Rank order 1985
London–Heathrow	25,734	−0.5	1	1
New York–Kennedy	15,395	−6.0	2	2
London–Gatwick	15,195	11.2	3	4
Frankfurt–Frankfurt/Main	14,544	1.0	4	3
Paris–Charles de Gaulle	12,899	−1.7	5	5
Amsterdam–Schiphol	11,602	2.7	6	6
Hong Kong–Hong Kong Intl	10,610	7.7	7	7
Tokyo–New Tokyo (Narita)	9,555	3.8	8	8
Singapore–Changi	8,912	2.5	9	9
Zurich–Zuerich	8,767	1.0	10	10

[1] Preliminary.
Source: ICAO. Statistical Yearbook 1986. Doc. 9180/12.

Table 4.5 The number of destinations from some major European airports – and Stockholm – in June 1987; regular embarkation of passengers.

To \ From	Amsterdam	Brussels	Copenhagen	Frankfurt/M	London	Paris	Rome	Zurich	Stockholm
Europe	86	57	66	85	109	83	44	67	29
Middle East	17	7	8	22	19	20	14	17	0
Other parts of Asia	18	8	7	16	14	18	14	12	2
Oceania	2	0	0	2	8	3	2	0	0
North Africa	9	7	2	11	9	20	5	8	0
Other parts of Africa	14	22	1	16	26	37	21	17	0
North America	10	8	7[1]	20	31	13	8	10	1
Central America[2]	7	0	0	4	9	7	0	0	0
South America	9	1	2	11	4	11	7	4	0
Domestic	(3)	(0)	(11)	(14)	(20)	(52)	(25)	(3)	(34)
Total international	172	110	93	187	229	212	115	135	32
of which intercont.	86	53	27	102	120	129	71	68	3
The number of non-European airlines	28	16	18	53	55	54	47	26	4

[1] Includes Greenland.
[2] Includes Mexico.
Source: Comén, L-G. (forthcoming).

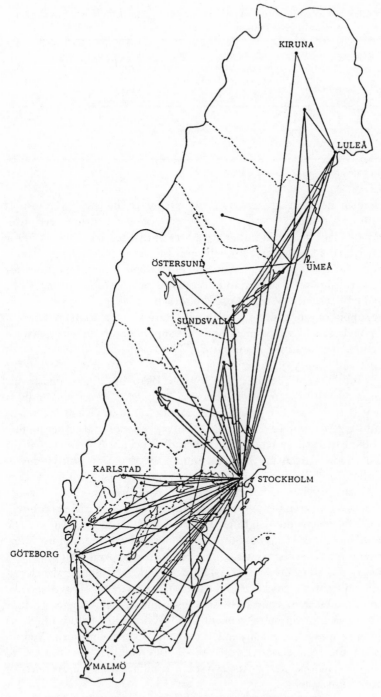

Figure 4.1 The routes within Sweden in 1987
Source: ABC World Airways Guide, June 1987

Table 4.6 Company control of private employment in the three metropolitan counties in Sweden in 1985 (percent; approx.)

County	Interdependence Employment by company and workplace located in the same area	Openness Employment by workplace located outside the region of the company
Stockholm	90	45
Gothenburg	70	20
Malmö	75	10

Source: *Storstaden som arbetsområde* 1987.

and circular model of city-system growth is used. The enlarged transport and communication services of high-speed railways, airlines and tele-communications described earlier are a part of that service complex in the model, which increases the attractiveness of large cities. These possible events in the global and the European context and their implications for Sweden are considered. The frame consists of Pred's (1977) theory of the national city-system, its structure and interdependencies.

The supply of jobs expresses a hierarchical feature in the city-systems. In metropolitan areas, there is an excess in the supply of jobs inducing inward interactions, for example, commuting. The relation between county of residence and county of workplace is thus a kind of interdependence. The absolute number of relations was low in 1980, but they indicate the existence of a subsystem. It is necessary to regard direct and indirect primary (largest number of relations county to county) and secondary (second largest number) relations. Primary relations are the joining elements of each subsystem. The county of Stockholm is the target for thirteen counties from northern to middle Sweden. The subsystems of Gothenburg and Malmö consist of three counties each. A fourth loose subsystem in middle and southeast Sweden contains four counties and a fifth contains two counties in the north. The interdependencies are roughly described related to the gainfully employed night-time population. The internal dependence is high in Stockholm, Gothenburg and Malmö, com-pared with the two other systems. The openness is generally high.

Migration is an example of an interaction with several dependencies. It indicates only three subsystems in the national city-system of Sweden. The Stockholm subsystem consists of sixteen counties, while there are four counties in each of the subsystems of Gothenburg and Malmö. Only the Gothenburg subsystem is truly independent of Stockholm. In the Malmö subsystem the secondary relations of all four counties are directed to Stockholm. The openness of these systems is higher than the openness of the Stockholm system.

Each metropolitan county is the center of its subsystem. On the county level the results are much the same. In this position the counties of

Stockholm and Gothenburg have high interdependencies, while the openness is high in Gothenburg and Malmö but low in Stockholm (see Table 4.7). Stockholm has, however, the highest openness towards the rest of the world (Forsström, 1987a).

The properties of the main Swedish subsystems can be roughly summarized. The city of Stockholm subsystem exercises a nationwide administrative control as the capital of Sweden. Both the Stockholm and county subsystems have a high interdependence concerning supply of jobs, migration and control of private companies. Its openness is high to moderate. It exerts control of employment in the rest of Sweden, but is less sensitive regarding migration. The Gothenburg subsystem resembles its capital counterpart, but it lacks control. On the other hand, the openness to migration is greater. In the case of Malmö, both county and subsystem interdependence varies from high to low, while openness is generally high.

One of the consequences of these system properties in the long run is a redistribution of population. There are two major features. Firstly, the metropolitan area of Stockholm exerts a strong centripetal power within the whole country, only interrupted in the years 1971–80. Secondly, there is also a steady, uninterrupted, stream to the whole subsystem of Gothenburg on the west coast of Sweden. The pulling force of Malmö, in southern Sweden, ceased in the mid 1970s.

Changing decision structure of organizations

BACKGROUND

The geographical pattern of decision making in Sweden has a unicentric structure (see above). Persons making and executing decisions are to a high degree concentrated in the capital Stockholm. The situation is similar in both central authorities, organizations and cooperations.

This structure has a long tradition regarding central government. Since King Gustav Vasa – the founder of modern Sweden – in the sixteenth century and even more during a line of absolute rulers, this hierarchical structure has extended to a lot of activities. The development has been similar within business companies. Large-scale production, motivated by

Table 4.7 Classification of central counties in the three migratory subsystems of Sweden 1986. (In- and outmigration per thousands of average population.)

County	Interdependence	System property Openness vs Sweden	Openness vs world
Stockholm	24	10	15
Gothenburg	24	23	10
Malmö	11	20	9

Source: Forsström, 1987a.

advantages of specialization and division of labor, among other things, has been an important factor influencing the build-up of new concerns and mergers. A consequence of this division and specialization is an increase in the dependence between different functions at work. At the same time certain areas have secured a strong position as centers of decision making (Engström, 1970; Sahlberg, 1970; Törnqvist, Wärneryd, 1968). The separation of the decision functions and production means that a lot of activities are governed by decision making far away. This relation between the center of decision making and the functions of production is important, as it can be disastrous to be located far away from the managing center (Godlund, 1972). In the Swedish parliament different proposals to decentralize have been discussed. One result is the reorganization of some of the activities controlled by the state. It is in this perspective that the interest of information technology as a factor influencing the decision structure of organizations should be seen.

INFORMATION TECHNOLOGY AND DECISION MAKING

The combined use of computer and communications technology is a means to realize a decentralizing organizational policy. This combination can be seen as a new break of structure in a similar way as the telephone in the 1880s was a socio-technical break of structure (Kuuse, 1984).

Information activities are usually concentrated in megalopolitan areas (Abler, 1974). In the case of Sweden the strong position of the capital should also be seen in connection with the central function of Stockholm as a communications center, reflected, for example by the hub organization of the air traffic (see above).

Possible geographical implications of IT are an important issue. There are different opinions including arguments for both centralization and decentralization (Hilz and Turoff, 1978; Marshall, 1984; Thorngren, 1972). Regarding decision making the use of IT can also mean possibilities of shaping new organizations characterized by a centralized or decentralized structure. Some tendencies can be seen as a consequence of technological developments.

The demand for expert knowledge and the existence of large-scale computers tended to strengthen the central level during the 1960s. More sophisticated techniques, including minicomputers with increasing memory capacity, developed during the 1970s, made it impossible to decentralize activities. During the 1980s the development of new tele-communications, the increasing installation of minicomputers and the spread of microcomputers meant better conditions for interaction between the central and the local level. Thus, regarding technical aspects it is possible to see the changes from the 1960s large-scale units to the small scale of the 1980s.

Are these tendencies also influencing the decision structure of organizations seen in a geographical perspective? Does the use of IT mean centralization or decentralization of activities? Some aspects of these issues from a Swedish point of view are given in two studies conducted at our department (Forsström, 1978b; Lorentzon, 1987). The result of the studies, mainly based on interviews with people employed in activities governed by the state, indicates that IT makes it easier to decentralize and delegate decisions. The traditional hierarchical structure makes it hard to push this issue as the flows of information are characterized by sending information from below. The needs from the central level have determined the design of the systems without regard to the needs of coordination at the local level. The knowledge of computerization has also been concentrated in the upper ranks of the organizations.

The need to be market-oriented, however, is a strong force towards an increase of the local influence. This also means that there is a demand for new channels of information. In this respect IT can be used to stimulate the building of local/regional nets of contact. This is a development leading to a stronger position of the activities at the bottom of the organizational hierarchy. At the same time the middle level tends to be weakened. This weakness is also caused by the centralization of important decisions. In the same way decisions are moving from the bottom of the organizations. Thus two main streams appear. One of them concerns the many routine

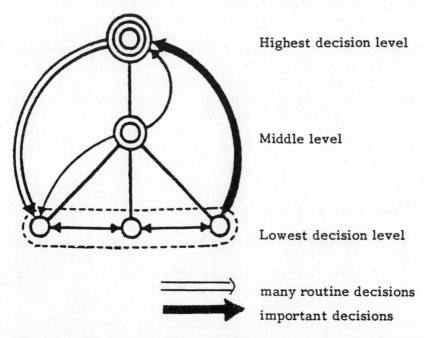

Highest decision level

Middle level

Lowest decision level

many routine decisions
important decisions

Figure 4.2   Illustration of how the use of IT is assisting the decentralization and delegation of decisions from the highest to the lowest level of organizations and the shape of new nets of contact. At the same time important decisions move upward in the hierarchy while the middle level is weakened.

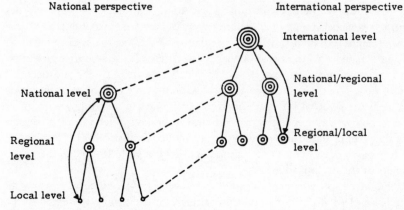

Figure 4.3   Illustration of how an increase of international relations in combination with an increase of the interaction between the top and the bottom of organizations can contribute to strengthen the regional level, while the national perspective indicates that the said level is weakened.

decisions that are moved from the top to the bottom of the organizations. The other stream concerns important decisions moving in the opposite direction. This development is illustrated in Figure 4.2.

Some geographical effects are suggested. The weakening of middle levels can cause a loss of jobs at regional centers now characterized by administration of state activities and wholesale trade. A development along these lines tends to strengthen Stockholm as the Swedish decision center.

It should be observed that the centralization seems to go hand in hand with delegation; important nonprogrammable decisions are centralized but routine programmable decisions are delegated. In making nonprogrammable decisions face-to-face communication is important. This means a greater need of personal contacts and a greater need of transport services. Here attention must be paid to the build-up of the infrastructure.

International geographic perspective – some reflections

Changing the perspective of a city-system from a national to an international angle implies new positions for the urban units. Regional centers in the Swedish city-system may take positions as local units from an international point of view. These units can be attractive alternatives to – in the Swedish case – Stockholm as places of location. A reason to choose a place outside the capital can be, for example, direct communication, to establish more precisely the market demands. Instead of a strong nation-based organization IT can contribute to include many local and regional decision centers in large international systems of activities. This means a stronger position for the regional level. See Figure 4.3.

A change of the perspective to the international level raises new issues concerning the connection between IT and decision making and the geographical impacts. One important issue is to observe the possibilities of contacts given by different types of communication systems (Hepworth, 1986; Kutay, 1986). The ability to satisfy the communication needs influences the design of organizations and the location of activities (Gillespie, 1985; Goddard, Thwaites and Gibbs, 1986; Markusen, Hall and Glasmeier, 1986; Wolf and Bördlein, 1986). The issues concern the location of new airports, airline routes, the functions of railways and roads or how systems for videoconferences should be built in the future.

Closing comments

In this chapter, the consequences of increased transport and telecommunication capacity are examined. Furthermore, the mutual interdependence between information technology and organizational policy is considered. Future geographical evaluations are suggested.

Classical geographical theory has for many years emphasized the role of distance for the settlement pattern, location of industry and land use. Implicit in the evaluations is a moderate traveling speed. Later, in other studies, the greater speed of railways, mass transit systems, cars and airplanes were reflected in redefined conceptions of cities, regions and city-systems. Examples are 'the urban field', 'the daily urban system', 'megalopolis' and 'the spatial implosion'. The megalopolis along the East Coast between Washington and Boston is partially motivated by a continuous settlement with a high population density and partially by jet airlines between the metropolitan cities. An urban system is imploding when the large cities tend to improve transport services and traveling time more than smaller cities. The larger cities converge, while smaller cities diverge in relative time–space.

What, then, is the suggested outcome of a sudden and many times increased speed in inter-city traveling? If a European high-speed train project is realized in the area of Paris to Cologne and Amsterdam, it is suggested that this city-system is converted to a city region. The area extension of the network, the distances between pairs of cities under 300 km and the speed indicate the possibility of commuting. Under these conditions, such cities will have overlapping locations from a temporal point of view (see Figure 4.4). Cities located more than 300 km away can be reached during the day and considerable working time will be available on business visits through the region. In a temporal perspective these metropolitan cities will be contagious. This speed- and time-defined city region is inhabited by 20 to 25 million people. A larger city-complex including London and the Ruhr–Rhein area would add another 18 million. The center with the highest accessibility would be the area containing Brussels and Lille.

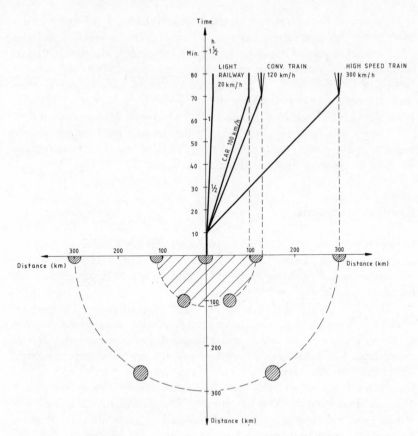

Figure 4.4   A spatially imploded city-system due to increased travelling speed by different means of transportation. (Various distances reached during identical time periods.)

Similar distance conditions prevail in the four largest city-complexes in the industrialized world (see Figure 4.5). The metropolitan cities are located within a range of 300 to 500 km, in California 600 km. In Japan these cities are linked by high-speed trains. In the USA airlines provide the inter-city services. In Europe high-speed train projects are emerging. The intercontinental contacts over the Atlantic and the Pacific are being facilitated by cheaper airline services and by tremendously increased telecommunicative capacity. Four of the six intercontinental air routes with the highest scheduled passenger traffic cross the Atlantic, while two bridge the Pacific. Thus it is likely that these city-complexes will strengthen their position of power as well as controlling and steering functions by means of an increasing share of high administrative organizations. The telecommunication network is essentially nodeless and thus without direct steering effects. The greater the capacity of the network, the easier organizational structures change.

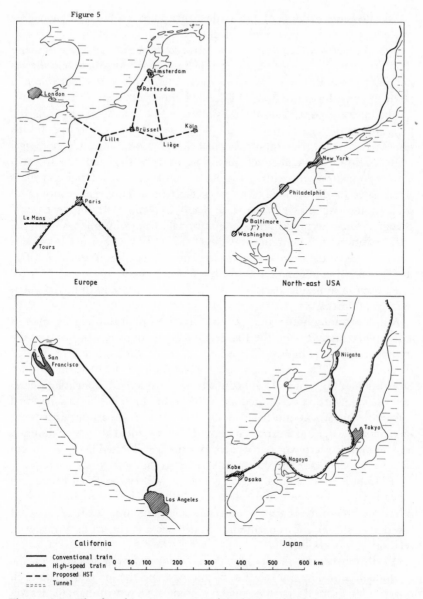

Figure 5

Europe

North-east USA

California

Japan

Conventional train
High-speed train
Proposed HST
Tunnel

0 50 100   200   300   400   500   600 km

Figure 4.5   The four big city-systems of the high-technology tripod – Europe,
North America and Japan in the industrialized world

The decision making of the large organizations is to a high degree concentrated in the upper ranks of the urban hierarchy. This often means location in metropolitan areas. The central cores of these areas are in many cases dominated by office buildings functioning as controlling and directing centers of both public and private activities. From a non-metropolitan point of view this concentration is looked upon with fear.

The interest of analyzing the possibilities of IT as a factor changing the prerequisites of decision making should be seen in this perspective. In this respect two tendencies should be mentioned. One is the technological tendency towards small-scale units. The other is the tendency towards flexible organizations with great local influence. This development is accentuated by the demand to be market-oriented. Thus the technological prerequisites are presently moving hand in hand with organizational changes. These technical and organizational changes imply a substantial differentiation of organizations. A direct effect of IT acceptance in organizations is suggested to be centralization and a delegation of decisions. The strategic decision making tends to cluster even more in metropolitan areas. The nonprogrammable decisions, on the other hand, tend to be delegated in large numbers.

A strengthening of the connections on the European continent implies an accessibility disadvantage for the Scandinavian countries. In Sweden the transport system is highly oriented to Stockholm irrespective of type of transportation. The passenger transportation system is formed with Stockholm at the center. Many contact-intensive activities, including decision making, are concentrated in Stockholm. In spite of this national concentration Stockholm has a weak position in comparison with other metropolitan areas in Europe. It should also be noted that Stockholm is located in the eastern part of Sweden. But the areas located in the south and west of Sweden have the shortest distances to the European core regions. Thus ambitions to achieve better connections between Scandinavia and other parts of Europe could result in a different layout of the Swedish infrastructure of communication. The southern and western parts of Sweden could, in a European context, achieve a fairly strong position in relation to Stockholm.

Another factor that may change the present position of Stockholm is the combination of new information technology and the internationalization of activities. From the decision-making point of view two main tendencies can be identified. One tendency is to centralize strategic decisions. The other tendency is to delegate routine decisions. A development according to these tendencies means an even stronger position for Stockholm. However, when regarding the decision making from an international instead of a national perspective, one perceives a different dimension of the geographical distribution of decisions.

The development of organizations working in the international market changes the competition surfaces. More activities are being integrated into international-based organizations. The administrative borders mean less. Instead, the functional aspects are stressed. This can result in the different hierarchical levels in organizations obtaining new functions. Instead of controlling the Swedish market via Stockholm, the needs of the market may be easier to satisfy by direct communication. Stockholm becomes a market just like other parts of Sweden. Thus internationalization can be a factor helping to counterbalance the present concentration in Stockholm. In this perspective, good connections between different areas of Sweden and also with different areas in the world seem to be a strategic competitive factor. The infrastructure should be built with more possibilities of inter-connections between activities located outside the capital. The route schedules should be less hub-oriented. The introduction of IT could be a factor stimulating the shaping of an outspread pattern of connections adapted more to real needs than to traditional organizational structures.

Some of the conclusions when analysing significant innovations and the development in the field of communications can be summarized as follows:

- there is an increase of the accessibility level all over the European continent,
- the central parts of the European city-system are growing in import-ance, and
- a focus is developing in the center of this system.

The importance of these tendencies for the pattern of localization in Sweden, situated in the periphery of Europe, is difficult to foresee. A logical consequence of the strengthening of the core regions of Europe should be a relative weakening of Sweden. Some factors, however, are counterbalancing these consequences.

Firstly, it is a question of relative changes. Sweden is already very strongly connected to Europe and the changes may be marginal. Secondly, it may – from a competitive point of view – be better to invest in the Swedish infrastructure to stimulate an integration of different parts of the country. The ambitions to raise the metropolitan area of Stockholm to the same level as the centers of Europe does not seem realistic. A more efficient way of competition should be to take care of the specific prerequisites of Sweden. To these belong the tradition in the field of telecommunications. This tradition may form a good base to meet the demand of one of the most strategic factors of the future: the possibility of different regions being integrated with international information flows. This does not mean only the physical connections in Europe but also the possibilities of communi-cating with other parts of the world. The developments of IT and also airway connections are of special interest.

Some further remarks should be made when regarding Sweden in global and European perspectives. The development towards an information society implies more attention to the human being as a factor influencing the preferences of location. Attention should, for example, be paid to the possibility of different regions being able to satisfy the wishes for a high-quality environment. Clean air and access to nature are important factors when describing the future of the location pattern. In this context the peripheral position of Sweden can be an advantage in comparison with heavily exploited core regions in Europe.

References

ABC World Airways Guide 1986, June 1987.

Abler, R. (1974), 'The geography of communications', in Eliot Hurst, ME (ed.), *Transportation Geography* (New York: McGraw-Hill), pp. 327–45.

Ahnström, L. (1973), *Styrande och ledande verksamhet i Västeuropa – en ekonomisk geografisk studie*. (Stockholm: EFI).

Amtrak Train Timetable 1986.

Bieber, A., and Coindet, J.-O. (1987), Les transports à courte distance: comparaisons et progrès récents, *Recherche Transports. Sécurité, revue de l'Inrets*, 13, pp. 29–32.

Comén, L.-G. (1988), *Air freight in Sweden and worldwide. Forecasts and trends* (Stockholm: Swedish Transport Research Board). Allmänna förlaget.

Engström, M.-G. (1970), *Regional arbetsfördelning*. Lund.

Eskola, A. (1985), 'Transmission systems using optical fibres. Sveriges tekniska attachéer', Conference on Japanese Technology for Information Systems, October 1985.

Felande länkar (1984), *Modernisering av den europeiska infrastrukturen för landtransporter över gränserna 1984*, A report from Roundtable of European Industrialists.

Forester, T. (ed.) (1985), *The Information Technology Revolution* (Oxford: Basil Blackwell).

Forsström, Å. (1987a), 'Lokaliserings- och bosättningsmönster – regional utveckling', in Andersson, L. (ed.), *Geografisk forskning i utveckling*, Högskolan i Karlstad, arbetsrapport 87:17.

——(1987b), *Kommunikationer och informationsteknologi i regionala system*, Department of Human and Economic Geography, University of Gothenburg. Occasional Papers 1987:11.

Giannopoulos, G. A. (1986), 'Intercity and regional passenger transport: policy objectives, organization and financing a European approach', *Proceedings of the World Conference on Transport Research. Vancouver, British Columbia, Canada, May 1986*. Vol. 2, 986–96.

Gillespie, A. (1985), 'Telecommunications and the development of the less favoured regions of Europe', *Le bulletin de l'IDATE*, No. 21, pp. 71–7.

Goddard, J., Thwaites, A. T., and Gibbs, D. (1986), 'The regional dimension to technological change in Great Britain', in Amin, A., and Goddard, J. B. (eds.), *Technological Change, Industrial Restructuring and Regional Development* (London: Allen & Unwin).

Godlund, S. (1972), 'Näringsliv och styrcentra, produktutveckling och trygghet, Förändringar beträffande struktur, ägande och sysselsättning inom industrin belysta med exempel från Norrköping', in *Regioner att leva i*. Uddevalla: Publica.

Hepworth, M. (1986), 'The geography of technological change in the information economy', *Regional Studies*, vol. 20, 5, pp. 407–24.

Hiltz, S. R., and Turoff, M. (1978), *The Network Nation. Human Communication via Computer* (Reading, Mass.: Addison Wesley).

Hägerstrand, T. (1970), 'Tidsanvändning och omgivningsstruktur', in *Urbaniseringen i Sverige – en geografisk samhällsanalys*, SOU 1970, 14, Stockholm.

ICAO Bulletin, June 1986.

ICAO Statistical Yearbook 1986, Doc 9180/12.

Ino, T. (1986), 'Speed-up on the Tohoku Shinkansen and its impact on regional transport and economy, *Proceedings of the World Conference on Transport Research. Vancouver, British Columbia, Canada, May 1986*, vol. 2, pp. 1082–9.

Kuhn, F. (1987), 'Les transports urbain guidés de surface', *Recherche Transports Securité* 13, pp. 19–28.

Kutay, A. (1986), 'Optimum office location and the comparative statics of information economies, *Regional Studies*, vol. 20, no. 6, pp. 551–64.

Kuuse, J. (1984), *Tele och data i ett historiskt perspektiv*. Datoriseringens inverkan på bebyggelsestrukturen. Centralisering – decentralisering? Department of Human and Economic Geography, University of Gothenburg, Occasional Papers 1984:2.

Lorentzon, S. (1987), *Informationsteknologi och beslutsfattande – med särskilt beaktande av samband mellan organisations- och lokaliseringsförändringar i Sverige*. Department of Human and Economic Geography, University of Gothenburg, *Choros* 1987:3.

Markusen, A., Hall, P., and Glasmeier, A. (1986), *High Tech America. The What, How, Where and Why of the Sunrise Industries* (Boston: Allen & Unwin).

Marshall, J. N. (1984), 'Information technology changes corporate office activity', *Geojournal*, September, vol. 9, no. 2.

Martin-Löf, J. (1987), 'Så utvecklas teletekniken internationellt', *Tele. Televerkets tekniska tidskrift*, no. 1, pp. 4–8.

Morellet, O. (1986), 'Une expérience de prévision de trafic menée en commun par quatre pays: L'etude des liaisons par trains à grande vitesse Paris–Bruxelles–Cologne–Amsterdam', *Recherche Transport Securité*, 11, pp. 7–16.

Näslund, B. (1986), 'Fortsatt utbyggnad av det digitala transatlantiska telenätet', *Digitalen 87*, no. 10.

Pred, A. (1977), City-Systems in Advanced Economies (London: Hutchinson).

Sahlberg, B. (1970), *Interregionala kontaktmönster. Personkontakter inom svenskt näringsliv*, Lund.

Storstaden som arbetsområde (1987), Department of Regional Planning, Royal University of Technology, Stockholm.

Thorngren, B. (1972), *Studier i lokalisering. Regional strukturanalys* (Stockholm: Beckmans).

Törnqvist, G. (1970), Contact Systems and Regional Development, *Lund Studies in Geography*, ser. B. no. 35.

——(1986), 'Frågor inför de kommande årens utveckling', *Conference on 'The New Information Technology – How can we use it to develop our local authority areas?'*, Lund, 20 October 1986.

Walrave, M., and de Tesslères, A. (1986), The French TGV system – achievements to date and future developments. *Proceedings of the World Conference on Transport Research Vancouver, British Columbia, Canada, May 1986*. Vol. 2, 1063–81.

Wolf, K., and Bördlein, R. (1986), *Dataenverarbeitungsbetriete im Rhein–Main–Gebiet–Kurzfassung der Untersuchungsergebnisse*. Presented at 'Rundgespräch Telekommunikation' 4–6 December 1986 in Koblenz Institut für Kulturgeographie der J. W. Goethe-Universität Frankfurt am Main.

Wärneryd, O. (1968), *Interdependence in Urban Systems* (Göteborg: Regionkonsult AB).

PART II

Telecommunications, Services and Economic Restructuring

5 Information flows and the processes of attachment and projection: the case of financial intermediaries

WILLIAM R. CODE

Mechanical means of communication: mechanical means of making and manifolding the permanent record, mechanical systems of audit and control . . . aided the rise of a vast commercial bureaucracy capable of selling in ever remoter territories. (Mumford, 1939, p. 266).

Introduction

Finance is the economy's elite umbrella function and, as such, it has been intricately dissected by the economic profession. However, the unequal distribution and growth of the financial communities at the core of the system have been largely ignored, or assumed to fall under an all-encompassing 'central place theory'. This is unfortunate for the financial function, whose central role is that of exchanging and processing information, provides an ideal, data-rich subject to explore the dynamics of information flows and location. As well, this function has a long history within which to examine the recursive processes dominating location.

The subject of this paper is the way in which the evolving communications systems have affected, at differing times and in differing ways, the distribution of financial communities. The following interpretation of how the financial system evolved differs considerably in its premises from that of the traditional explanations rooted in the concept of threshold fulfilment. It argues that, with higher-level financial activity, the demand by the consumer for a minimization of the distance to purveyors of a service has never been strong and has become weaker through the period with which we are concerned. It suggests that such demand on the part of consumers of higher-level financial services has been greatly overshadowed by the need for access to information on the part of the financial intermediaries themselves, a need which they found best satisfied in compact financial communities. Moreover, in any given area, the forces influencing the extent of higher-level financial decision making have been more exogenous than endogenous. The requirements of preexisting financial communities influence the distribution of financial decision making in a developing area more than the demands of that area itself.

Understanding the geography of quaternary functions such as finance requires not only an examination of the spatial patterns of information availability but also functional variations in the demand for information. The spatial biases in the availability of information of various qualities acts as a screening mechanism segregating functions with varying demand characteristics. Thus, in a space characterized by sharp variations in the availability of information and the rate at which it becomes available, functions such as finance are spatially arranged according to their demands for information, which in turn is frequently a function of the levels of uncertainty in their operations.

The relationship between intelligence flows and the geography of financial intermediaries cannot be studied in a static, solely contemporary frame. More than most economic activities, the distribution of financial communities is characterized by locationally constraining recursive processes operating through a long time span. The present distribution is a reflection of prior patterns of information supply and demand maintained by cumulative growth processes. What is generated is what Vance called a 'process of attachment' which is on occasion disrupted by a need for functional projection.

The venue for this interpretation of the development of systems of financial communities is Canada during the period between the early nineteenth century and the middle of the twentieth century. Canada has advantages for such a study. Among these is that it possesses a relatively small economy with a fairly simple financial history and yet still encompasses an immense physical extent. If the friction of space has influence, it will be here. As well, Canada experienced major expansion of her financial system during a period when important communications innovations were transforming the time and costs of transmitting financial intelligence. Finally, the spatial arrangement of Canada's finance, particularly banking, has not been molded by governments to the degree it has in the United States. In that country, the lingering Jacksonian ethic of decentralized finance not only resulted in unit banking but also in that unusual form of central banking – the Federal Reserve. These strongly distorted the market-based spatial growth processes of finance.

Financial intermediaries and financial communities

The difficult businesses of transferring funds from those with a surplus, or a debt, to those who are more able to use them, or to a creditor, is the primary function of financial intermediaries. Financial intermediaries obviate the need for each group of savers to seek out and choose among the wide variety of capital users, and conversely, for each group of capital users to seek out and choose among the wide variety of savers (Kuznets, 1961). While the exact nature of the service may vary among different types, the fundamental role of the intermediaries is facilitating this transfer.

The central functions of a financial community are performed by a relatively small number of people, but a number which is so vital that their performance determines the size and health of the entire community. This group comprising what Robbins and Terleckyj once called the 'money market core' is made up of the specialists, researchers and analysts and their immediate staff who largely manage the market for financial instruments and thereby adjust supply and demand (Robbins and Terleckyj, 1960). A second group of financial functions such as bank and insurance company investment departments are closely tied to the information environment of the money market core. The back offices, where much of the employment occurs, are concerned with the routine functions connected with processing the paper and electronic symbolism representing real and imagined wealth. Despite some recent suburbanization of some parts of this function, much still remains in the towers of the world's financial communities. This is largely because of the dual role of the elite who are responsible for not only keeping on top of the market but also administering large organizations. Finally there are the street offices of the banks, insurance companies, brokers and dealers and other financial organizations which are found wherever a teller stands or an insurance salesperson keeps his/her desk. For the most part they are distinct from the financial communities.

Whether in London's 'City' or Wall Street or Bay Street, financial communities have displayed unusual levels of compactness and locational conservatism over protracted stretches of urban history. Even today's advances in organizational control and the technology of communication are not significantly weakening the compact financial communities (Code, 1983; Pye, 1979; Tauchen and Witte, 1983).

The prime centralizing factor in the financial community is the continuous interplay of intelligence among the specialists and their research staffs. This exchange could be made electronically from one section of the country to another, for much of the exchange of information is made by this medium within the financial community. But most of the electronic exchange of information is made on the basis of prior face-to-face contact. When stripped of its electronic mask, one finds that at the center of the communications system is a method of information exchange not substantially different from that of medieval European financial centers. One reason for this is the need for trust. Vast amounts of money are moved daily on the basis of advice, and equally vast amounts are transferred without the benefit of a written agreement but on the basis of understandings and custom. It would appear that there is a difference in the nature of the information flow within large financial communities and the nature of the flow into and out of it. Within the community, there is an intense interchange of information connected with the decision-making process. As yet, technological and organizational advances have not been sufficient to allow the community's dispersion, even within the urban area. On the other hand, technology and organizational improvements appear to have increased the rate at which relevant information is fed into and disseminated outward from the community.

The growth of large financial communities and their spatial cohesiveness through a hundred years of dramatic communication advances lends support to Eliot Chapple's comments:

> so highly developed is the communication network that one might suppose that the more 'primitive' face-to-face interaction is no longer necessary. Certainly, messages can be transmitted from person-to-person with remarkable speed, but they contribute very imperfectly to the nature of the emotional (and behavioral) adjustment of the individuals concerned. . . . Distance is still a barrier to be overcome. (Chapple, 1970, pp. 195–6).

Uncertainty, information flows and organization

The pervasive view of the spatial evolution of the financial system is still that popularized by N. S. B. Gras in the 1930s. Here, the metropolitan economy proceeds, almost organically, through several phases, the fourth and last being the financial phase (Gras, 1939). He states that the developing metropolis is usually under the tutelage of some fully developed metropolis. Gradually, a large measure of financial independence is secured and then the new and old are equal, at least in function. No reason is provided, however, as to why this financial independence should come about. Why, with improved communications, would the 'more mature' metropolis not increase its dominance of the 'less mature' ones? What disrupts the whole process of circular and cumulative agglomeration in the larger center, a center which would possess a far-flung information gathering and disseminating system, and greater access to profitable information, greater scale economies, large specialized labor pool, more extensive research facilities and more prestige? Gras was writing over fifty years ago, but this view of the evolution of finance has remained dominant to the present, carried forward by the logic of central place and the metropolitan school of economic history (Christaller, 1966; Innis, 1939). Market-threshold based logic implies that the level of decision making in a financial community is a direct function of the size and wealth of the city and region in which it resides. While the dimensions of a local or regional market are important, this factor does not adequately account for the many incongruencies between financial center size and the magnitude of the regional economy. Nor does it account for why, as many regional economies have grown and financial thresholds been fulfilled, the local financial communities atrophy.

Central to an alternative view is the concept of uncertainty reduction through information acquisition and the manner in which information flows in organizations. The uncertainty which those involved in the financial markets are attempting to limit 'has a spatial dimension . . ., not only because the inherent stability of phenomena may vary from place to place, but also because communication channels diffuse information unevenly in space, and because perception varies' (Wolpert, 1964, p. 547).

The bases of an alternative theory can be found in the pioneering works of Marshak and Isard and Tung (Isard and Tung, 1963; Marshak, 1959). Thomas Marshak's information-system model centered on the key role of the ntaure of information demand and the communications mechanism in determining the relative efficiency of centralized or decentralized allocation of decision making. Marshak stressed the importance of the decision-making capability of the center and the communications system as vital factors influencing organizational centralization or decentralization. He observed that if the transmission time of sending data from dispersed observers to a central agent is too great, no capacity to use information at the center can compensate for the lag in transmission (Marshak, 1959, p. 426). This is important in explaining some aspects of the location of finance, for when the quantities of information are great (such as with certain aspects of mining finance or with large low-value transactions), decision making tends to shift from the centralized solution to a decentralized one. On the other hand, in Marshak's model placing of sufficient computing capacity at the disposal of the central agent may reverse the preference for the decentralized over the centralized system. Isard and Tung's conceptual model formalized decision-making costs relating the spatial structure of organizations to the major costs of decision making. The roots of an information-based explanation of the spatial evolution of the financial system can also be found in the Swedish (and later British) contact linkage literature (e.g. Warneryd, 1968) in Myrdal's and Pred's circular and cumulative causation models, (Myrdal, 1963; Pred, 1966, 1973) and James Vance's mercantile model (Vance, 1970).

Attachment and projection in finance: a conceptual model

The central element in the evolution of financial communities is the type and quality of information required to make decisions and the way in which this intelligence has been communicated over space at various time periods. Six variables operating in concert underlie movement towards concentration or replication within the financial system. These factors are:

1   The relative size of individual financial communities
2   The financial intermediary's need for uncertainty reducing information
3   The degree of localization of the investment opportunities
4   The mass of information required in relation to its value
5   Regional market size
6   The state of the communications system.

The first variable, that of the relative sizes of financial communities, is a surrogate measure of the advantages which a financial community is able to impart to the units working within it. This is primarily through greater access to internal information exchange systems in which intelligence, frequently semiprivate, nonroutine and of high but transitory economic value, is exchanged on the basis of face-to-face contact and personal relationships built up prior to its transmission (Goddard and Morris, 1976). Because of the relationship between agglomeration size and information quality this is unlikely to produce an S curve productivity function in the agglomerations and a consequent tapering off of advantage with size (Code, 1986).

A factor which varies among financial functions is the need for uncertainty-reducing information. Some, such as insurance, have had relatively little need for up-to-the-minute information because they provide a service in which there is relatively little uncertainty (the risk in providing the insurance service can be discounted) and their investments are by statute relatively stable.

A satisfactory investment decision involves intensive gathering of information, and the quantity of information required may be greater than the capability of the communications system to transfer information, in unprocessed form, with sufficient speed to a distant financial community. The result is an organizational shift in decision making to the origin of the information. Such a shift is warranted only where the source of the information is not only remote but also clustered, such as in a large mining field or oil region. With wide dispersion of the information sources, the inducement for certain financial functions to shift to a lesser financial community are reduced.

The fourth variable, that of the relation of the mass of information to the return per unit of that information, can also affect the spatial distribution of financial decision making. If the particular business involves the lending of small amounts of money to individuals (such as is the case with most branch-bank operations prior to computerized operations and ATMs) the mass of information required about those individuals is of relatively low value in comparison to the return on the loan. In many communication environments, this sort of information does not warrant the cost of moving it to a central financial district. On the other hand, large loans to few individual customers may justify the movement of required information.

The magnitude of the local or regional market has received almost exclusive attention as a causative variable in the development of commercial and financial activity. As a reason for the development of financial communities, the sustaining power of regional markets takes on the character of a special case, one which is parochial in time and space and varies in its importance with the protection offered by spatial friction. Nevertheless, within this important constraint, the size of the local market for financial services can be important in the development of financial communities. It becomes relevant to the growth of a new regional financial center if spatial friction is sufficient to provide the center with protection from distant larger financial centers with superior information environments. On the other hand, in an environment of low-cost, high-capacity and nearly instantaneous communication, the size of the regional market becomes irrelevant as far as the potential regional center is concerned.

The final variable, and one which must be taken into account with all the others as a locational criterion, is the adequacy of the communications systems to move with the required speed not only given quantities of information but also information of particular qualities (for example, highly valuable information which could only be transferred on a basis of trust). Centralized decision making in the financial system can only take place if there is an adequate system of moving the information on which those decisions are made.

Attachment

The foregoing locational factors tell little about financial communities as a constantly evolving system and the interacting processes producing spatial change in the system. Dominating the spatial dynamics of the financial system is the recursive process diagramed in Figure 5.1. Here the only exogenous variable is an assumed improvement in the ability of the

Model of Growth Process of Financial Communities

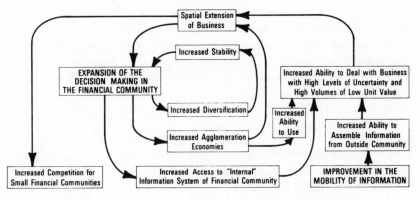

Figure 5.1 Model of growth process of financial communities

communications systems to move information. Such an improvement results in an elevated ability of many of the functions in the financial community to invest funds under their control or direction over greater distances and encroach on the hinterlands of smaller financial centers. This in turn, represents a net spatial expansion of the hinterland of the financial district and an intensification of the interaction within that hinterland. This increased interaction, in turn, results in expansion of the decision making taking place in the financial community, and its expanded size enhances the possibility of the exchange of information through the internal communications system. This, in turn, results in an improved capability for making satisfactory investment decisions, which improves the competitive advantage of the financial center *vis-à-vis* other centers. Coupled with this cycle based on information flows are other advantages resulting from the larger size of the financial community. The increased scale also results in improved access to external business services, and the ability to pool labor and capital resources. As well, with the increased size and diversity of the financial community, it becomes less dependent upon any one particular type of financial activity and less likely to decline rapidly with the demise of that function, which has been a particular failing of many small regional financial centers.

The improved competitive position of the larger financial community may push smaller financial communities into a downward spiral, depending upon the degree to which they are protected by distance and other barriers. Improvements in communications reduce the relative advantages of regional or local financial communities and expose them to larger financial communities which have greater sources of information emanating from within, and increasingly equal access to information about investment opportunities in the more distant regions. Carried to its logical conclusion, with communications steadily improving, distance from a financial community becomes irrelevant as a factor influencing the quality of service which it is able to provide.

Why should the external communications of the financial community be more affected by communications improvements than internal communications? The answer lies in the nature of the messages being transmitted, the external ones being more standardized, less spatially concentrated, and less dependent upon face-to-face contact. Once the messages can be rapidly transmitted, marginal increases in distance become of negligible importance in comparison to the difference in the quality of service between large and small financial communities. The medieval-like communications system used within the financial community has been less readily affected by technology (Pye, 1977; see also Daniels, 1987). The effect is concentration of information flows on the financial community without a corresponding fragmentation of that agglomeration beyond the metropolis.

Clearly, this model is insufficient to account for those times when small financial communities arise, survive and even become the equal of large prior communities. There have obviously been important exogenous influences which affect the velocity of this spatial growth process even to the point of reversal. It is therefore necessary to expand this model into one reflecting both the overall trend towards the establishment of points of attachment for financial decision making and the forces which can create new points of attachment.

Projection

Generating the geography of finance is the interplay of the foregoing ideal-typical growth process of financial communities and the constraints imposed on this locational process by spatial variations in the quantity, nature and mobility of information emanating from outside the financial community. For example, the development of a new financial community and its becoming the equal of a previously existing one (e.g. New York–London; Toronto–Montreal) can be envisioned as following one of the sequences presented in Figure 5.2. In this diagram the right side assumes that the communications system is able to transmit adequate information to make satisfactory investment decisions while the left assumes that the communications system is unable to transmit adequate levels of information to an established financial center.

The simplest manner in which information and the transfer of capital could take place is on a one-to-one basis, such as in Figure 5.2, A1, where the custodian of an investment opportunity in a remote area espouses a particular project to the potential investor in the metropolitan financial center. This case does not require an intermediary in the remote region. Here, the intelligence needed for allocation of funds flows in diverse streams to the metropolis where it is digested and funds are allocated and forwarded. This represents the ultimate in centralization and dominance.

At the other extreme, uncertainties arising from impediments to communications (such as the North Atlantic in the mid nineteenth century) demand the creation of mechanisms which act as a credit bridge. This mechanism takes the form of a system which establishes trusted intermediaries proximate to the potential investment opportunities, logically in the largest commercial center of the remote region (Figure 5.2, A2). Information on those investments is preprocessed and decision-making authority given to intermediaries nearer the remote investment opportunities. Here, the incipient financial community holds more decision-making power. Knowledge of various opportunities is acquired by intermediaries through inquiry on their part or by being approached by promoters. On the basis of such information, the intermediaries inform the established financial community on the advisability of particular investment opportunities. With this advice, the established center can then forward money to the distant intermediary and its ultimate recipient. Alternatively, intermediaries in the remote center can endorse investments, and with a favorable

Hypothetical Sequence of Financial System Development.

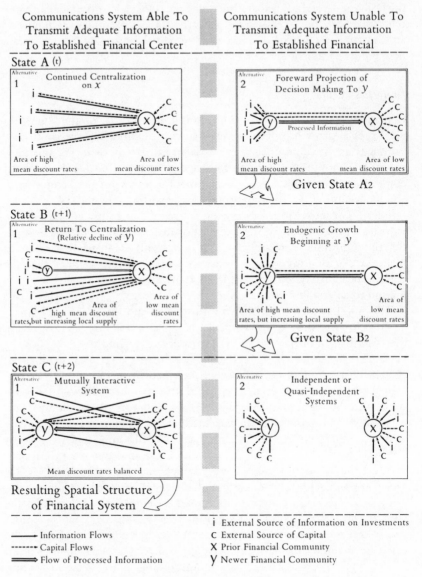

Communications System Able To Transmit Adequate Information To Established Financial Center

Communications System Unable To Transmit Adequate Information To Established Financial

State A (t)

Alternative 1 — Continued Centralization on x

Area of high mean discount rates — Area of low mean discount rates

Alternative 2 — Foreward Projection of Decision Making To y

Processed Information

Area of high mean discount rates — Area of low mean discount rates

Given State A2

State B (t+1)

Alternative 1 — Return To Centralization (Relative decline of y)

Area of high mean discount rates, but increasing local supply — Area of low mean discount rates

Alternative 2 — Endogenic Growth Beginning at y

Area of high mean discount rates, but increasing local supply — Area of low mean discount rates

Given State B2

State C (t+2)

Alternative 1 — Mutually Interactive System

Mean discount rates balanced

Alternative 2 — Independent or Quasi-Independent Systems

Resulting Spatial Structure of Financial System

⟶ Information Flows
------ Capital Flows
⟹ Flow of Processed Information

i External Source of Information on Investments
C External Source of Capital
X Prior Financial Community
y Newer Financial Community

Figure 5.2 Hypothetical sequence of financial system development

record of reliability gain funds from the established center. In either event, decision-making power develops on the distant shore.

This latter scenario, typical of periods with inadequate communication, does not necessarily hold in perpetuity. The conditions presented in A2 could become the conditions in B1 if the relationship between information demand and supply should shift so that satisfactory information movement occurs. If this should not occur a system of dual financial nuclei will ensue with each playing an intermediary role within its domain and with funds flowing in response to information preprocessed by the distant intermediaries (Figure 5.2, B2).

If the situation in B2 continues to be reinforced by poor communications and capital sources in the remote region develop, the need for long-distance funds transfer will decline. This results in two independently operating financial centers. If communications should improve, as is the case in C1, the result is two equal financial centers with high levels of interaction and shared market areas – increasingly the norm in the modern world.

In summary, projection is a rational geographic response to an imbalance between demand for uncertainty-reducing information in established financial centers and the capacities of the organizational and communications systems to deliver satisfactory levels of information with sufficient speed.

The development of the Canadian financial system

There are three central questions in the development of financial geography of Canada. The first is why Montreal developed as a center of intermediary activity during the first half of the nineteenth century. The second is why Toronto was able to overtake Montreal by the mid twentieth century as a financial center. The third is why simultaneously both centralization and decentralization of the Canadian financial system occurred after the country's post 1870 expansion.

Transatlantic capital flows and the development of Montreal 1821–71

The wealth which once entered Montreal in the fur packs was replaced in the first two decades of the nineteenth century by a new income connected with the production of squared timber and deals, wheat and potash in the peninsular country between Lakes Ontario, Erie and Huron. Montreal reentered the commercial rivalry of North America's eastern seaboard cities in an attempt to capture the commerce of the expanding agricultural frontier. It built canals and railways and entered the bitter struggles to capture hinterlands.

Between 1821 and 1871, the Canadian economy evolved from an elemental state, almost exclusively dependent upon the primitive extraction of export staples to a more rounded, integrated condition. The economy was nevertheless characterized by a shortage of fixed capital and money. There were two areas of demand for financial services. The first related to the flows of investment capital, particularly that relating to the insatiable demand for expansion of the transportation infrastructure. The most continuous direct demand for funds emanated from the commercial sector however. Commercial credit constituted the primary source of capital for agriculture, lumbering and manufacturing during the first half of the nineteenth century. It is impossible to give an accurate quantitative value to the amount of commercial credit outstanding throughout the period in Canada, but it is evident that the supply coming through the merchants' credit chain was insufficient. This is indicated by the fact that the merchants found it necessary to pool their resources and form banks. Typically, banks were creations of the commercial system and in North America they 'always arose from a shortage, not from a surplus, of funds' (Pentland, 1950).

The supply of capital in Canada between 1821 and 1871 came mainly from three sources – Britain, the United States and domestic savings. The British and the domestic sources were by far the most significant and of these two the imperial source was paramount, for a considerable portion of the domestic savings was a consequence of the sometimes massive infusion of British funds. In Britain, investors and particularly the 'country investors' – those who had benefited from the industrial processes taking place in Lancashire, Ireland, the Midlands and the Eastern Counties – were actively seeking to put to productive use the sizable amount of funds at their disposal. The large quantity of funds was a major reason for the relatively low rates of interest on Lombard Street throughout most of the nineteenth century (for example, between 1828 and 1832, excellent commercial paper could be discounted in London for as little as 1.5 percent). The interest rate gradient outward from the industrializing British Midlands induced a large flow to London, some to be returned as discounts on commercial paper and some to be sent overseas. Britain became by far the prime source of investment capital for the world for most of the nineteenth century.

A considerable portion of this pent-up capital flowed to North America. Initially, the vast majority of it went to the United States but Canada increasingly obtained a share. There were differing characteristics in the movement which involved varying needs for intermediaries, differing levels of skill and knowledge on the part of the middlemen, and ultimately different locations for them on either side of the often treacherous North Atlantic.

The barrier of the North Atlantic

The mobilization of funds from Britain to North America involved not only a movement of capital from an established economy to the new. It also involved risk. The money was entrusted by its owners to borrowers separated by a vast expanse which caused a large uncertainty discount. Before the advent of the submarine cable in 1866 and regular steamship service, communications were slow and erratic. In the early nineteenth century, Lord Dalhousie bitterly complained about the official (and least efficient) postal delivery, noting that his dispatches which left England in November did not reach Quebec until February and his February dispatches lingered until May (Short and Doughty, 1914, p. 734). The more efficient private system of depositing mail with friendly captains of private vessels was somewhat faster but also uncertain for regular communications. Even as late as 1852, when a regular line of screw steamers was initiated between Liverpool and Quebec, with departure only twice a month in summer and once a month in winter the speed of service was still slow and irregular. The sailing time in 1853 averaged fourteen days to Montreal and twelve days for the return voyage for the fastest boat of the line, and twenty-two days out and eighteen home for the slowest.

The situation scarcely improved until 27 July, 1866 when telegraphic communication was established between Valencia, Ireland and Heart's Content, Newfoundland. The price to send twenty words across the Atlantic in 1866 was £20 dropping to £5 in 1867 and leveling off by 1872 at $1 or 4s a word (Brown, 1898, pp. 536–43). This rate structure limited information transmission to the most vital kind, or to information of relatively broad interest such as quotations from the various markets, and press reports. While the accessibility to important up-to-the-minute information did reduce the risk involved in certain kinds of European investment in North America, for the most part it had little effect, for the investment fraternity had already evolved a mechanism which permitted an efficient movement of funds. This solution involved a shift of decision-making to North America.

The transfer of the intermediary function

The organization and operations of the firm of Baring Brothers illustrates the trans-Atlantic allocation of decision making. The importance of information in the distribution of investment and loan funds did not escape these powerful merchant bankers. As Hidy recounts:

> The wealth and credit of Baring Brothers and Company . . . rested upon the integrity and judgement of hundreds of remote correspondents as much as upon the decisions of the partners themselves. (Hidy, 1949, p. 94).

Typically, Thomas Baring in 1828 made an extensive tour of transatlantic business centers to acquire correspondents and information preliminary to the establishment of a system of operation. In most of the major ports he established connections. They even brought Americans into the London offices to help construct the credit bridge. The House of Baring was not an isolated case. Fletcher, Alexander & Co., W. and J. Brown & Co., Overend, Gurney, Glyn Mills and other British houses had similar arrangements. In effect, a system which minimized the flow of information across the North Atlantic was being implemented.

Similarly in Canada, the slow communications links in the commercial credit system demanded that intermediaries exist in Canada despite the colonial link. However, even with intermediaries functioning in Canada, the flow of commercial credits was restricted by a distance-based uncertainty discount. Over such a distance, it was often difficult for British merchants to discriminate among Canadian applicants for credit. Moreover, the inadequacy of Canadian bankruptcy laws meant that British merchants on occasion took heavy losses, thus impairing the credit of the Montreal community as a whole.

This weakness in the transatlantic credit flow was a root cause of the development of Montreal merchant banking organizations such as Forsyth, Richardson & Company and the early chartered banks such as the Bank of Montreal. Such banking organizations in Montreal were a logical response to the communications related restrictions in transatlantic credit flow and is typical of a response which recurs throughout the development of the Canadian financial system. In effect, inadequate communications between a developing area and a previously existing financial community necessitated the development of financial intermediaries in the developing area both to tap substitute local savings and provide a more secure basis for the transfer of funds.

An inland competitor

In general, the expansion of the financial network into North America involved an organization adaption which projected financial decision making into the 'hinge' cities of the eastern seaboard of North America, thereby decreasing the amount of uncertainty-reducing information required to traverse the North Atlantic. While the cities of the United States seaboard and Montreal generally shared in this process of forward projection they differed in the degree to which financial decision making was transferred. As intermediaries, the American financial institutions were powerful enough to attract money to themselves from Europe, and British financial houses found it advisable to allocate a significant portion of their organization to the listening posts of New York, Boston and Philadelphia. In general, Montreal was in a less powerful position with respect to Canada. The role played by financial intermediaries resident in Montreal in organizing British financing of major investment projects was considerably less than that played by their American counterparts.

By 1871 Montreal had acquired a substantial body of financial activity and certainly the largest such grouping in Canada, but unlike New York, the size and complexity of its financial community was not great enough to provide it with overwhelming information and agglomeration advantages. The stage was thus set for the appearance of a small but increasingly successful upstream rival. Toronto's financial community had by the end of the 1870s acquired a body of experience, an information gathering system, and a loyal clientele which, with the important political support of Canada West (to become Ontario), permitted its financial functions to compete with Montreal from its Toronto base. Spatial competition for most functions whose activities involved limited uncertainty and low levels of information acquisition was based on level of service. This level was less dependent upon the friction of space than on the efficiency with which a particular financial function could be performed in each of the financial communities.

If Montreal's commercial fortunes had been more firmly established when Toronto arose as a city; if Montreal had benefited to a greater extent from a more extensive overseas capital inflow; if Montreal's commercial fortunes had not waned during the 1840s; if Toronto had not been protected by the parochial political power of Canada West; and if communications between Montreal and western Upper Canada had been better in the early part of the century, then St James Street might well have caused Bay Street to languish as the side street it once was.

The spatial relationship of the financial communities of Montreal and Toronto to the Canadian economy after 1871

Between 1867 and 1871 the Canadian state expanded to encompass half a continent. Land was engulfed which was soon to be exploited to feed the industrial demands of Europe and the eastern United States. These developments resulted in an explosive expansion in the spatial demand for financial services. As well, it was not a regional zero-sum game: the great investment in the new areas increased the demands for manufactured products and services from the older areas of Quebec and Ontario and compensated for the migration of capital, labor and displacements in eastern agriculture.

Again, the financial system confronted a problem of how to provide service over vast space, but now with a more developed communications system. The system of communications which developed between central Canada and the Canadian frontier was in marked contrast to those affecting the financial communities of Canada before the mid nineteenth century, despite the great increase in the distances involved. The American railroads, the Canadian Pacific Railway (CPR) and the later developmental railroads permitted the rapid delivery of mail. The CPR, for example, enabled Montreal mail and passengers to be delivered in Winnipeg in three days. From the 1870s telegraphic communication was possible from the Canadian west through the United States. Between 1880 and 1886, construction by the Great North Western Telegraph Company and the Canadian Pacific provided a thorough coverage throughout almost all the Dominion. Even in remote areas of the Northwest Territories, where private companies would hardly be justified, the Dominion government operated lines, and by 1897 it had 2,548 miles of line serving the frontier. At roughly the same time, Melvin Bell's Bell Telephone Co. of Canada extended services to every part of the country except British Columbia, providing comprehensive service affordable by business to the larger towns.

The second half of the nineteenth century witnessed a dramatic change in the pace of time–space convergence in Canada. The telegraph produced a quantum reduction in the time required to move impersonal messages, and the railway generated a smaller but still sizable reduction in the time required to institute personal contact. In the financial system, these new conduits reduced the need for preprocessing information and projecting decision making forward towards the frontiers of the expanding economy.

After 1871, Montreal and Toronto came to dominate the expanding frontiers of the new country. The bipolar center totally absorbed banking head offices. The thirteen banks in the two metropolitan centers constituted only 41 percent of the Canadian total in 1871. By 1910 the twenty-two banks in these cities constituted 71 percent of the Canadian total. By 1940 all banks were headquartered in Toronto or Montreal. They created a vast branching system which literally followed the railhead across the prairies, serving the navvies and the immigrants from sawhorse tables until their prefabricated classical 'edifices' arrived by boxcar. The branches were controlled in a military-like organization and used rule books as a substitute for communications, although the rail and telecommunications were making organizational flexibility easier as the twentieth century progressed.

Securities trading continued to be dominated by Montreal and Toronto, with Toronto surpassing Montreal's exchanges by the early 1930s. The lesser security markets in Canada have been largely concerned with mining and oil securities, and especially those of a speculative nature. Of the exchanges outside Toronto and Montreal only Vancouver's met with success.

After 1900, insurance became less centralized than the other major components of the financial system, reflecting its concern with statistically manageable risk rather than uncertainty, and its governmentally imposed conservative investment policies. Companies located in small cities such as Kitchener, Winnipeg, London and Ottawa grew disproportionately through the first half of this century.

The Trust and Mortgage Companies (Savings and Loans) and their populist parallel, the Credit Unions and the Caisses Populaires, were more dispersed. These were descendants of the building societies and originated as a means of mobilizing mortgage money predominantly (but not exclusively) from local sources. Initially the Trust and Loan companies were oriented to the developing rural markets in Ontario and the west, but over time they too became focused on the major cities. By 1960, Toronto and Montreal Trust institutions were clearly dominant.

In many fields, such as banking and the securities trade, the level of penetration of regional markets by Montreal and Toronto headquartered institutions varied little with distance. The constraint of distance to markets as an influence on the distribution of higher-level financial decision making steadily decreased. In its stead, the locational factor of greatest importance became access to profitable information which increasingly was concentrated in the two large financial communities. This information more and more originated within the financial community and was increasingly concentrated there even if it emanated from outside.

Projection in Canadian mining finance

The financing of Canadian mining and oil development is of unusual importance in the growth of the Canadian financial communities. For over a hundred years it constituted a large segment of the activities of the Canadian financial centers. As well, mining finance in Canada (as in the United States and Australia) clearly reveals the relationship among uncertainty, information demand, the structure of the communications system and the need for projection of the financial intermediary function.

Mark Twain's definition of a mine as 'a hole in the ground with a liar at the top' has had some validity, both a hundred years ago in the Comstock and in many mining (and by extension, oil) fields of more recent vintage. Uncertainty is particularly characteristic of mining and oil field investment, particularly in those types of development characterized by a great many small companies operating in a localized area. In dealing with the worth of small mines, one is confronted with a market featuring high price volatility, a factor which can be of great value to the speculative investor (if the situation is understood). However, such investment requires hard-to-get information, frequently of the 'inside' variety, even to begin to understand the movement of the value of one security, let alone the shift of hundreds. The uncertainty comes from many sources: three main ones being the property (or the ability of the prospector, if an exploration company), the promoter and the strange events surrounding the rise and fall of 'worthless' stocks. In addition they are subject to all the other vicissitudes which characterize more normal financial exchanges. In most of these mining fields, not only was a large amount of information needed to sift even a minimum of facts about investment opportunities, but also it was required 'in a hurry' for, in this field, opportunities are particularly ephemeral.

The high degree of uncertainty involved in financing mining areas (only where there is a diffused pattern of control) and the considerable information required to reduce this uncertainty has resulted in development of financial communities at nearby locations. With the opening of a sizable mining area to exploration and early development, there normally exists a large number of individuals, syndicates and small companies all seeking financial backing. Usually there was insufficient local financing and the financial markets were distant. In addition, financial organizations were usually reluctant to extend money for an unknown commodity where they had insufficient information. As a result, the pioneering mine developers in Canada were required to seek out friends and acquaintances willing to put forth money on the basis of confidence in the developer. At this early stage there is no one center of financial intermediaries but a scattering dependent upon the personal information fields of the participants in the mining development. In Canada these linkages were oriented to a wide range of towns and cities, particularly in the northern tier of American states.

As the Canadian mining fields developed, they required greater amounts of exploration and development capital. The ability of individuals and small syndicates to risk enough money for even the initial work grew increasingly more limited. They needed broader-based financing. The most common way of acquiring these funds was the sale of equity shares to a range of investors. With these increasingly impersonal linkages, the uncertainty connected with investment increased. Financial intermediaries began to play a more important role, and tended to locate near the mining area where the maximum amount of information was available.

As mining fields in Canada developed, consolidations typically took place, either through mergers of small companies working the field or through acquisitions by already established corporations. Because of the security of size and diversity, uncertainty was reduced. The structuring of the information flows through the internal organizations of these large corporations made information as readily available in more distant financial communities located near the large mining companies' head offices. Less information was needed for satisfactory investment and it could be moved more readily through the internal communications system of the corporations thereby decreasing the advantages of nearby financial intermediaries. However, if numerous profitable internal linkages developed in the financial community located near the mining field, then mining finance remained there. If they did not, this function shifted over time to the more distant but larger financial community which could provide those linkages.

One can see this process in many of the world's mining areas. (A particularly good example is the development of San Francisco's financial community as a base for investment in the Comstock.) In Canada, the clearest example is the development of Toronto's financial community as a base for Cobalt and Noranda-Rouyn (ultimately tipping the scales in Toronto's favor in its rivalry with Montreal) and Vancouver as a base for mining in the Canadian Cordillera.

When the financing of mining development could no longer be handled by individuals and small syndicates, Toronto had a distinct advantage over rival cities as a center for the expanded intermediary function. It was the closest city to the mining fields and private information and knowledge of local conditions on the shield were more prevalent here than in rival centers. Moreover, it possessed an already existing financial community which could provide sufficient external economies for the limited needs of mining finance. Trading of securities took place here which attracted shares and funds from other centers, which in turn increased trading. An upward growth spiral resulted.

The growth cycle could have ended with the success of the mining field, the increase in capital requirements and the evolving supremacy of large mining corporations funded from large international financial communities. With reduced uncertainty, and information systems controlled and internalized within the corporation, the spatial constraints promoting nearby locations disappear. However, in Toronto's case (and unlike comparable situations such as Spokane and Colorado Springs), the city possessed a broader financial community which was comparable to Montreal's and able to provide continuing service to many of the larger mining corporations when they needed large-scale financing.

The financing of speculative mining and oil ventures in the Canadian Cordillera and adjacent high plains did not centralize on Toronto. Again the imbalance between uncertainty-driven information demand and the capability of the communications system induced projection. It is more than accidental that the largest financial center in Canada, outside of the old centers of Montreal and Toronto, is Vancouver's, focused on the vicissitudes of speculative mining, oil drilling and the world's lumber markets. In the modern world it is only with such conditions of high uncertainty that a new financial center is likely to develop.

Conclusion

During the time period covered by this paper there has been a staggered but one-directional movement in the time and cost of moving information, resulting from the sometimes spectacular innovations in the technology and improvements in the organization of communications. Higher-level financial activity, dependent upon intelligence and dealing with information of sometimes inestimable value, has been one of the first functions to be affected by the reduction in the constraints of distance. The lowering of the walls of distance has been selective in its impact, however. In a truly frictionless space, information may easily flow to the center, but there would be no greater advantages in this center than elsewhere if the ties which bind it together were to be infinitely stretched or broken. As the friction of distance has declined, its first impact appears to have been that of enabling the financial communities with particular advantages to exercise them more easily over greater distances. The backflow of information from the large financial centers through public sources and through the corporate communications system, has permitted less information-dependent financial functions such as insurance to operate in small, sometimes remote, financial centers. However, we have not reached the point where universal immediacy creates a financial version of Webber's 'non-place community'.

References

Brown, W. J. (1898), 'Canada and the Atlantic Cables' in J. Castell Hopkins, ed., *Canada an Encyclopedia of the Country* (Toronto: The Linscott Publishing Co.), pp. 536–43.

Chapple, D. (1970), *Culture and Biological Man* (New York: Holt Rinehart & Winston).

Christaller, W. (1966), *Central Places in Central Germany* (Englewood Cliffs: Prentice-Hall).

Code, W. R. (1983), 'The strength of the centre: downtown offices and metropolitan decentralization policy in Toronto', *Environment and Planning A*, 15:10, pp. 1361–80.

——(1986), 'Information redundancy, information quality and the agglomeration economics of large office communities', *Modeling and Simulation*, 15:1, pp. 289–93.

Daniels, P. (1987), 'Technology and metropolitan office location', *The Service Industries Journal*, 7:3, pp. 274–91.

Goddard, J. B., and Morris, D. (1976), 'The communications factor in office decentralization', *Progress in Planning*, 6:1, pp. 1–80.

Gras, N. S. B. (1939), *An Introduction to Economic History* (New York: Harper & Brothers).

Hidy, R. W. (1949), *The House of Baring in American Trade and Finance* (Cambridge, Mass.: Harvard University Press).

Innis, H. A. (1939), 'Toronto and the Toronto Board of Trade', *The Commerce Journal, Annual Review* (Toronto: University of Toronto Press).

Isard, W., and Tung, T. H. (1963), 'Some concepts for the analysis of spatial organization' (parts 1 and 2), *Papers and Proceedings of the Regional Science Association* 11 and 12, pp. 17–40 and 1–28.

Kuznets, S. (1961), *Capital in the American Economy: Its Formation and Financing* (Princeton: Princeton University Press).

Marshak, T., (1959). 'Centralization and Decentralization in Economic Organizations', *Econometrica* 27, pp. 339–430.

Mumford, L. (1938), *The Culture of Cities* (London: Secker & Warburg).

Myrdal, G. (1963), *Economic Theory and Underdeveloped Regions* (London: Methuen).

Pentland, H. C. (1950), 'The role of capital in Canadian economic development before 1875', *Canadian Journal of Economics and Political Science* 16:4, pp. 457–74.

Pred, A. (1966), *The Spatial Dynamics of U.S. Urban-Industrial Growth, 1800–1914* (Cambridge, Mass.: MIT Press).

——(1973), *Urban Growth and the Circulation of Information, The United States System of Cities, 1790–1840* (Cambridge, Mass.: Harvard University Press).

Pye, R. (1977), 'Office location and the cost of maintaining contact', *Environment and Planning A*, 9, pp. 149–68.

Robbins, S. M., and Terleckyj, N. E. (1960), *Money Metropolis* (Cambridge, Mass.: Harvard University Press).

Shortt, A., and Doughty, A. G. (1914), *Canada and its Provinces* (Toronto: Glasgow Brock & Company).

Tauchen, H., and Witte, A. D. (1983), 'An equilibrium model of office location and contact patterns', *Environment and Planning A*, 15, pp. 1311–26.

Vance, J. E. (1970), *The Merchants World: The Geography of Wholesaling* (Englewood Cliffs: Prentice-Hall).

Warneryd, O. (1968), *Interdependence in Urban Systems* (Göteborg, Regionhonsult Atlliebolag Goteborgs Universitet Geografiska, Meddelanden, Series B. NR. 1, 1968).

Wolpert, J. (1964), 'The decision process in a spatial context', *Annals of the Association of American Geographers*, 54, pp. 537–58.

6 Information technology and the global restructuring of capital markets

MARK HEPWORTH

Introduction

In a relatively short period of time, only a decade or so, the world's capital markets have undergone a major transformation in terms of their geography, technological structure and offering of financial products. This paper discusses recent processes of global capital market restructuring, with specific reference to the internationalization and integration of secondary markets for securities trading, and their relationship to flows of information and capital and the growth of multinational financial intermediaries. At the heart of this market transformation are the new information and communications technologies, which have effectively removed the spatial and temporal constraints on twenty-four-hour global securities trading and created pressures for 'deregulation' in all countries across the world.

We begin with a description of the global hierarchy of equities markets, dominated by the major stock exchanges of London, New York and Tokyo. This is followed by a brief overview of the changing technological structure of global capital markets, focusing on the electronic interconnection of national markets through computer networks and the role of financial on-line information vendors. In the next section, these global information flows are interrelated with international capital movements, including the growth of foreign portfolio investment and foreign direct investment, in the emerging context of financial-services deregulation. Finally, we suggest the geopolitical implications of the global restructuring of capital markets, especially its effects on national sovereignty, regulatory practices and domestic economies.

The 'three-legged stool'

Global capital markets are increasingly interconnected by computer-network technology, common financial products and investors prepared to

Table 6.1 The size of the world's stock markets, 1985, equity market value and turnover, £ million.

		Market value	Share (%)	Turnover	No. companies quoted		Foreign share (%)
					Domestic	Foreign	
1	New York	1,302,241	42.3	671,280	1,487	54	3.5
2	Tokyo	648,674	21.1	271,481	1,476	21	1.4
3	London	244,711	7.9	52,777	2,116	500	19.1
4	Germany (Association of Exchanges)	123,783	4.0	68,825	451	177	28.2
5	Toronto	108,867	3.5	21,823	912	54	5.6
6	Zurich	76,161	2.5	n.a.	131	184	58.4
7	Basle	69,151	2.2	n.a.	320	523	62.0
8	Paris	58,477	1.9	14,522	489	189	27.9
9	Geneva	55,061	1.8	n.a.	271	537	66.5
10	American	43,740	1.4	18,214	732	51	6.5
11	Australia (Association of Exchanges)	41,611	1.3	22,455	1,069	26	2.4
12	Amsterdam	41,200	1.3	14,181	232	242	51.0
13	Hong Kong (All Exchanges)	40,792	1.3	n.a.	247	22	8.2
14	Milan	39,377	1.3	10,885	147	0	0
15	Johannesburg	37,579	1.2	1,671	501	26	4.9
16	Singapore	27,918	0.9	2,480	122	194	61.4
17	Stockholm	25,820	0.8	7,578	164	7	4.1
18	Brussels	14,014	0.4	2,482	192	144	42.8
19	Madrid	13,730	0.4	2,248	334	0	0
20	Barcelona	12,298	0.4	377	458	0	0
21	Kuala Lumpur	11,181	0.3	1,754	223	61	21.5
22	Copenhagen	10,451	0.3	224	243	6	2.4
23	Luxembourg	9,106	0.3	55	n.a.	n.a.	n.a.
24	Oslo	7,060	0.2	2,914	156	7	4.1
25	Tel-Aviv	5,290	0.2	408	267	1	0.4
26	Kuwait	4,716	0.2	190	38	4	9.5
27	Helsinki	4,244	0.1	209	50	1	2.0
28	Vienna	3,182	0.1	567	64	38	37.2
29	Athens	527	—	11	114	0	0

(Total) (3,080,562) (100.0)

Source: The Stock Exchange, London.

trade twenty-four-hour positions. Nevertheless, stock exchanges across the world are national institutions with separate identities, and the emergence of a so-called 'Global Capital Market' is still some way off. The relative dominance of capital markets, at a world scale, by the New York, London and Tokyo exchanges accounts for Hamilton's (1986) description of this financial system as a 'three-legged stool'.

In the secondary (rather than new issues) market for equities trading, the New York and Tokyo stock exchanges account for about two-thirds of world market value (Table 6.1); London (7 percent share) follows as a 'distant third' but it far outranks other financial centers which comprise the lower order of a global hierarchy of national equity markets. Although the turnover (trading volume) of all major stock exchanges has grown considerably – the world average rising 37 percent in 1985 and 39 percent in 1986 (*Financial Times*, 17 March 1987, p. 23) – the rank order of this market hierarchy has been relatively stable through the 1970s and early 1980s (Bergstrom, Koeneman and Siegel 1983).

This decade has also witnessed the expansion of equity markets in Third World countries (Table 6.2). Between 1982 and 1987, total market capitali-

Table 6.2 The size of equity markets in selected developing countries, 1980 & 1985: Market capitalization and trading volumes, $ US billion.

	Number of listings		Market capitalization		Trading volumes	
	1980	1985	1980	1985	1980	1985
Africa						
Morocco	78	77	0.4	0.2	—	—
Nigeria	90	95	3.1	2.6	—	—
Zimbabwe	62	55	1.5	0.4	0.2	—
Asia						
India	2,265	5,751	10.4	29.4	2.8	5.7
Korea	352	342	3.8	7.4	1.9	4.1
Malaysia	249	283	12.4	16.5	2.6	2.3
Philippines	196	138	2.1	0.7	0.6	0.1
Singapore	261	316	24.4	33.5	3.7	2.9
Thailand	77	100	1.2	1.9	0.3	0.6
Europe						
Turkey	314	373	0.5	1.0	—	—
Western Hemisphere						
Argentina	278	227	3.9	1.4	1.1	0.5
Brazil	1,040	1,144	9.2	42.9	5.4	13.4
Chile	265	227	9.4	2.0	0.5	0.1
Colombia	193	102	1.6	0.6	0.1	—
Mexico	271	162	13.0	4.2	2.7	4.4
Venezuela	98	116	2.7	2.0	0.1	0.1

Source: International Capital Markets, International Monetary Fund, Washington, DC, 1986, page 73. Compiled from International Finance Corporation data.

zation of about thirty-five equity markets in developing countries rose
from $US 55 billion to $US 130 billion; further, before the debt crisis,
eight of these smaller exchanges surpassed the world average of stock
market performance (Behrmann, 1987). The key factor underlying stock
market expansion in the developing countries is the growth and multinational diversification of portfolio investments made by European, Japanese and US pension funds.

In addition to size distribution, the world's equity markets are also
differentiated by their location in three main time-zones – American,
European and Far East – which dictate official hours of trading activity (see
Figure 6.1). Competition between financial centers, at a very broad geopolitical scale, tends to be analyzed in terms of this zonal partitioning of the
international equities market. This is illustrated by the following exem-
plary comments from different sources:

New York's position is assured because the players are there, the
communications are in place and the markets are already huge. Other
centres of the USA and of Canada – which is having to debate the role
of its three main markets in Toronto, Montreal and Vancouver – are
going to have to survive on a speciality that they can internationalise
or, at least, become internationally famous for. (Hamilton, 1986,
p. 129).

One of the strengths of London itself is that it is ideally situated in a
time zone which permits dealing with the Far East in the morning,
New York in the afternoon and the European capitals for most of the
day. (Whiticar, 1985, p. 19).

Thus, apart from being a gateway to China, Hong Kong is centrally
located in the Asian–Pacific region. Its time-zone position enables it to
attract transactions in the forex and gold markets when London and
New York are closed (p. 35). . . . Most bankers feel that, in the long
run, the real competition (other than Singapore) may come from
Tokyo. (Jao, 1985, pp. 35, 45).

We are now witnessing the 'collapse' of these temporal boundaries in the
international equities market. The superficial expression of this 'collapse' is
recent changes in the trading hours of different stock markets: the London
trading floor now opens thirty minutes earlier (at 9.00 a.m.), and the
American and New York stock exchanges are examining the feasibility of
twenty-four-hour or round-the-clock trading (see 'Can New York's stock
exchanges afford to sleep', The Economist, 8 October 1984, pp. 67–8). These
changes have, however, resulted from new and different forces operating in
the world equities market, which relate directly to the use of information
and communications technology by financial institutions.

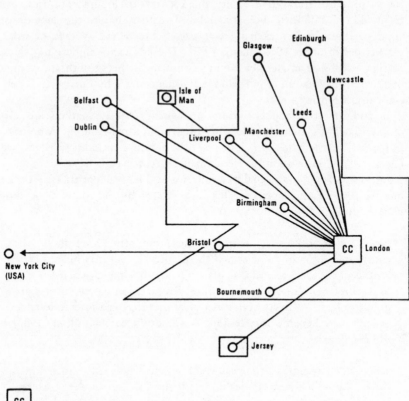

Glasgow
Edinburgh
Newcastle
Belfast
Isle of Man
Leeds
Dublin
Manchester
Liverpool
Birmingham
Bristol
CC London
New York City
(USA)
Bournemouth
Jersey

CC   Computer Centre (incl. on-line data bases)

O   Multiplexer/s supporting 12 Topic terminals each in local sub-network

———   Telecommunications link

Source : Author.

Figure 6.1   The London Stock Exchange topic network

First, equities markets are becoming increasingly global, as multinational corporations obtain listings on several different stock exchanges. By 1984, 236 companies were traded daily in international markets, and the incidence of these multinational listings of shares is growing rapidly.[1] The simultaneous trading of these equities depends on the new technology, particularly for the around-the-world transfer of trading positions among branches in multinational securities firms.

Second, national stock exchanges are now competing with computer trading systems, whose equity transactions are not physically confined to a central trading floor but are carried out on electronic dealing screens. By

1985 turnover in equities trading on NASDAQ (National Association of Securities Dealers Automated Quotation system) amounted to £179,580 million, making it third in rank to the New York and Tokyo stock exchanges at a global scale (see Table 6.1). As national equities markets evolve into communication systems, rather than physical locations, trading activity is becoming less constrained by the spatial and temporal parameters of stock exchange floors and the regulations governing their hours of operation.

And third, the new technologies have been used to create a widening array of financial products (futures, options, swaps, and so on), which compete directly with traditional securities as debt instruments. In particular, 'securitisation', or the 'conversion into paper of more and more borrowing than can be bought and sold in a secondary market' (Hamilton, 1986, p. 64), has further reduced the institutional and regulatory barriers which traditionally differentiated equities trading from other areas of the global capital marketplace. With the 'collapse' of these market barriers, the geography of securities trading threatens to be transformed: 'Globalisation and securitisation are tearing down most of the national barriers. All the exchanges are still testing their defences' (Hamilton, 1986, p. 77).

Information and the New Technology

The new information technologies, particularly transnational computer networks, play a critical enabling role in the globalization and integration of the world's equity markets. Throughout this century, new media of communications have effectively closed the 'information gap' between international financial centers – that is, the time delays in communicating price information which largely contribute to inter-market price differentials (Garbade and Silber, 1978). The impact of this process of technological change in world capital markets is clearly illustrated by the following real-life example of a New York dealer's experience:

> Mr. Cass started as a trader in 1965, a few years after the *telex* replaced *cablegrams* as the means for handling trans-Atlantic transactions. The telex meant that the time required to complete a transaction went from a few hours to several minutes. By 1970, *telephone* service replaced the telex as the preferred medium for international dealing. Then about four years ago, the Reuters *network and video screens*, allowing virtually instant access to dealing rooms worldwide, became the international standard (*New York Times*, 23 March 1987, p. 25).

In addition to the 'Reuters network', international securities trading is now supported by several on-line financial information services – for example, Telerate and Quotron in the US and Extel and Datastream in

Europe. A new entrant to this burgeoning information market, US-based Automatic Data Processing (ADP), offers real-time stock price quotations, together with company, financial and international news and information, from a wide range of databases; all of this information, collected from different stock exchanges, is disseminated to dealer workstations from three large computer centers located in Mount Laurel, New Jersey (serving US investors), Farnborough, Hampshire (Europe) and at a planned site in the Far East/Pacific Rim area.[2] The growth of this on-line financial information market led Walter Wriston, Chairman of Citicorp (1970–84) to remark: 'Information about money has become almost as important as money itself' (cited in Hamilton, 1986, p. 32).

In a recent report prepared for the European Commission, Information Dynamics (1985) identifies four main types of on-line financial data services which are currently distributed over private and public computer networks: trading-related data (e.g. share prices), company-related data (e.g. annual reports), financial news (e.g. interest rate changes) and economic and political information. The principal users of these network information services are major international banks, securities trading houses and multinational corporations. Present technical trends are towards the merging of variegated services at user workstations, where special decision-support software can be applied to integrate and manipulate financial data from internal and external sources, and the consolidation of separate computer networks into broadband, high-speed transmission systems with open access to larger user groups, including 'market-makers' in equity trading (see, for example, Nisse, 1987).

The emergence of new vendors of financial information poses a considerable threat to the traditional market-making role of the stock exchanges. Recognition of these competitive pressures led the Chief Executive of the London Stock Exchange to comment:

> Capital markets are all about information and the dealer will turn to the organisation which can provide the information he wants. In the days of domestic trading and before the technological era that information was best provided by people meeting in one place at particular times, using facilities provided by a stock exchange floor. The telephone and telex now allow international trading to take place without a physical presence. It is clear that the international capital markets lie not so much with those who provide a physical floor, where trading can take place, but with those who control the information systems. In order to maintain their competitiveness, the Stock Exchanges will have to become high technology companies, with the facility to transmit information to each other. (Knight, 1984, p. 15).

In addition to its Topic viewdata network for disseminating market information from the London trading floor, the Stock Exchange now operates a computerized trading system, Stock Exchange Automated Quotation (SEAQ) International, on which prices are displayed by forty market makers for a total of 460 non-UK stocks (Hepworth, 1987). The recent expansion of the SEAQ network is central to the Exchange's plans to make London the world's leading center in the burgeoning market for internationally traded shares. Market makers outside London are increasingly contributing prices to SEAQ: the trading system is being extended worldwide to include the US, Canada, Japan, Hong Kong, Australia, South Africa, France, Germany and Scandinavia.

Other examples of the electronic interconnection of stock exchanges include 'link ups' between Chicago and Toronto and Tokyo and Amsterdam (Maranoff, Tate and Whitehouse, 1987). In addition to the London Stock Exchange's transatlantic interconnections, UK-based NMW Computers now operates a satellite link-up with the Toronto Exchange's Computer Assisted Trading Systems (CATS), which enables British investors and stockbrokers to carry out automated transactions in Canadian stocks (see *Financial Times*, 21 May 1987, p. 17). Full-screen dealing within Europe, under the aegis of the Committee of Stock Exchanges in the European Commission, is expected to come about by 1990.

The London Stock Exchange's SEAQ network, as a computerized system that replaces the trading floor, is similar to the US-based NASDAQ network. The latter, established in 1971, is effectively an electronic 'over-the-counter' marketplace for equities trading, and the size of the computer network (centered on data centers in Turnbull, Connecticut) has grown from a few thousand terminals in 1972 to 40,000 in 1978, to more than 120,000 in 1985 (Hamilton, 1986, p. 43). About 10 percent of NASDAQ terminals are located outside the US, the main locations being Canada, Britain and Switzerland (Walton, 1985, p. 14). The entire transnational network depends on leased telephone lines for simultaneously communicating share prices across the world, recording transactions and supporting other market-making functions.

In order to keep pace with this process of market internationalization, investors and securities firms are applying the new technology (computers and data communications networks) to keep track of their exposure in different stock markets at a global scale. For example, I. P. Sharp Associates, the Toronto-based computer services bureau (see Hepworth, 1987) which was recently taken over by Reuters, markets a value-added network for worldwide currency trading (see *The Banker*, March 1987, pp. 49–53). These types of twenty-four hour, computer network-based trading systems are increasingly used to transfer 'books' of trades (portfolios of financial instruments) between different regional branches of multinational financial houses:

From London the book would be sold to New York, which in turn would sell it to Tokyo eight hours later, and in another eight hours,

Tokyo would sell it back to London. . . . Some observers warn that this around-the-world (book passing) could prove to be an economic nightmare of global proportions to tax collectors, currency regulators, and the whole trading industry. (Maranoff, Tate and Whitehouse, 1987, p. 77).

Whilst the new technologies have provided investors and brokers with a real-time 'window' onto the global securities market, through transborder flows of variegated financial information, they have also contributed to 'an international struggle for territory between communications corporations, between the exchanges and between the non-floor trading systems' (Hamilton, 1986, p. 47). The basis for this 'territorial struggle', however, is not only a new technological context for equities trading but also the deregulation of capital markets across the world, including the relaxation of national controls on the movement of financial capital.

Deregulation and the 'Big Bang'

The global restructuring of equity markets has depended on both the timely flow of information (enabled by new computer network technology) and the ability to move very large volumes of capital across national boundaries, as well as the growth of foreign direct investment in financial services to support multinational securities transactions. With the removal of foreign exchange controls (or their equivalent) in different countries over the past decade, there has been a massive increase in international capital movements. For example, new issues of Eurobonds, securities sold outside the home country of the borrower, grew from $26.7 billion in 1981 to $200 billion in 1986 (see *The New York Times*, 22 September 1986, p. 34). A recent study carried out by the US Securities and Exchange Commission (SEC) estimated that Americans purchased a record $102 billion in foreign stocks through 1986, whilst foreigners invested $277 billion in US stocks (*Financial Times*, 14 May 1987, p. 36). In the two years to the end of 1985, Japan's outflow of portfolio capital rose from $16 billion to $60 billion, of which 78 percent went to the US (*Financial Times*, 7 May 1987, World Banking Supplement, p. II). The year following the British government's removal of foreign exchange controls (1979) witnessed a dramatic increase in the share of (net) overseas investment made by UK pension funds (8.7 to 22.6 percent) and insurance companies (6.0 to 12.9 percent (figures from Coakley and Harris, 1983, p. 39)). The scale of this internationalization of capital, as it is expressed by bank and pension fund holdings of foreign securities, is illustrated by the selected figures presented in Tables 6.3 and 6.4.

Parallel and linked to these transborder flows of portfolio capital is the globalization of the financial services sector. The largest 'players' in the

Table 6.3 Estimated size and non-domestic exposure of private pension funds in selected countries, 1980 and 1986, $ US billion.

	1986 Total assets	Holdings of foreign securities	Percentage of total	1980 Total assets	Holdings of foreign securities	Percentage of total
United States	1,250.0	45.0	3.6	330.0	3.3	1.0
Japan	145.0	14.5	10.0	40.0	0.4	1.0
United Kingdom	243.0	56.6	23.3	108.0	9.7	9.0
Netherlands	85.0	8.5	10.0	38.0	1.5	3.9
Canada	62.0	5.6	9.0	29.6	2.0	6.9
Switzerland	65.0	3.3	5.1	33.0	1.3	3.9
West Germany	48.0	1.9	4.0	25.0	0.5	2.0
Australia	12.0	1.8	15.0	5.0	—	—
France	12.0	0.2	1.7	4.0	0.7	17.0

Figures include both equity and debt: equity at market value and other asset at book value. Amounts converted to US dollars for each year at the then-effective exchange rates.
Source: Financial Times, London, March 17 1987, p. 23. Compiled from Goldman Sachs data.

Table 6.4 Bank holdings of international bonds and other long-term securities, 1981–June 1986 (in billions of US dollars).

	Outstanding amounts at end of period					
	1981	1982	1983	1984	1985	June 1986
Estimated total holdings[1]	46.7	59.2	76.7	99.5	157.7	—
Holdings of banks in United Kingdom[2]	18.1	23.5	32.0	40.7	62.0	68.1

Source: Bank of England.
[1] Estimates based on the holdings of international securities by banks in Belgium, Canada, France, the Federal Republic of Germany, Italy, Luxembourg, the Netherlands, Sweden and the United Kingdom (see note 3), as well as the consolidated holdings of Japanese banks booked at head office and at all domestic and foreign branch offices plus their holdings at merchant banking subsidiaries located in London net of possible double counting.
[2] Including holdings of short-term certificates of deposit.

Table 6.5　Presence of selected financial institutions in the London, New York and Tokyo stock markets, 1986.

	London		New York		Tokyo	
	GBD	SE	GBD	SE	GBD	SE
Citicorp	✕	✕	✕	✕	✕	✕
Bank America			✕	✕	✕	
Chase Manhattan	✕	✕	✕	✕	✕	
Bankers Trust	✕		✕		✕	
Security Pacific	✕	✕		✕	✕	
Morgan Guaranty	✕		✕		✕	
Merrill Lynch	✕	✕	✕	✕	✕	✕
Salomon Brothers	✕		✕	✕	✕	
Morgan Stanley	✕		✕	✕	✕	✕
Goldman Sachs	✕		✕	✕	✕	✕
American Express	✕	✕	✕	✕		
Barclays Bank	✕	✕			✕	
National Westminster	✕	✕			✕	
Midland Bank	✕	✕	✕			
Kleinwort Benson	✕	✕	✕			
Mercury Int. Group	✕	✕		✕	✕	✕
Paribus		✕			✕	
Deutsche Bank	✕	✕		✕		
UBS	✕	✕		✕		
Swiss Bank Corp				✕		
Credit Suisse	✕	✕			✕	
Nomura		✕		✕	✕	✕
Daiwa				✕	✕	✕
Nikko				✕	✕	✕

GBD – Government Bond Dealers; SE – Member of Stock Exchange
Note: Many banks are represented through part-owned subsidiaries. Most US banks are members of the New York State Exchange through limited service discount broking subsidiaries.
Source: Financial Times, London, April 2 1986.

world equity market are establishing regional/branch offices in all of the major financial centers. In the wake of the 'Big Bang' in London (27 October 1986), overseas institutions have purchased minority or 100 percent stakes in Stock Exchange member firms, such that 'about 60 percent of this last "English" enclave of the City is now foreign-owned' (Goodhart and Grant, 1986, p. 7). The internationalization of Japan's domestic market in securities trading has occurred more slowly: by 1983, only six foreign securities firms were licensed to open a total of seven branches, although some seventy-seven companies operated resident offices to collect and offer information on securities markets; at the same time, and in sharp contrast, Japanese securities companies had established a total of ninety-six branches or subsidiaries in twenty-three major cities distributed across seventeen countries (Japan Securities Research Institute, 1984).

The timing of financial services deregulation, or 'Big Bangs', differs significantly between countries:

> That the bans should all be going off at once is partly coincidence: the politics of deregulation in Australia, the UK and Japan are quite different, and in the US there has conspicuously been no deregulation at all: the Glass-Steagall Act which prevents banks underwriting corporate securities looks almost as solid as ever, though this has not prevented US institutions taking the lead in foreign and offshore markets. But there is a common fuse: the wish to stimulate competition and, in many countries, attract more international finance business (Lascelles, 1986, p. 17).

In addition to restrictions on foreign membership of stock exchanges, which are generally overcome through the use of local brokers, non-tariff barriers affecting the internationalization of capital markets include statutory restrictions on the securities trading activities of banks, which relate to the US in particular. The Glass–Steagal Act (1933) states that affiliates of banks can not be 'principally' engaged in securities business. Thus foreign securities firms, which are normally affiliated to banks whose parent organizations are allowed to undertake a full range of investment activities in their domestic markets (for example, Britain), are effectively confronted by a significant barrier to US market entry.

A plethora of other non-tariff barriers currently limit the degree to which national capital markets can be effectively integrated. For example, the Japanese Ministry of Finance applies an administrative system ('red tape') to foreign firms seeking branch status, which is generally regarded to be 'extremely slow and cumbersome'; the procedures imposed by the SEC in the US to ascertain the competence of foreign management have resulted in 'some confusion'; and, the Italian system for registering and settlement is regarded as 'archaic, inefficient and wholly unsuited' to catering for the increasing volumes of stock traded.[3] Although the European Commission has targeted 1992 as a deadline for the complete unification of the Community's financial services market (see Pastre, 1986, p. 29), progress in harmonizing different trading and settlement procedures is slowed down by international competition between London and the fast-growing continental 'bourses' (for example, Paris and Milan), once regarded as 'poor relations' by the London and New York stock exchanges (see *The Economist*, 21 March 1987, p. 81).

The 1980s have, indeed, witnessed a continuous stream of policy changes – such as the abolition of withholding tax or stamp duty – which have reduced equity transaction costs, as part of a general thrust towards capital market deregulation. International competitive pressures have acted to create a 'domino effect', such that different countries, regardless of their political ideologies, have rapidly succumbed to a 'Big Bang' of global proportions. As noted by Hamilton (1986, p. 181):

Socialist governments in Australia and France might believe that world recession and industrial restructuring signalled the need for greater direct control of banks in the interests of domestic markets. But once exchange controls on most of the major currencies had been abolished in the 1970s and once the big financial centers started to deregulate in earnest in the 1980s, it became impossible for individual countries to stand out against change.

The 'fallout' of the 'Big Bang' in the world's security markets, and capital markets in general, does raise a number of important political questions, including the implications of deregulation, internationalization and technical change for national economic sovereignty. In the last section of this paper we will briefly review these questions, which range from international trade issues to concerns over the appropriate institutional mechanisms for regulating trading activity in a global equities market where the electronic 'convergence' of capital and information has increasingly rendered national and market boundaries 'invisible'.

The geopolitics of capital

As stated above, the globalization of capital markets, including equities trading, acts to limit the scope for the autonomous pursuit of national economic policy (Llewellyn, 1980). This applies most obviously to *monetary policy*, where higher international capital mobility may undermine government attempts to tighten or ease financial conditions, due to the greatly expanded role of the 'Rest of the World' as a residual source of demand for excess domestic liquidity and a residual source of supply of funds. As a result, national governments are increasingly forced to depend on fiscal policy for the stabilization of domestic economies at the expense of more flexible and (in the short run) politically insulated monetary policy instruments (Cooper, 1972).

National controls over *exchange rate policy* are similarly affected. Today, exchange rate movements are dictated more by rates of return in different world financial centers than the relative prices of goods exchanged between countries, such that they may operate to generate, and not to correct, balance of trade and payments problems at the national level. Thus, even if foreign exchange controls were to be reimposed, it is unlikely that they would be effective under current conditions of electronic methods of global information and capital transfer.

In the area of *business taxation and regulation*, national governments are confronted by the more effective concealment of interest income and repatriated profits through the use of global electronic funds transfer systems within multinational corporations. For example, Touche Ross International (TRI), accountants, markets a computerized 'World Tax

Planner' to multinational corporations, which selects electronic communication routes for repatriating funds designed to 'slip the tax net' of different countries:

> (TRI) built a data base of information on the tax systems of more than 185 countries, together with an analysis of corporate and withholding tax rates. To this was added information on tax treaties and other news and developments, all up-dated monthly. Then came the big challenge: to design a system to use this information in planning international money flows, with the aim of helping a client to pay the minimum amount of foreign tax on dividends coming back to Britain. Now using an IBM or compatible desk-top personal computer, TRI's tax experts worldwide can gain, within minutes, detailed print-outs of up to 10 alternative recommended routes for repatriating money, some via several countries. The programme – contained on hard disc, with monthly updates on a floppy disc mailed from TRI's London office – is capable of carrying out 5,000 calculations every two minutes (*The Sunday Times*, 22 October 1985, p. 42).

In addition to the new and different issues raised by transborder data flows of financial information and capital for national economic policy, the internationalization of equities markets presents a major challenge for the institutions which regulate and supervise dealing activity, particularly in the areas of market surveillance and enforcement. In Britain, 'policing' securities markets, following deregulation or the 'Big Bang', has been left to the so-called 'Self-Regulatory Organizations' (SROs), under the overall aegis of the Securities and Investment Board. Although a new organization of foreign securities houses, the International Securities Regulatory Organization (ISRO), has been established recently, the essential problem is that

> no single SRO has sufficient surveillance information or broad enough disciplinary jurisdiction fully to regulate conduct affecting its market, nor (in the US case) does the SEC have the information or jurisdiction to regulate conduct occurring beyond the country's national borders that has consequences in the US securities markets (Scribner, 1986, p. 24).

With the globalization of capital markets, some British observers have argued that domestic industry has been 'starved' of investment funds needed for the renewal of plant and machinery (Minns, 1982).In addition to these putative effects of the relative growth of foreign portfolio investment (Coakley and Harris, 1983), particularly investments made by UK pension funds and insurance companies (see Table 6.3), it is suggested that the 'Big Bang' has encouraged 'short termism' amongst financial intermediaries, such that British industry has needed to invest for short-term profits rather than for long-term trading and production. According to Goodhart and Grant (1986, p. 8), for example:

> Industry is also worried that Big Bang will encourage the short-term investment horizons of many institutions – now even criticised by the Chancellor himself. This 'short-termism' is caused by the competitive monitoring of fund managers, and has helped contribute to the 'merger mania' of the past few years; fund managers are unable to resist a bid premium. Big Bang will further loosen the link between shareholders and companies by increasing turnover in the equity market and by spreading the shares of larger companies all over the world, thus placing them in the hands of investment managers who will have no concern for their long-term interest. The fear of takeover will thus concentrate management attention on keeping the share price up and artificially boosting profits – instead of making the long-term investment that everyone agrees is desirable.

In Britain, at least, there is little academic research which might illuminate the social and economic issues raised by the generalized impact of capital market internationalization on domestic industries, the financial base of government policies, employment and social investment programs. The state of our current knowledge of how transborder data flows affect the balance of trade in financial services and on-line information services is similarly underdeveloped. Yet, both of these types of services trade, supported by transnational computer networks, represent the fastest-growing elements of the world economy, and financial services have now entered the General Agreement on Trade and Tariffs (GATT) negotiations.

Conclusion

The new information technology, and transnational computer networks in particular, have played and continue to play a critical enabling role in the global restructuring of equity markets. However, the accelerated 'collapse' of space and time in the world's capital markets has also derived from major institutional changes in different countries, which have eroded national barriers to the movement of portfolio capital and foreign direct investment in the financial services sector. Thus, the role of information and the new information technology in global capital market restructuring must be examined in this broader context of the internationalization of capital.

Against this background of global information and capital transfer, whose unprecedented scale far outweighs the current value of international merchandise trade, national governments are confronted by major challenges to their economic sovereignty. Indeed, the converging financial and information 'revolutions', presently underway across the world, raise new and different issues for social and economic policy, which would have yet to be thoroughly explored by researchers from different disciplines, including geographers. The urgency of this research effort has been underlined by recent events in the world's capital markets, which 'collapsed' through October 1987 and virtually precipitated a financial crisis leaving no country unaffected.

Notes

1   Annual estimates of the multinational listing of shares are published in *Euromoney* magazine.
2   ADP example is cited in *The Financial Times*, 17 February 1987, p. 12.
3   All examples taken from the *International Securities Digest*, No. 10, January, 1986, The London Stock Exchange.

References

Behrmann, N. (1987), 'World Bank body sees pension billions going to smaller markets', *The Melbourne Age*, 2 July, p. 6.
Bergstrom, G., Koeneman, J., and Siegel, M. (1983), 'International securities market', in Fabozzi, F. (ed.), *Readings in Investment Management* (Homewood, Ill.: Richard Irwin), pp. 210.36.
Coakley, J., and Harris, L. (1983), *The City of Capital* (Oxford: Basil Blackwell).
Cooper, R. (1972), 'Towards an international capital market', in Dunning, J. (ed.), *International Investment* (Harmondsworth: Penguin), pp. 220–40.
Garbade, K., and Silber, W. (1978), 'Technology, communication and the performance of financial markets', *Journal of Finance*, 3, pp. 819–32.
Goodhart, D., and Grant, C. (1986), 'The internationalisation of Capital', *New Statesman*, 24 October, pp. 6–9.
Hamilton, A. (1986), *The Financial Revolution* (Harmondsworth: Viking/Penguin).
Hepworth, M. (1987), 'Information services in the international network marketplace', *Information Services and Use*, vol. 7, pp. 167–81.
Information Dynamics (1985), *Financial Information Services in Europe*. Report to Directorate-General for Information Market and Innovation, Commission of the European Communities, Brussels.
Knight, J. (1984), 'The interconnection of European stock exchanges', *The Stock Exchange Quarterly*, December, pp. 14–18.
Jao, Y. (1985), 'Hong Kong's future as a financial centre', *The Three Banks Review*, 145, 35–53.
Japan Securities Research Institute (1984), *Securities Market in Japan* (Tokyo: JSRI).
Lascelles, D. (1986), 'The Stampede to become global players', *The Financial Times*, 2 April, p. 17.
Llewellyn, D. (1980), *International Financial Integration* (London: Macmillan).
Maranoff, J., Tate, P., and Whitehouse, B. (1987), 'Around the world in 24 hours', *Datamation*, 15 January, pp. 75–7.
Minns, R. (1982), *Take Over the City* (London: Pluto Press).
Nisse, J. (1987), 'Wiring for profit', *The Banker*, June, pp. 87–93.
Pastre, O. (1986), 'Now for Europe's Bigger Bang', *The Financial Times*, 5 November, p. 29.

Scribner, R. (1986), 'The technological revolution in securities trading: can regulation keep up?', in Saunders, A., and White, L. (eds), *Technology and the Regulation
   of Financial Markets* (Lexington, Mass.: D. C. Heath), pp. 19–29.
Walton, P. (1985), 'NASDAQ: up and over the counter', *The Accountant*, 2 October,
   pp. 14–15.
Whiticar, J. (1985), 'The UK capital market', *The Treasurer*, February, pp. 17–21.

7 Internationalization, telecommunications and metropolitan development: the role of producer services

P. W. DANIELS

The significance of producer services: some generalizations

Structural change in advanced economies are symbolized by the continuing growth and diversification of producer services as employment in other sectors, most notably manufacturing, has continued to contract (Elfring, 1989; Gershuny and Miles, 1983; Riddle, 1986). It may be helpful to begin with some generalizations about producer services derived from the results of recent research: first, producer services are tradable both within organizations and on a commercial basis on the open market; second, the output of producer services is exportable; they are not dependent only upon local demand; third, these services are part of the supply side of a metropolitan or regional economy; fourth, they therefore have a role to play in adjustment to structural change; fifth, producer services are not necessarily footloose with respect to locational choice (even though they are tradable); sixth, large producer service organizations are seeking to maintain market share through active market-making and integrated global strategies; small producer service firms are oriented towards market niches allied with specialization in their output (global strategies are less significant); seventh, internationalization is a major part of the growth and development strategies of large producer service organizations; eighth, agglomeration and urbanization economies are significant for the location of many producer services with telecommunications helping to sustain rather than to dilute the importance of these economies.

Advanced economies have become much more 'open' as rapid advances in telecommunications technology, growing mobility of labor at international levels, the innovativeness of producer service firms and the growing pressure for liberalization of trade in services (see, for example, Siegel, 1987; UNCTAD, 1988, 1989) has made it more difficult for service organizations and metropolitan areas to survive purely on the basis of trade/interaction within the confines of their own national boundaries.

These circumstances, together with some of the attributes listed above, have increased competition at the global scale; metropolitan areas are invariably the key nodes in international communications networks and are therefore also the foci for the globalization strategies of producer service organizations (see for example, Rimmer, 1990).

As internationalization of services continues to gain momentum it poses some new research questions. These include its effects on, and contribution to, trade in service; (McConnell, 1986; Noyelle and Dutka, 1986; UNCTAD, 1989); its consequences for the division of labor (Bertrand and Noyelle, 1986); for methods of delivery; and access to the specialist knowledge embodied in many of these services. The spatial consequences of these processes are also important both at the global level (Dunning, 1989; Dunning and Norman, 1987; Noyelle and Dutka, 1987) and within national space economies (Daniels, Leyshon and Thrift, 1986; O'Connor, 1987), especially for the relative growth of the urban areas that are usually the focus of locational choice by services choosing to operate offshore. Such choices will clearly be associated with employment effects, the reinforcement of information and related linkages and the stimulation of commercial office and residential property markets (see for example, Noyelle, 1986, 1987a).

Meaning of internationalization of services

Internationalization may be viewed, therefore, from a number of perspectives: as the transborder flows of business travellers, foreign currency transactions, information and data generated by the business activities of service, manufacturing and government organization; as a response to domestic regulations restricting services trade; as part of acceptance of and rise in international lending; as a by-product of the expansion of multinational enterprises (MNEs) and their service affiliates in accounting, advertising, insurance as product development to meet the particular needs of markets outside the national context, i.e. localization as part of the globalization strategies of large corporations; as the expansion overseas of business organizations in order to achieve direct representation in foreign markets (this may take the form of merger/takeover, agency or direct establishment of a new office/manufacturing plant).

Schwamm and Merciai (1985) enumerate five ways in which services can be provided internationally: first, exchange of data, images or people between countries; second, granting licensing contracts to suppliers in overseas markets; third, consumers can travel to another country to purchase/benefit from a service; fourth, service producers travel to another country; fifth, sale of services through a foreign affiliate. Whichever approach is used (and some may be used in parallel by the same organization) it is important to note that while a manufacturer can export a product without necessarily being present in the buying country this is not the case with services (Schwamm and Merciai, 1985). Since services are intangible it is often not possible for them to be effectively supplied without the physical presence of the producer or an agent/representative. The legislative or regulatory environment in individual countries may prevent cross-border transactions, therefore necessitating direct representation. Finally, successful marketing of services requires proximity to prospective and existing clients if suppliers are to attract a larger share of a market.

World trade in services

The growth of world trade since 1945 has been accompanied by much greater global economic interdependence. This has been substantively reinforced during the last decade by widespread relaxation of constraints on capital flows and the dramatic developments in communications, computers and electronics that have created a global market for financial and other advanced services (Sauvant, 1987; OECD, 1983, 1984). It has been estimated that global service exports in 1980 accounted for some 20 percent of world trade valued at $350 billion (US Government, 1983). The UK is only marginally behind the US as the leading exporters of services ($37 billion) followed by West Germany, France, Italy and Japan.

It remains extremely difficult to measure the share of producer services in national exports; one US study quoted by Noyelle and Dutka (1986) estimates the foreign revenue from business services to be 25 percent with about half of this proportion coming from sales of major producer services such as advertising, management consultancy, accountancy and legal services. Only some 10.4 percent of Canadian Current Account receipts in 1984 were contributed by services (Siegel, 1987) largely composed of trade with the US. The situation might improve following ratification of the new open trade agreement between Canada and the US (signed in mid October, 1987) (Harrington, 1989). The agreement, the first to include comprehensive treatment of services, will ensure that each side will provide treatment to each other's citizens that is not less favorable than that granted to its own citizens with respect to all new measures affecting services. Financial institutions in each country will be able to compete with each other, there will be fewer restrictions on cross-border investments and steps will be taken to develop an open and competitive telecommunication and computer services market.

It is hoped that the ratification of this agreement will lend more weight to US efforts to obtain worldwide liberalization of trade in services which are still not included in GATT. Vast sums of money are now being invested worldwide in the internationalization of advanced service operations and a large share of the investment is being placed in telecommunications hardware and software. Gottman (1983, p. 19) notes how telecommunications has altered 'considerably the significance of distance . . . so that for . . . the transmission of information across space, distance has become a secondary consideration. What matters is the organization of the network installed to transmit, not the physical fact of distance and the extent of space.'

The international provision of services has become a dynamic process involving both trade (even though services are often thought of as untradable) and foreign direct investment (FDI) (Noyelle, 1987b). Air transport, shopping and related travel are predominately trade oriented while producer services such as banking, other financial services, professional services and telecommunications are largely oriented towards FDI as the preferred method of international provision. The latter is encouraged by the tight domestic regulation of many services which have important social and political roles in national economies. Thus the services share of FDI outflows from the UK rose from 41 percent in 1970–1 to 49 percent in 1971–80 and in Japan from 20 percent in 1975 to 67 percent in 1984 (United Nations, 1983). Producer services' contribution to foreign FDI is not easy to quantify but seems to range between 25 and 50 percent (OECD, 1981).

It is likely that the majority of international trade in services is accounted for by multinational enterprises (MNEs) (Enderwick, 1988; Stern, 1985) with a good deal of the activity controlled by transnational corporations (TNCs) (Clairmonte and Cavanagh, 1984). TNCs can be further subdivided into Transnational Service Conglomerates (TSCs) that operate in two or more service sectors and Transnational Integral Conglomerates (TICs) which cover industrial and service sectors. During the last decade a number of major firms in accountancy and advertising, for example, have become TSCs through merger or takeover of overseas firms or through the establishment of agency or representative office managements at offshore locations (see, for example, Leyshon, Daniels and Thrift, 1987a). Control of these conglomerates remains highly centralized with respect to administration, finance or corporate planning but this is combined with a decentralized system of service distribution to clients throughout the world. Telecommunications is clearly vital for consolidating such networks, ensuring effective two-way exchanges of information relating to many aspects of conglomerate business activity. TICs are, in many respects, even more significant than the TSCs and their predecessors, the manufacturing conglomerates. They permit the amalgamation of almost all the stages in the production process from design through assembly of a product to marketing and promotion for sale to a consumer or client. Because of such vertical integration the 'internationalization of services is being prodigiously speeded up' (Clairmonte and Cavanagh, 1984, p. 217).

Some of the trade generated by these trends is visible, such as business travel, but a large proportion is not visible and includes business services and foreign direct investment. The internationalization of services must therefore be examined in relation to the growth of multinational enterprises which have developed to overcome international transaction costs (the legal and other impediments in the way of individuals and corporations crossing frontiers to provide services) (Caves, 1982; Dunning, 1989; Rugman, 1981, 1987). The structure of an MNE permits control of overseas subsidiaries where production and distribution of a service which it owns can take place while still retaining ownership of any proprietary information (Dunning and Norman, 1983, 1987). Service activities are driven by their ability to market their services and successful MNEs are able to adapt quickly to changes in consumer requirements and identify and occupy niches in their markets; economies of scope are of much greater importance than economies of scale. The extent to which MNEs produce services in-house rather than contracting-out is also important. Three factors are relevant to this decision: the frequency of the transaction, the uncertainty of the transaction, and the asset specificity of the transaction (Williamson, 1986). MNEs can consider producing more services in-house because they are large enough to benefit from scale economies if they centralize production of the service. Subsidiaries purchasing these services from their parent firms will therefore be engaging in international trade. Finally, MNEs will also want to purchase services of known quality and standard for their operations around the world and this has encouraged service firms to meet demand for such global services by creating global service networks that act as efficient channels of trade for such services (Feketekuty, 1985).

Internationalization of large accountancy firms

Thus far large producer service firms have been developing as TSCs rather than TICs. An example is provided by accountancy; the leading twenty firms in the UK employed more than 300,000 people worldwide in 1986 (International Accounting Bulletin, 1986, Table 7.1). The internationalization of these firms has only recently received detailed attention but it is not a new phenomenon; it has evolved over several decades. Some firms have indeed only established international links since 1975 but a number of others had gone multinational before 1900.

There are two clearly distinguishable historical periods in the internationalization of large UK accountancy firms. The 'early' period began in the 1890s and continued until 1939 while the second or 'late' period commenced in 1945. During the early period the impetus for the development of an international operating capacity came entirely from the UK as accountancy firms based in the City of London followed the movement of manufacturing capital into overseas markets. The second phase has involved a more structured extension of multinational activities via the formation of international partnerships. A large proportion of these have

Table 7.1   The ten leading accountancy groups worldwide, 1985.

Firm	Fee income ($ m)	No. countries	No. offices	No. partns.	No. prof. staff[1]	All other staff	Total staff
Arthur Anderson	1,574	47	191	1,630	21,336	6,836	29,802
Peat Marwick Mitchell	1,445	89	335	2,533	20,482	6,849	29,864
Coopers & Lybrand	1,410	98	519	2,850	n.a.	n.a.	36,000
Price Waterhouse	1,234	94	378	2,113	20,656	7,603	30,372
Ernst & Whinney	1,185	77	359	2,199	17,201	5,600	25,000
Arthur Young International	1,060	68	370	2,560	17,640	6,600	26,800
Touche Ross International	973	90	463	2,550	17,500	5,950	26,000
Deloitte Haskins & Sells	953	63	433	2,125	16,621	5,266	24,012
Klynveld Main Goerdler	n.a.[2]	73	487	3,215	19,300	12,817	29,766
Grant Thornton International	479	57	449	1,419	n.a.	n.a.	14,800
Total: Top Ten	10,313	—	3,984	23,194	150,936	57,521	272,416
Top Twenty	12,910	—	6,230	30,716	187,262	71,818	333,544

Notes:
[1] Excluding partners.
[2] 1983 the income was estimated as $1m (*International Accounting Bulletin*, 1983).
   Source: International Accounting Bulletin (1986a).

been between British and American firms. The purpose of each merger is to strengthen the overseas representation of the participants and to facilitate the penetration of hitherto unentered national markets. The increasing scale and diversity of MNEs and the requirement for them to produce consolidated accounts based on returns from each of their operating units has also encouraged accounting firms to extend their international office networks. By doing so, the large accounting firms increased their market share of the world accounting market.

  In consequence the total number of offices operated by the twenty largest firms increased by some 115 percent between 1975 and 1985, from 2,323 to 4,991. In both 1975 and 1985 the United States, Canada and Europe accounted for over two-thirds of the total number of offices operated worldwide but Europe now contains more offices in absolute terms than North America following a 160 percent increase in the number of offices operated there by the leading twenty firms; more than 1,000 offices were opened in Europe by the twenty firms between 1975 and 1985. Some 32 percent of the European office growth has taken place in the UK and the Netherlands alone. Nevertheless, the USA has been the location for almost 25 percent of all large accountancy conglomerate offices in 1985. The largest relative increases have occurred in Asia (290 percent) and in Australasia (138 percent). More than 50 percent of the new offices in Asia have been located in Japan, Malaysia and the Philippines. Remaining regions in the world, such as Latin America, have experienced a decrease in their share of accountancy offices.

Internationalization of banking

Financial centers around the globe are vying with each other for a share of the expanding international trade in financial and other services. London's share of international banking, for example, declined between 1980 and 1986 as New York, Tokyo and various Offshore centers became more competitive (Table 7.2) (see also Lamb, 1986). The number of foreign banks in London has stabilized at around 400 but has risen to 356 in New York; however, the number of staff has grown by 25.9 percent (1985–86) to 53,883 in London compared with 15 percent, to 29,221, in New York (*The Banker*, 1987).

The internationalization of financial markets has been encouraged by a number of interrelated trends. There has been a steady contraction in the regulatory environment created by governments to control their national financial markets (Noyelle, 1988; White and Vitas, 1986). This has been encouraged by the ability of financial services to find ways of minimizing the effects of such restrictions (by the formation of Euromarkets in which the Eurodollar is preeminent, for example) although the Wall Street crash (1929), the crisis in Secondary Banking in the UK (1973–4) or the sharp fall in share values in stock markets around the world in late 1987 shows why governments are anxious not to allow financial services to operate in totally unregulated environments. Thus, the Glass–Steagall Act (1933) ensured a highly regulated environment in the US (see, for example, Holly, 1987) which, in the early 1960s, gave London an opportunity to become the center of the Euromarkets and allowed it to maintain its role as the world's leading international financial center. By 1981, however, the US had introduced International Banking Facilities (IBFs) which allowed national banks to circumvent domestic banking restrictions and to develop Euro-

Table 7.2 Market shares in international banking, by center, 1980 and 1986.

Center	1980	1986
London	27	23
EMS Center[1]	35	23
Offshore Centers[2]	11	21
New York	13	15
Tokyo	5	10
Other	9	8
	100	100

Notes:
[1] Belgium, Luxembourg, Denmark, France, Germany, Ireland, Italy, Netherlands.
[2] Bahamas, Caymans, Hong Kong, Singapore, Netherlands, Antilles and US branches in Panama.
Source: Data produced by Bank for International Settlements, *The Banker*, 157, 1987 (March).

markets within New York, which had recaptured nearly 10 percent of the Eurocurrency banking market. Some countries, such as Canada, continue to enforce a rigid separation between commercial and investment banking and this undoubtedly inhibits the development of Toronto as an international money center (Courchene, 1986).

Internationalization of financial markets has also been encouraged, secondly, by significant reorganization of the way in which international debt is arranged and the principals involved as intermediaries. Euromarkets remain the medium through which debt is funded but the emphasis has shifted from loans to bonds and from commercial banks to investment banks and securities houses. The banks have moved into much shorter-term financial commitments with exposure limited to the time that it takes to redistribute the issued bonds. Therefore it is of vital importance that the banks possess effective electronic distribution systems and are able to operate them at the global scale. As debt has become more securitized so the opportunities for trading in instruments of debt have increased; many are now traded globally for twenty-four hours. This requires participant firms to have representation in the appropriate centers, London, Tokyo and New York, and to be linked by highly sophisticated telecommunications systems. Continuous trading enables firms to respond immediately to sudden changes in the market and to eliminate overnight risks (Lambert, 1986).

British merchant/investment banks are beginning to establish branches in overseas markets. Before World War II there was little, if any, internationalization with overseas activities consisting mainly of independent banks. Schroder's had an effective New York office but Warburg's office failed and other UK merchant banks had low profiles there. Morgan Grenfell had no offices overseas twenty years ago; they now have offices in twenty different countries. Their prime purpose is to service multinational clients but without indigenous business they would not be viable and it is considered important to expand the local portfolio. Some UK-based banks have devised strategies based on maintaining a strong London office servicing branches overseas; others, such as Schroders (Travers, 1987) have gone for a more dispersed strategy which involves building separate investment banking bases in New York and in the Far East (Hong Kong and Tokyo) employing more than 60 percent of Schroder's total staff. The number of branch and representative offices of seven merchant banks has increased from forty-six to seventy-seven (including a number outside London but within the UK) between 1975 and 1985.

The internationalization of accountancy conglomerates, merchant/investment banks or property consultants (Leyshon, Thrift and Daniels, 1987b) is a process which involves a good deal of spatial discrimination. An emerging 'elite' comprising global financial/corporate complexes are the principal beneficiaries (see Friedmann, 1986). Although London is still acknowledged as the leading international financial center, New York is exerting more pressure. It is not only at the hub of the massive US domestic market but is at the heart of the twenty-four hour global securities market (Fairlamb, 1987) and is the headquarters location for some of the massive financial conglomerates expected to dominate the banking world by the end of this decade (Citibank, Merrill Lynch, Morgan Guaranty). All the professional producer services operating in the international arena must therefore be alert to the dynamic competition between global money centers and adopt spatial configurations that reflect the fluctuating relationships between them.

Significance of telecommunications

There can be little doubt that internationalization and expansion of trade in services has been made possible (if not accelerated) by rapid advances in telecommunications. While technology has revolutionized the economics of the telecommunications industry for the manufacturers of equipment and for the suppliers of the services, this has coincided with more diverse and testing demands from users as they become familiar with the real and potential value of telecommunications for enhancing business efficiency and competitiveness. Much of the growth in the installation and application of telecommunications is attributable to services, especially business services (*The Economist*, 1987; Langdale, 1989). In addition, the impact of telecommunications on the spatial development of services is irretrievably intertwined with questions of public policy: the public service monopoly that has long dominated telecommunications services in many parts of the world has become much more vulnerable since governments have begun to question its efficiency; improved services through deregulation leading to increased competition will likely bring costs down for users and thus encourage them to make full use of the power of telecommunications in their global strategies. It does remain to be seen, however, whether deregulation will actually mean that customers receive better services from a tightly organized monopoly or a free-reined competitive system (Crandell and Flanam, 1989; Moss, 1986a). The greater variety of services and equipment now available, for example, possibly causes confusion amongst businesses trying to identify the best services for their needs.

An example of the effect of deregulation is provided by a British company, Mercury, which was licensed in 1986 to provide services that compete directly with British Telecom's main network. The company is aiming to get a 5 percent share of the UK telecommunications market by 1990 and has been concentrating, initially at least, on long-distance traffic. Many of its customers are business service firms in the City of London, where there has been an explosive growth in demand for telecommunications in the period leading up to Big Bang and subsequent to it. Mercury believes that it has three principal advantages over British Telecom: price (long-distance calls up to 24 percent cheaper), quality of service (modern digital equipment, itemized billing), and diversification (companies have been transferring all or part of their business to Mercury as an insurance against breakdowns in British Telecom services). This is especially important for international banks, stockbrokers, foreign-currency dealers or insurance brokers for whom a breakdown in telecommunications and computing services would be disastrous.

The telecommunications industry is itself becoming more international through cross-border joint ventures, minority stakes in companies with strong local market positions or cooperative research and development projects. One reason is the sudden globalization of world markets. In the public switch market, for example, most of the large telecommunications equipment manufacturers were secure in their own home markets until the development of digital equipment; domestic markets are now barely sufficient to support the cost of continuously enhancing the efficiency of the present range of products and services using this equipment. The recent convergence of telecommunications and computer technology has been encouraged by user firms who no longer want data processed on computers in different locations, often thousands of miles apart, treated in isolation. Linking together remotely located computers, printers or files into systems (networking) is primarily a telecommunications task (see, for example, Hepworth, 1986; Holly, 1987; Langdale, 1989; Marshall and Bachtler, 1984). Likewise, telecommunications companies are themselves identifying network possibilities for computer users as a way of offering a complete package of services to their customers. IBM, for instance, has links with MCI, a long-distance telecommunications carrier, and with Rolm, a manufacturer in private telephone exchanges. Some companies are forming alliances with overseas partners as a swift way of filling gaps in their own products and services, e.g. British Telecom's takeover of Mitel (Canada) or Cable and Wireless, historically an operator with lines linked to British overseas interests, which has recently completed a number of overseas arrangements that will allow its 'global highway' concept based on a fiber-optic cable network from Western Europe to North America and on to Japan to be realized.

Value added network services (VANS) have been invaluable for electronic mail, financial information and network management services. By using these systems companies can save on the capital and staff they would otherwise need to provide the same services in-house. At the same time small companies can start their own VANS for a small start-up cost by 'piggybacking' them onto established systems. Furthermore, international telephone lines make it easy to develop international VANS such as EDINET, which carries documents across the North Atlantic in a joint operation between British Telecom and McDonnell Douglas. Information services based on computer database retrieval are good examples of VANS; the value of the economic data, text abstracting, securities trading information, or assessments of risk and financial implications facing organizations trading overseas, must outweigh their cost to subscribers. These services tend to be expensive, therefore, and only large companies (able to spread the benefits as widely as possible) can afford to purchase them, only large companies can set them up, and such services must be expensive to be viable. This suggests that the major users and suppliers will be multinational service firms, especially financial and business services. One difficulty with these systems is that millions of pounds worth of information cross international borders, traveling on wire or via satellite, without any means of monitoring by customs and excise.

There is another dimension to the impact of telecommunications and information technology on producer services. Instead of just being users of these services they are also increasingly involved in developing and marketing them. Specialist software packages for accountancy, banking or brokering are marketed by the organizations initially responsible for creating the demand for them. They may also sell access to their own databases or trade access for increased business while other specialist services sell information updated daily (if not hourly in some cases) on, for example, property deals, firms most likely to be seeking additional office floorspace, the locations likely to be required, and so on. There is therefore a massive market for computer systems and office and communications equipment. It has been estimated that 5,650 firms in banking and finance in Britain had over \$2 billion in installed computer equipment at the end of 1983 and that this will have increased to \$3.7 billion by 1988. Equivalent figures for insurance are \$1.5 billion in 1983, rising to \$2.8 billion by the end of the subsequent five-year period. With expenditure per employee averaging some \$1,400 in 1984 British service firms would seem to be heavily committed to information technology but expenditure per head compares unfavorably with US firms with an average of \$3–4,000 per employee and is less than the average for French or West German firms (*The Economist*, 1985).

A large proportion of these expenditures are made in the information intensive corporate complexes such as London, New York, Chicago, or Toronto and such differences in the investment might be interpreted as giving a competitive edge to the firms in high-spending locations. A recent survey of telecommunications (*The Economist*, 1987, p. 1), notes that there are 'more telephones in New York than the whole of Black Africa' and that the waiting list for telephones in India is close to the 1 million mark. But if technology is a necessary condition for success in the global market place it is not, on its own, a precondition. Much depends on the way it is used, for what purposes, and the human resources deployed to make decisions using the large volumes of detailed information available at high speed in fast-moving markets. The hardware is getting cheaper as user volumes increase but the software requirements are becoming more sophisticated and demanding; the cost of the human resources to produce the software is therefore rising. The problems are compounded by the need to link the word processors, facsimile machines, personal computers, mainframe computers and telex machines in order to derive maximum benefit from using them.

Technology and internationalization of service firms

The way in which technology has liberalized the development of service firms to incorporate an international dimension is illustrated by foreign exchange dealing. The development of this activity can be simplified into three stages. First, in the mid 1970s the emphasis was on large and expensive centralized computer systems dedicated to the automation of back-office accounting. During the late 1970s the first branch-based systems were installed, again mainly to handle accounting, and foreign exchange dealers became used to obtaining the dealing and related information they needed from video screens. Second, subsequent to the development of minicomputers it became possible to distribute computer power to the branches, especially branch offices overseas, and foreign exchange dealing was a natural candidate for early automation. Organizations such as Reuters and Telerate also recognized that substantial profits could be made by bringing the latest prices and price changes directly to foreign exchange dealers' desks. The third stage has involved a more integrated approach to banking software in which foreign exchange dealing can take place with ready access to other money-market information or facilities for making internal rate calculations, for example, which make it possible for customers to keep ahead of the market by inserting their own specialist rates. Further introduction of digital information-switching should reduce the number of separate screens that dealers must scrutinize to reach the information they need to make decisions while greater use of decision support systems (computer systems that massage basic data in such a way as to make it easier for the dealer to make an informed, sensible decision) is also now possible.

The symbiosis between telecommunications and advanced services operating at a global scale has been explored by Langdale (1984, 1985). The internationalization of financial services, for example, has encouraged the development of electronic funds-transfer systems while, in turn, advances in information technology have stimulated demand for such services from financial institutions. Langdale (1984) identifies three groups of telecommunications services used by banks and other financial institutions: public switched services (telephone, telex, data transmission), international leased communications networks, and cooperative systems such as SWIFT (Society of World Interbank Financial Telecommunications) for interbank transactions. It is difficult to know the share of total international traffic using switched networks accounted for by financial services because the lines are shared by numerous other business and non-business users. But this results in unacceptable levels of security risk for many users who prefer to operate private leased networks which offer greater reliability, flat-rate charges allowing a higher volume of use, links tailored to specialist requirements between offices located in the world's major financial centers and an ability to 'piggyback' information services on their network of leased lines (Langdale, 1984). In order for these leased networks to function they require computer switching and data processing centers; typically, these are located in key financial centers so that Singapore, for example, is the switching center for Citibank offices in Australia, New Zealand, Sri Lanka, Indonesia, Malaysia and Thailand (Langdale, 1984; see also Langdale, 1987, 1989). Indeed, telecommunications now ranks third after salaries and accommodation in the operating overheads of major MNEs such as Citicorp (Moss, 1987a). The Hong Kong and Shanghai Banking Corporation has an extensive leased network covering more than a hundred offices in sixty countries organized around six switching centers in, amongst others, the United States, Sydney and Brussels. SWIFT involves more than 1,000 banks in a comprehensive system for foreign exchange confirmations, bank transfers and transmission of statements but not payments, which still rely on linkages through correspondence. The SWIFT network is dominated by United States and European Banks (well over 50 percent of the membership), who also account for a large proportion of the traffic.

Services, telecommunications and metropolitan development

Business corporations therefore no longer view telecommunications as simply an adjunct to the conduct of their activities; it is necessary for them consciously to allow for them in their corporate strategies. Some firms not only rely on major telecommunications carriers but are also involved in developing leased systems that fit in with their corporate strategy (Langdale, 1989). It also seems that deregulation (as in the United States) or privatization and deregulation (as in Britain) is not necessarily making the most advanced telecommunications services more widely available; if anything, the infrastructure is increasingly focused on the information processing complexes in major metropolitan areas since these represent the major market (although not necessarily the only one) for the use of advanced telecommunications. Thus, the development of Mercury's fiber-optic network in Britain is predicated on the assumption that the bulk of the traffic handled will be generated by the major cities linked by a network sharing the same routes as those used by the railways linking the same cities. Another reason for the disproportionate focus of advanced telecommunications on information-intensive environments is the removal, following deregulation, of the extensive cross-subsidies used to standardize prices across a state or country irrespective of variations in user density and the costs of linking users to the system (rural costs for installation being clearly much higher when compared with urban areas).

Moss (1987c, p. 35) has therefore, advanced the view that telecommunications is 'creating a new urban hierarchy, in which certain cities will function as international information capitals, with the most extensive electronic infrastructure and richest opportunity for economic interaction' (see also Moss, 1987b). In other words the 'emerging telecommunications infrastructure is an overwhelmingly urban-based phenomenon' (Moss, 1986a, p. 37). This is increasingly underpinned by a move away from universal provision and costing of telecommunications services (whereby urban users subsidized rural users and business users subsidized residential users) towards user-related rates and investment. The disparity between large cities and smaller cities will therefore be reinforced. These cities will be able to act as control points for extending the geographical limits of their markets using telecommunications but the growth associated with such expansion (employment, incomes, expenditure on retail services) will not be dispersed as markets expand to the global scale. Castells (1985), in a review of the relationship between economic restructuring and high technology in the United States, observes that 'new technologies also enhance, simultaneously, the importance of a few places as locations of those activities that cannot easily be transformed into flows and that still require spatial continuity' (Castells, 1985, p. 18). The recent expansion of professional producer services both within and between countries seems to accord well with this prognosis (Leyshon, Daniels and Thrift, 1987a, 1987b; Leyshon, Thrift and Daniels, 1987). In certain respects it is paradoxical that the enhancement of personal interaction using a variety of communication channels from the telephone to the videoconference has been accompanied by a continuation of the importance attached to direct face-to-face interaction in, for example, the dealing rooms of securities firms, on the floor of commodity or futures exchanges, or in the offices of major insurance underwriters such as Lloyds of London.

What kind of empirical evidence is there to illustrate the validity of the arguments advanced by Moss, Gottman, Noyelle and others? It is possible to chart the growth of certain services in particular metropolitan areas, to outline the characteristics of recent developments in telecommunications services and, conversely, to trace the internationalization of producer services headquartered in countries such as Britain. The expansion of international banking in London between 1975 and 1985 (Lamb, 1986) provides an example of the impact of internationalization on a major financial center. London was already a key center for international banking in the 1870s with lending largely linked to the finance of overseas trade. Hence, the governments of Russia, Chile, Spain and Sweden (amongst others) were issuing securities on the London market before 1875 (Lamb, 1986, after Jenks, 1971). Since that time London's historical importance, a tradition of creating an environment conducive to international banking activities, and the advantages of its time zone position in relation to the world's two other major financial centers (Tokyo and New York) and the operation of related foreign exchange markets have sustained its attractiveness for overseas bank representation. These three major centers engaged in almost 50 percent of all international banking activity in 1985 and have the largest capitalized stockmarkets in the world (Price, 1986). Each acts as a regional financial center but their global status results from their strategic position within different time zones (Clarke, 1986). Transactions can take place throughout the working day in each center by passing on deals from one market to the next. Telecommunications permits a London-based Eurobond dealer who starts work at 6.00 a.m. to catch the end of trading on the Tokyo exchange, to trade all day in London and to catch four to five hours on the New York exchange, depending on when he decides to stop work. Only foreign exchange dealing currently operates on this basis but other markets in Eurobonds and international futures are increasingly operating in this way.

A measure of the status of London as a banking center is that at the end of 1985 British-owned banks and the branches and subsidiaries of foreign banks held just under 25 percent of all international claims booked in countries reporting to the Bank for International Settlements. This market share was twice that of banks in the United States, followed closely by Japan. London also accounted for 30 percent of all Eurobanking activity (cross-border lending in foreign currencies, mainly in dollars) at the end of 1985. With all the top hundred banks in the world represented, sixty-three countries had direct banking representation in London in December 1985. Japanese banks are prominent and almost 40 percent of their international business worldwide was booked through their London branches and subsidiaries (almost more than the total international business of banks in Japan) compared with 25 percent for US banks even though the latter have larger absolute representation. London is less significant for the international business of European banks headquartered outside Britain but Swiss, German and Italian banks do have significant operations. London is also the principal center for Eurobanking and international operations of foreign banks from countries not reporting to the Bank of International Settlements (Lamb, 1986). From the point of view of evidence for highly selective location of foreign bank operations it is notable that business between banks dominates the total international business conducted in London; some 75 percent of outstanding claims at the end of 1985 were interbank loans. Japanese banks, for example, have been prominent borrowers from other banks in London and overseas in order to lend mainly to their own head offices in Japan. The confidence derived from geographical proximity and scope for face-to-face interaction for negotiations therefore encourages locational convergence of international banks but telecommunications permit cross-border transactions or transfers of foreign currency to head, branch, or subsidiary offices in parent countries or elsewhere.

Both facets of bank behavior are supported by long-distance tele-communications systems which have identified large and expanding markets based on business and other services in the information-intensive environments of cities such as London and New York. It has been sug-gested that the telecommunications infrastructure of New York for example, should be viewed as an economic development asset; it provides a wide choice of sophisticated services in a highly competitive environment in which medium-sized and small firms can also benefit because they can occupy space in large 'wired up' or 'intelligent' office buildings without the need to invest heavily in equipment and facilities (Moss, 1986b). Since New York is the principal generator of long-distance international telephone traffic in the US it has been the focus of investment in communication systems based on fiber optics. These are invaluble for telecommunications services because of their large capacity compared with copper wire or satellite systems, their high security, signal strength (fewer repeater stations for signals) and declining costs. In 1986 there were five long-distance fiber-optics systems on-stream or planned for New York (Moss, 1986b). New York also has comprehensive satellite facilities including the Tele-port on Staten Island with fiber-optic links to New York City and New Jersey. The complex includes satellite earth stations providing advanced voice and data transmission services between locations in other large cities around the world. Buildings in Manhattan such as the Empire State Building and the World Trade Center are nodes in the network served by Teleport. Communications between service and other businesses within New York is also being enhanced by fiber-optic systems such as the InterBorough Optical Network with its capacity for 5.8 million voice or data conversations between thirty-one office buildings in Queens, the Bronx, Brooklyn, Westchester and Nassau Counties and Manhattan. Facil-ities of this kind, together with microwave and mobile telephone systems, give New York an unrivalled telecommunications infrastructure which, as a result of competition and deregulation, is constantly being extended and updated with the latest facilities. The result is that business services as well as other economic activities can 'extend their geographic reach and market new products and services on a global basis' (Moss, 1986b, p. 395) while retaining their key control functions in the information-rich New York environment.

We should not lose sight, however, of the importance of other factors for sustaining the dominance of metropolitan areas on the location of producer services. Agglomeration economies undoubtedly exist and are important, otherwise why would MNEs and the supporting complexes of business and other advanced services accept the high accommodation, staffing and related costs of operating from large metropolitan areas? It may also be the case that the corporate resources present in large cities are actually more productive: one estimate suggests that productivity (value added per worker) for services rises by 10 percent as city sizes are doubled

(Henderson, 1986). Specialized labor requirements are increasingly important for successful producer services and the possibilities for obtaining such labor are most numerous around the largest corporate complexes (Vernon, 1963). Agglomeration not only makes investment in the most advanced telecommunications infrastructure attractive, it also makes public sector investment in highways, for example, both necessary and feasible. This provides indirect subsidies for the operations of firms within the agglomeration. Goldberg, Helsey and Levi (1987) also suggest that investment decisions by financial services reinforce the status of the major cities. The risks in investment projects in large cities is lower because the default value of diverse but immobile assets is higher. The expected value of office development investment projects, for example, rises with city size and this will have a diversionary effect on the allocation of capital. Large metropolitan areas will benefit disproportionately and will tend to be the most stable part of the urban system.

Conclusion

International markets for producer services will continue to grow in at least three ways. Firstly, through expansion into new geographical areas, primarily outside Europe and North America where market consolidation is already taking place. Some areas such as the Middle East or Latin America are unlikely to be attractive (stability, scale of existing debts) but the newly industrializing countries (NICs) such as Singapore, Hong Kong, Malaysia, or South Korea do offer largely undeveloped market opportunities. Secondly, growth will continue as producer service firms devise new services in response to customer demand and their own identification of market needs and opportunities. The major accountancy conglomerates have been particularly effective at diversifying into services that allow them to provide clients with a fully comprehensive 'bundle'; chartered surveyors, advertising services and legal firms are adopting similar strategies. Finally, growth will also take place as producer services targeted at a broader range of customers, such as medium-sized manufacturing firms, who have tended to use specialist services only occasionally or in order to comply with legal requirements for certified audits, for example.

While technology will undoubtedly be a key intermediary in this growth process its effect will be curtailed by the regulatory environment in existing and targeted areas. The overseas expansion of any business requires the deployment of special expertise (often from the parent country) and the combination of this with other expertise in the new market in ways that do not fit easily into existing practices and regulations. Noyelle and Dutka (1986) identified several types of restrictions encountered by multinational business service firms when trying to extend the geographical range of their markets: restrictions on the mobility of professional personnel, restrictions on international payments, on technology and information transfers, on market access through local procurement policy and on the business scope of firms.

Note

Parts of this contribution are derived from research conducted jointly with Dr N. J. Thrift (University of Bristol) and Dr A. Leyshon (University of Hull) on the growth and location of large professional producer service firms. This research was made possible with a grant from ESRC (D00232198).

References

Bertrand, O., and Noyelle, T. (1986), *Changing Technology: Skills and Skill Formation in French, German, Japanese, Swedish and US Financial Services Firms* (Paris: OECD).

Castells, M. (1985), 'High technology, economic restructuring and the urban–regional process in the United States', in Castells, M. (ed.), *High Technology, Space and Society* (Beverly Hills: Sage), pp. 1–18.

Caves, R. E. (1982), *Multinational Enterprise and Economic Analysis* (Cambridge: Cambridge University Press).

Clairmonte, E. and Cavanagh, J. (1984), 'Transnational corporations and services: the final frontier', *Trade and Development*, 5, 215–73.

Clarke, W. M. (1986), *How the City Works* (London: Waterloo).

Courchene, T. J. (1986), *Some Perspectives on the Future of Banking* (Victoria, BC: Institute for Research on Public Policy), Discussion Paper Series on Trade in Services.

Crandall, R., and Flanam, K. (eds) (1989), *Changing the Rules: Technological Change, International Competition and Regulation in Communication* (Washington, DC: The Brookings Institution), p. 233.

Daniels, P. W., Leyshon, A., and Thrift, N. J. (1986), 'UK producer services; the international dimension', *Working Papers on Producer Services, 1*, St David's University College, Lampeter and University of Liverpool.

Dunning, J. H. (1989), 'Multinational enterprises and the growth of services: some conceptual and theoretical issues', *The Services Industries Journal*, 9, pp. 5–39.

Dunning, J. H., and Norman, G. (1987), The location choice of offices of international companies, *Environment and Planning A*, 19, pp. 613–31.

Elfring, T. (1989), 'The main features and the underlying causes of the shift to services', *The Service Industries Journal*, 9, pp. 337–46.

Enderwick, P. (ed.) (1988), *Multinational Service Firms* (London: Routledge).

Fairlamb, D. (1987), The New York connection, *The Banker*, 137, March, pp. 54–5.

Fekeketuky, G. (1985), Negotiating strategies for liberalizing trade and investment in services, in Stern, R. M. (ed.), *Trade and Investment in Services* (Toronto: Ontario Economic Council), pp. 203–14.

Friedmann, J. (1986), The world city hypothesis, *Development and Change*, 17, pp. 69–83.

Gershuny, J., and Miles, I. (1983), *The New Service Economy: The Transformation of Employment in Industrial Societies* (London: Frances Pinter).

Goldberg, M. A., Helsley, R. W., and Levi, M. (1987), The evolution of international financial centres. Final Report to Ministry of Finance and Corporate Relations. Vancouver: UBC Faculty of Commerce and Business Administration (mimeo).

Gottman, J. (1983), *The coming of the transactional city* (College Park, Md: University of Maryland, Institute for Urban Studies).

Harrington, J. W. (1989), Trade in services between the US and Canada: status and prospects', Buffalo: Canada–United States Trade Center Occasional Paper No. 2, SUNY, Buffalo.

Henderson, J. V. (1986), Efficiency of resource usage and city size, *Journal of Urban Economics*, 19, pp. 47–70.

Hepworth, M. (1986), The geography of technological change in the information economy, *Regional Studies*, 20, pp. 407–24.

Holly, B. P. (1987), Regulation, competition and technology: the restructuring of the US commercial banking system, *Environment and Planning A*, 19, pp. 635–52.

International Accounting Bulletin 1986, 'Arthur Andersen first through a new barrier', March, 3.

Jenks, L. (1971), *The Migration of British Capital to 1875* (London: Nelson).

Lamb, A. (1986), 'International banking in London, 1975–85', *Bank of England Quarterly Bulletin*, 26, pp. 367–78.

Lambert, R. (1986), '24 hour markets: more products now need round-the-clock trading', *Financial Times*, 27 October.

Langdale, J. (1984), Computerization in Singapore and Australia, *The Information Society*, 3, pp. 131–53.

——(1985), 'Electronic funds transfer and the internationalization of the banking and finance industry', *Geoforum*, 36, pp. 1–13.

——(1987), 'Telecommunications and electronic information services in Australia', in Brotchie, J. F., Hall, P., and Newton, P. W. (eds), *The Spatial Impact of Technological Change* (Beckenham: Croom Helm), pp. 89–103.

——(1989), 'The geography of international business communications: the role of leased networks', *Annals, Association of American Geographers*, 9, pp. 501–22.

Leyshon, A., Daniels, P. W., and Thrift, N. J. (1987a), Internationalization of professional producer services: the case of large accountancy firms', *Working Papers on Producer Services 3*, University of Bristol and University of Liverpool.

——(1987b), Large commercial property firms in the UK: the operational development and spatial expansion of general practice firms of chartered surveyors', *Working Papers on Producer Services 5*, University of Bristol and University of Liverpool.

Leyshon, A., Thrift, N. J., and Daniels, P. W. (1987), 'The urban and regional consequence of the restructuring of world financial markets: the case of the City of London', *Working Papers on Producer Services 4*, University of Bristol and University of Liverpool.

McConnell, J. E. (1986), 'Geography of international trade', *Progress in Human Geography*, 10, pp. 471–83.

Marshall, J. N., and Bachtler, J. (1984), 'Spatial perspectives on technological changes in the banking sector of the United Kingdom', *Environment and Planning A*, 16, pp. 437–50.

Moss, M. L. (1986a), 'Telecommunications and the future of cities', *Land Development Studies*, 3, pp. 33–44.

——(1986b), 'Telecommunications systems and large world cities: a case study of New York', in Lipman, A. D., *et al.* (eds), *Teleports and the Intelligent City* (New York: Dow-Jones Irwin), pp. 379–97.

——(1987a), 'Telecommunications: shaping the future', paper presented at a Conference on America's New Economic Geography: Nation Region and Central City, Washington, DC: 29–30 April.

——(1987b), 'Telecommunications and international financial centres', in Brotchie, J. F., Hall, P., and Newton, P. W. (eds), *The Spatial Impact of Technological Change* (Beckenham: Croom Helm), pp. 75–88.

Noyelle, T. J. (1986), *New Technologies and Services: Impact on Cities and Jobs* (Washington, DC, University of Maryland, Institute for Urban Studies).

——(1987a), 'New York City in an era of global financial markets', Contribution to the New York Region paper prepared under the editorial responsibility of the Regional Plan Association for the World Association of Major Metropolises, Mexico City, May 1987.

——(1987b), 'International trade and FDI in services: a review essay', *The CTC Reporter*, 23, Spring, pp. 55–8.

Noyelle, T. J., and Dutka, A. (1986), *Business Services in World Markets: Lessons for Trade Negotiations* (Washington, DC: AEI Press).

——(1987), *Business Services in World Markets: Accounting, Advertising, Law and Management Consulting* (Cambridge, Mass.: Ballinger).

Noyelle, T. J. (ed.) (1988), *New York's Financial Markets: The Challenge of Globalization* (Boulder, Colo.: Westview Press).

O'Connor, K. (1987), 'The location of services involved with international trade', *Environment and Planning A*, 19, pp. 687–700.

OECD (1981), *Recent International Direct Investment Trends* (Paris: OECD).

——(1983), *International Trade in Services: Insurance* (Paris: OECD).

——(1984), *International Trade in Services: Banking* (Paris: OECD).

Price, K. A. (1986), *The Global Financial Village* (London: Banking World).

Riddle, D. (1986), *Service-Led Growth: The Role of the Service Sector in World Development* (New York: Praeger).

Rimmer, P. J. (1990), 'The global intelligence corps and world cities: engineering consultancies on the move', in Daniels, P. W. (ed.) *Services and Metropolitan Development: International Perspectives* (London: Routledge), (in press).

Rugman, A. M. (1981), *Inside the Multinationals: The Economics of Internal Markets* (London: Croom Helm).

——(1987), 'A transaction cost approach to trade in services', *Series on Trade in Services* (Victoria, B. C.: The Institute for Research on Public Policy).

Sauvant, T. (1987), *International Trade in Services: The Politics of Transborder Data Flows* (Boulder, Colo.: Westview Press).

Schwamm, H., and Merciai, P. (1985), *The Multinationals and the Services* (Geneva: Institute for Research and Information on Multinationals).

Siegel, B. (1987), *The Internationalization of Service Transactions: The Role of Foreign Direct Investment on International Trade in Services* (Victoria, BC: The Institute for Research on Public Policy, Series on Trade in Services).

Stern, M. (1985), *Trade and Investment in Services: Canada/US Perpectives* (Toronto: Ontario Economic Council).

The Banker 1987, p. 137, March.

The Economist, 1985, City of London survey, 6 July.

The Economist, 1987, Telecommunications, 17 October.

Travers, N. (1987), 'City revolution 6: Flemings takes the lead in market making', *The Banker*, May, pp. 18–19, 22–3.

UNCTAD (1988), *Services: Issues Raised in the Context of Trade in Services* (Geneva: UNCTAD Secretariat).

——(1989), *Trade in Services: Sectoral Issues* (New York: United Nations).

United Nations (1983), *Salient Features and Trends in Foreign Direct Investment* (New York: United Nations).

United States Government (1983), *US National Study on Trade in Services* (Washington, DC: Office of the US Trade Representative).

Vernon, R. (1963), *Metropolis 1985* (Garden City, NY: Doubleday Anchor Books).

White, B., and Vitas, D. (1986), 'Barriers in international banking', *Lloyds Bank Review*, 161, pp. 19–31.
Williamson, O. (1986), *Economic Organisation* (Brighton: Wheatsheaf Books).

8   *The re-internationalizing of apartheid: information, flexible production and disinvestment*

JOHN PICKLES

Whose future is it anyhow?

As we suddenly confront a world of computers and satellites and video imagery, as familiar industries decline and strange new ones arise, as neighborhoods, businesses and family lives are transformed, painful political questions become inescapable.

Every civilization has its own characteristic distribution of power as between classes, sexes, races and even regions. Today as industrial civilization passes into history, powerful voices demand to know whether the emergent civilization will find room in it for the millions, indeed, billions on earth who today are discriminated against, harassed or oppressed because of their racial, ethnic, national or religious backgrounds. Are the poor and powerless of the past going to stay that way, viewing the future, as it were, through bullet-proof glass – or will they be welcome in the new civilization we are creating?

Tough questions. And perhaps the most dangerous of all.

Toffler, 1983, p. 139

Introduction

In his recent works, *The Third Wave*, *Previews and Premises* and *The Adaptive Corporation*, Alvin Toffler writes about the nature and consequences of the wholesale restructuring of industry brought about in part by the rise of new industrial forms and products. Although Toffler (1983, p. 13) locates his explanation for this phenomenon only in the changing *forms* of production: in the breakdown of the old smokestack industries (the Second Wave industrial era economy) and the emergence of new forms (the Third Wave economy), and although he explicitly rejects deeper structural processes such as crises of redistribution, and over- or underproduction, he does point to important consequences of the development of new technologies of production and the shift from Fordist approaches to flexible specialization. In particular he makes the connection between the flexibility of the new industrial processes – and their increasing use of space as a factor in organizing production – and the increasing manipulation of information both in the interests of profitability and in the interests of legitimizing new practices and modes of operation. Thus in *The Adaptive Corporation* (his report on the AT&T Corporation) Toffler (1985, p. 2) argues that:

> Instead of constructing permanent edifices, today's adaptive executives may have to *de-construct* their companies to maximize maneuverability. They must be experts not in bureaucracy, but in the coordination of ad-hocracy.

In this newly flexible corporate environment even the type and relative signficance of factors of production have changed.

> In the past, land, labor and capital were the key elements of production. Tomorrow – and in many industries tomorrow has already arrived – information will be the crucial ingredient. (Toffler, 1985, p. 29).

Flexibility of production, ad-hocracy and de-construction in organization, the manipulation of space and information in the interests of restructuring the labor market on the one hand and the broader ideology of industry on the other (including enticing the state to partake and foster such a changing ideology): this nexus of concepts raises particularly interesting questions about contemporary South Africa. The structural changes in South African manufacturing, agriculture and mining, the broadening of opposition struggles within the country, and the increasing effectiveness of international pressure on the South African government have resulted in the loss of foreign capital (Table 8.1) and the attendant restructuring of production within South Africa.

The pressures currently affecting the South African economy are far

Table 8.1 Capital movements in and out of South Africa (net inflow of capital R million).

1980	− 2,282
1981	846
1982	3,085
1983	− 331
1984	− 19
1985	− 10,418
1986	− 6,139

Source: South Africa Reserve Bank, *Quarterly Bulletins.*

from clear, and do not necessarily represent a single coherent strategy on the part of the business community. First, although the past six years have seen a general trend towards increasing levels of net outflow of capital from South Africa, this is not reflected evenly across the South African space economy nor in the investment strategies of all nations. One area which runs counter to this trend is the Ciskei, which has seen a large increase in the inflow of capital between 1981 and 1985. One nation which has increased its investment in South Africa is Taiwan, whose industrialists invested $20 million in the first six months of 1987 to establish sixty-eight factories in the 'independent' homelands of Ciskei, Transkei, Bophuthatswana and Venda (*IDAF News Notes*, May 1987, 37, p. 3). More broadly, there appears to have been a progressive drift in foreign manufactures from core areas to peripheral locations (Rogerson, 1986). Thus, beneath the aggregate pattern of reducing levels of international and local corporate investment there appears to be another pattern emerging of internal restructuring of the space economy, in which capital, attracted by high levels of government and local incentives, is investing in certain homeland areas.

Current economic circumstances are by no means ideal for a general restructuring of capital and labor location, and what follows should not be taken to imply that conditions now favor this restructuring on a broad scale. Rather, the restructuring of the space economy is taking place *despite* generally adverse economic conditions.

Second, having said this it must also be pointed out that the withdrawal of foreign capital is itself not a straightforward process. The South African government already exerts tight controls on the repatriation of asset sales, and even with such mechanisms for evading these restrictions as transfer pricing, funding of subsidiaries in the country, indirect routing of funds, the manipulation of financial transactions and the running-down of productive assets over time, international corporations still have difficulty in liquidating their assets. In part this explains the proliferation of creative arrangements for selling to local companies when disinvesting.

Third, the transfer of assets to local companies also is not without its disadvantages for South Africa. The generally held view of withdrawals of foreign companies is that South Africans are able to buy plant and stocks at very low prices. The buying of grossly undervalued companies has indeed resulted in windfall profits for South African investors, as the recent purchase of Barclays South Africa to form the First National Bank of South Africa illustrates. On the negative side, however, an appreciable part of domestic savings has to be invested in buying such companies and their existing plant; savings which otherwise might be available for investing in new plant and the renewal of worn-out machinery (Moorsom, 1986, pp. 56–7).

To these three factors: the outflow and partial counterflow of foreign capital, the creative restructuring of foreign companies within South Africa and the long-term cost of local takeover of older plant and machinery, must be added the changing technologies of production and organization which underpin and explain some of these changes. The ability of capital to move and to take advantage of regional differences in tax and incentive levels has been greatly enhanced by the revolution in communications and information-processing. It is these technologies and the new forms of organizing production which are causing a new 'ad-hocracy' to form, permitting a largely hidden restructuring of the industrial geography of South Africa, even in the face of the much publicized withdrawal of corporate investments. This chapter is concerned with this trend, one which runs counter to the general impression of capital flight, and it asks: under what conditions is international capital investing in South Africa and what effect does this have on the margins? That is, what is the impact of a new regional restructuring on the territorial structure of apartheid?

I will raise questions about the role of increasing flexibility of organization and production and the use of space as a means (seized by the apartheid state and accepted by foreign investors) to legitimize operations in South Africa and to restructure production in the face of opposition from internal and external groups. In this context, information technologies and communication industries play a dual role: they are the mechanisms which underpin the current restructuring of industry, the move to flexible specialization, and the attendant global penetration of space, *and* through the manipulation of media reports and the creation of ideology they are the means by which such activities are legitimized. Production and ideology are linked through the power of technology. No longer does the industrialist rely solely on the general ideology of the day (if *he* ever did) to permit the appropriation of wealth from a social process. The industrialist of the Third Wave forges the ideology in consort with the state, and transforms the informational environments within which the restructuring takes place. In what follows I will concentrate (a) on the role of communications technology in the current transformation of apartheid, and (b) on the linkage between structural changes in the mode of production and organization *and* the apparent withdrawal of capital from South Africa as a result of the disinvestment and divestment campaigns in the West.

Communications technology, information, and the restructuring of apartheid

Communications technology, computers and information play at least three roles in the articulation and current restructuring of apartheid.

1 They have historically enhanced the oversight and control mechanisms of the state at local, regional and national levels. In this way the state has been able to provide attractive and stable conditions for foreign investment. Through strict control of labor markets cheap labor has been

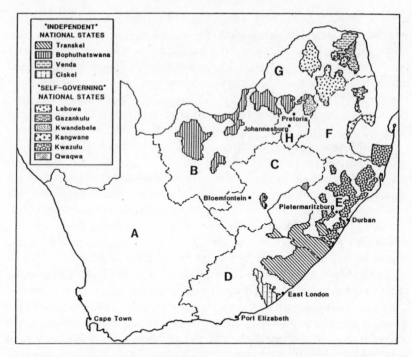

Figure 8.1 Homelands and the new development regions. The development regions A–H replace the Bantu Administration Boards.

made available, and through import controls internal markets have been protected.

2   These technologies have fundamentally altered the nature and relations of production. These are particularly significant because of their linkage with the explicit attempts by the government to shift from formal to informal controls of the majority of the population. At the simplest level this is seen in the recent abolition of laws relating to the infamous pass system and its replacement by a dual system of informal influx control: a universal identity document for all South Africans and an upgrading of the scope and sophistication of the labor bureaus which control access to employment and hence to the urban centers. The development since 1982 of a revitalized policy of industrial decentralization organized around functional regions ('development regions') also, when linked to the policy of 'orderly urbanization' (a system of informal, economic controls of movement to the urban areas), provides for a restructuring of economic and administrative regions within which the 'logic of the marketplace' (albeit heavily structured by the state) will replace the 'artificial boundaries' of classical Verwoerdian apartheid (see Pickles, 1987a, 1987b) (Figure 8.1). Technological change can thus be linked directly to the modernizing of racial domination in South Africa, even as classical apartheid is dismantled.

3   In a broader context, the regional restructuring fostered by the South African government and local and international business sits within a restructuring of international business operations, of production and sales and of financial markets. Ironically the latter, while being the dynamo behind these changes, has also produced its own opposition. The internationalizing of information and communications works both ways: while it permits the increasing flexibility of operations for business, it also generates an international constituency which may weigh against such operations. The South African case is instructive, information flows have been instrumental in adversely affecting financial markets, in raising the pressure for institutional, local and state *divestment* from companies doing business in South Africa, on companies to *disinvest* from South Africa, and from federal and international *sanctions* on the Pretoria government.

The significance of information flows in these regards is illustrated by two events in recent years. The first case is the Muldergate scandal in which the Department of Information was found to be investing large amounts of money in groups and individuals in America to counter negative reporting of events in South Africa (the conservative *Washington Star* and the *New York Daily Tribune*, the right-wing Christian League, Gerald Ford's Presidential campaign, the challenger to Dick Clark in the Senate elections who, as it turned out, successfully unseated the chairperson of the Senate Africa Sub-committee, and Senator Hayakawa all being mentioned as recipients of funds (M., 1979, pp. 10–12)). The Muldergate scandal undermined B. J. Vorster's position as Prime Minister and wrecked the chances of Connie Mulder, his heir apparent, in favor of the 'military coup' of the then Minister of Defence, P. W. Botha. The second case is the 1986 extension of the State of Emergency to clamp down on all foreign news reporting from South Africa. The net result of this clampdown, and the reason the government was so adamant in its imposition, was that it was effective. The media ban resulted in the almost immediate reduction of coverage of the South African situation in the American media. For example, coverage of South Africa on ABC's evening news in the month after the media ban was only 3.2 minutes, compared to 10.8 minutes in the month prior to the ban. Similarly, coverage of South Africa in the *New York Times* in the month after the ban was only 402 column inches, compared with 727.5 column inches in the month prior to the ban (*Harpers*, February 1986, p. 13). The subsequent dormancy of the issue, and perhaps the apparent loss of momentum in the anti-apartheid movement in the USA during 1987, particularly the divestment/disinvestment campaign, can in part be attributed to the success in controling information and removing it from the national agenda. The extension of South Africa's public relations efforts to the international scene is a topic that has yet to be considered in detail. The South African government has long exerted similar control over information inside South Africa itself, and in August and September of 1987 substantially increased yet again its control over the press, the courts and the universities.

Changes in the international economy

In recent years South Africa has experienced the effects of four major shifts in the operation of the international economy.

1   The crisis of underconsumption/overproduction stemming from the 1970s on, and the attendant search for means to reduce the cost of production. In particular, this has given rise to a searching and differentiation of labor markets, with the consequent development of international, regional, and local divisions of labor (Smith, 1986; Stohr, 1985c, 1987). On the other hand, this search to reduce costs has brought the South African government even further into the arena of labor relations as a provider of incentives and controler of labor (exercising itself through the training programs, its strict control over the 'Bantu education' system, incentives for location, controls on organized labor activities, etc.), and the detention of those members of organized labor who 'stray' away from workplace reform to more political, community issues.

2   The ability to increase competitiveness in a glutted market has been aided by the adoption of new technologies of production and communication. These have permitted the segregation of production processes across space. The result is an increased flexibility of production and a contemporary restructuring of industry to take advantage of the regional division of labor. In regard to South Africa Trevor Bell (1984, p. 9) has expressed this position bluntly:

> The essence of this argument, thus, is that the observed tendency towards industrial decentralisation has occurred largely spontaneously in response to the pressures of competition in world markets for manufactured goods. The geographical deconcentration of industry has largely been the result of a striving after lower labour costs through the substitution of Black labour obtainable at low wage rates in the less industrialized areas, for more expensive White, Coloured, Asian and, indeed, also Black labour available in the larger industrial centres.

Furthermore, as Soja (1987, p. 290) has pointed out in a more general context, the sweeping changes in the technology of production and communication have been attended by 'unexpected shifts in long-standing patterns of cultural expression, economic development, and political power' as well as 'equally unanticipated revisions in the immediate social relations of class, gender, kinship, and allegiance' (and, we might add, race).

3   The enhanced ability of financial markets and industrial production to move investments willy-nilly through space has been facilitated by the changing nature of the unit of production. The development of massive, fully integrated transnational corporations has increased both flexibility and bargaining power *vis-à-vis* local, regional and national governments,

which have become increasingly involved in the business of 'industrial inducements' to encourage firms to locate 'here' not 'there'. In the South African context this has taken on even greater significance. The need to maintain the high levels of foreign capital investment that South Africa has traditionally enjoyed has become even more crucial as the threat and reality of withdrawal have arisen. Moreover, the structural changes in the economy, particularly the developing tendency towards capital deepening and the substitution of technology for labor, increases South African industry's reliance on foreign capital.

At the same time, on the home front South African companies, faced by increasing economic and political uncertainty have pursued major campaigns of reorganization and diversification through foreign expansion. Notable in this regard is Anglo-American Corporation, which over the years has established and purchased a number of companies whose specific purpose is to build up Anglo-American interests outside South Africa (see Innes, 1984, Appendices 2 and 3; Kaplan, 1983).

4   These three structural changes in the organization of production have permitted a fourth change in corporate practice. With increasing vulnerability of large corporations to widescale consumer boycotts and opposition, and with their need to maintain customers in these markets, the divestment, disinvestment and sanctions campaigns have achieved perhaps surprising success in affecting corporate trading and investment practices (see Smith, 1987; Adam, 1987). The results of these campaigns have been several key changes of operations, and these are instructive for two reasons. On the one hand they illustrate the power of coalitions to affect business decisions in a depressed marketplace. On the other hand they illustrate the extent to which the new flexibilities of production and corporate organization permit publicity coups to be achieved, such as those achieved by the pullout of Coca-Cola, IBM and General Motors, which were achieved by a reorganization – in Toffler's terms a *de-construction* – of the firm *in the interests of greater flexibility*.

If the restructuring of production and communication systems, and the flexible specialization it allows, is the harbinger of a rearticulation of spatial and social arrangements, then as these processes affect South Africa the geography of apartheid will require a fundamental reassessment.

The spatial division of labor in the South African context

During the 1980s the South African government has begun to move away from the Verwoerdian conception of apartheid, predicated on the territorial fragmentation of the state. Recognizing on the one hand its inability to control rural–urban migration or to stem the economic collapse of the rural homelands, and on the other hand its dual need to stem the rising tide of opposition in the townships to the apartheid laws (such as the Pass Laws) and the parallel needs of the economy for greater flexibility in the operation of labor markets, a range of new policies has been adopted.

These policies move away from the Riekert Commission recommendations and the territorial basis of apartheid it presupposed, and embrace the notion of functional economic regions within which labor markets operate according to 'economic' principles. Several advantages follow:

1   'Minor' frictions and oppositions are overcome, and as the state is seen to move away from racial territorial segregation (towards a differentiation based on economic factors) the state hopes to regain legitimacy.
2   The needs of capital in the metropolitan centers for an increasingly skilled, permanent workforce, and of the state for a rising black middle-class can be accommodated.
3   At the same time under the industrial decentralization policies and new tax schemes of the Regional Service Councils employers in metropolitan areas will have to pay for the benefits that accrue from such a location, while industries willing to relocate in the industrial deconcentration areas will receive large capital incentives to do so.
4   One net effect of this restructuring is the undermining of the power of urban black organized labor. This is reflected in the recent changes in policy towards disinvestment adopted by the Congress of South African Trade Unions (COSATU) at its first annual congress, held in Johannesburg in 1987. At this congress COSATU modified its stand on disinvestment as a result of GM's withdrawal from South Africa in 1986, which was cited at the congress as highlighting the need for workers to ensure that their interests are not harmed by disinvestment.

COSATU's initial position was that it supported intensified disinvestment subject only to the condition that the 'social wealth of South Africans remains the property of South Africans'. Angered by foreign companies selling to white-controlled South African companies COSATU is now demanding that disinvestment companies negotiate the conditions of withdrawal with representative trade unions. Correspondingly the congress adopted a resolution that the 'wealth created by workers must remain in South Africa and be controlled by workers' (*The Guardian*, 16 July 1987, p. 10).

Such an arrangement was, at the time, under negotiation between the National Union of Metalworkers and Ford Motor Company. The latter agreed to place part of its 42 percent share in the South African Motor Company (SAMCOR) in a workers' trust. Such an arrangement would give workers a stake in SAMCOR and provide Ford with an opportunity to withdraw from South Africa with considerable favorable publicity (*The Guardian*, 16 June 1987, p. 8). The final arrangement gave 24 percent to SAMCOR employees, while Ford sold 18 percent to Anglo-American (*The Guardian*, 15 June 1987, p. 8).

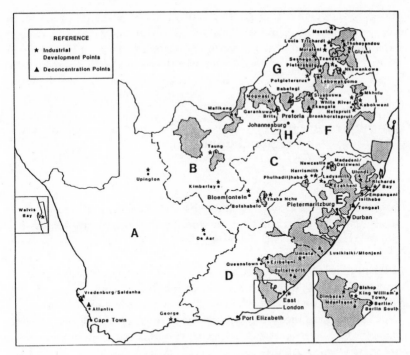

Figure 8.2 The development regions and industrial development and deconcentration points in South Africa

A second effect is the economic entrenchment of the fundamental patterns of the territorial organization of the South African state *in spite* of the state's ostensible move away from Verwoerdian territorial apartheid. Ciskei provides perhaps the best current illustration of these changes. According to the *Sunday Times* (2 June 1985, p. 20) Ciskei is experiencing an economic boom brought about by the declaration earlier in the year of a tax-haven on the model of Hong Kong: 'there has been an influx of foreign businessmen eager to set up new factories, while Western governments, which refuse to recognise Ciskei's sovereignty, are showing a keen but discreet interest.' Since 'independence' in 1981 the industrial base of Ciskei has grown by 500 percent. In 1982 there were thirty-two factories, in 1984 there were 144, and during 1985 a further fifty-seven were being built. Companies from Israel and Taiwan are the largest investors, but there are also investing companies from the USA, Britain, Holland and Spain. Under the incentives package which attracts these investors foreign firms are not taxed. They receive state subsidies of 80 percent on the rental of factories, 50 percent on other assets, 60 percent rebate on transport costs, and 60 percent mortgage subsidies on housing for managerial staff. Moreover, such companies receive a tax grant of 95 percent of wage bills up to a maximum of R 110 a month for each worker (Figure 8.2).

Policies of regional restructuring and the political concessions which accompany them only slowly reform the more obviously repressive aspects of apartheid. Thus from the government's point of view, if these reform measures are to have a chance to succeed they must be accompanied by increasing control over the daily lives of their opponents.

Communication technologies and the control mechanisms of municipal and regional government

More than any other single technological advancement, the computer has fostered the concentration of administrative power in the hands of South Africa's white elite. (NARMIC, 1984, p. 14).

In 1985 the Anti-Apartheid Act passed the US Congress and was signed by President Reagan. Among other things the Bill introduced a series of sanctions on contact with South Africa.[1]

One element of the 1985 sanctions was the ban on further sales of computers to South Africa. This perhaps unusual sanction makes much more sense when situated within the recent trend towards the computerization of the control mechanisms of apartheid. From influx control, the courts, the police, to local and regional planning under apartheid, operations have been rapidly computerized since the mid 1970s.

One example of the importance of the computer is the Department of Plural Affairs (formerly the Bantu Affairs Department) in its administration of the influx control laws. The department stored fingerprints and personal records on the black population (Department of Plural Affairs, *Annual Report*, 1977), and through its network of fourteen regional Bantu Administration Boards one of its main tasks was the administration of the influx control system. In November 1980 plans were announced to expand the scope of this system by establishing a national network linking the Administration Boards and the police to a central computer. The new system would rationalize the influx control and labor procurement systems 'by providing instant information on where jobs are and where workers are who can do the jobs' (quoted in NARMIC, 1984, p. 19). Since the consolidation of the Bantu Administration Boards (BABs) into the new development regions and the apparent dismantling of influx control legislation during 1986 and 1987, it is not yet clear what role the Department of Plural Affairs continues to play in efforts to control population movement.

Since 1970, when few local and regional government bodies used computers, there has been a rapid computerization of government functions. Not only has the number of government computer installations increased, but the range of applications has been extended. P. E. Claasen

(1980, p. 7) has described this trend as the 'total computerization of local government'. Moreover, the central government's policy of Total Strategy has brought with it the increasing militarization of civilian institutions, as military and civil operations are increasingly integrated (South African Defence Force, 1979). This has been particularly noticeable in urban planning, where the 'technologies of political control' exercised in Soweto (hippos, sneeze machines and military sweeps through the township) demand a system of straight roads providing easy access to houses and neighborhoods, and a detailed and accurate knowledge of the area. Informal squatter settlements are extremely difficult to control in this manner. Black townships have always been designed and laid out with the ends of controlling them in mind. Thus the introduction of Control Data Corporation's simulation software packages: 'Urban Planning Package' and 'Perspective', in the 1980s permits civilian and military authorities to plan and control new urban areas even more efficiently.[2]

In 1981 Control Data Corporation supplied computer components to the South African Police to enhance their surveillance abilities for controlling the pass system, and it supplied computers to South Africa's National Institute for Telecommunications Research, some of whose work 'is classified as it relates to defense'. Digital Equipment Company participated with a British company in supplying the South African Air Force with a radar surveillance system. IBM provided technical data to the agency which operates project Konvoor: a system designed to streamline support for South African military posts in Namibia. General Electric actively provided a satellite system which would permit the South African military to keep track of troop movements inside and outside the country (Anderson, 1981, E51).

The debate about the effect and value of sanctions has been a particularly muddy one, especially in the United States where inconsistencies in the claims made for or against sanctions seem to be ignored: in this regard the US performance at the United Nations has not been stellar. Under Jean Kirkpatrick the USA has argued that sanctions are an ineffective means of influencing policy at the very time that US sanctions against the Soviet Union, Cuba, Libya, Nicaragua and Poland were in operation. But the biggest fear is that '[i]n the short term . . . disinvestment will probably drive white South Africans further into the laager. . . . Sanctions would only deepen Boer stubbornness, harden white attitudes, and make progress all but impossible' (*Atlantic*, February, 1987, pp. 31–2).

On the other hand, the ability of the South African government to control opposition groups (through more effective means of social control) and to articulate the terrain of debate (through its control over the media and communication services generally) forces us to ask about the kind of change which would be likely without sanctions of some sort on key information technologies. In other words, what would be achievable with ready and cheap access to modern information and communication technologies?

Controlling the operation of the labor market

The significance of foreign companies in the South African computer and communication industries is not just a matter of the use to which these technologies are put, but is also a function of the dependency of South Africa on these sources of technology. South Africa is highly vulnerable to restrictions on the supply of equipment. Despite reserve stocks of about one year and the development of local production, the denial of access to foreign technology could have major repercussions.

The recent history of industry and labor in South Africa has been one of increasing demands for skilled and semi-skilled labor in production, capital deepening and the adoption of new technologies to overcome shortages and costs of labor, offsetting the increasing militancy of organized labor, and increasing the positional power of semiskilled and skilled workers. In these circumstances increased automation and capital intensity has been commonplace in industries from agriculture and mining to manufacture. Central in this process has been the automation and computerization of production. Thus, the loss of a single computer could necessitate 'having to find hundreds of bookkeepers who are not available on [the] labor market' (US Embassy, Pretoria, Cable to Secretary of State, 13 October 1978).

Not only would the loss of computers and other automated machinery cause a disruption in the recent trends in production and changes in labor markets (within which context organized labor has achieved substantial positional power – an argument used by some to oppose sanctions and disinvestment), but it would have major repercussions for labor policy generally, and along with that for the new regional and urban policies of the state (see Pickles, 1987b). Since these policies are predicated on the increased regional differentiation of the labor market (see Hindson, 1985), and on the development of a skilled and semiskilled black labor force, any constraint on the acquisition of technology would indirectly affect these broader state policies.

Local subsidiaries and the disinvestment movement

The rise of the divestment and disinvestment movements in South Africa and the West, particularly in the United States, has substantially altered the climate for doing business in South Africa. In January 1985, 300 US companies operated in South Africa, but by November 1986 this number was down to 250 (*Facts on File*, 1986, p. 827). By 1987, 144 Sullivan Signatory Companies had withdrawn from South Africa. Some companies disinvested from South Africa in part as a result of these pressures, in part because of threats to markets at home from local opposition, and in part from the deteriorating economic conditions of doing business in South Africa. Prudential Corporation announced on 3 September 1986 that it was selling its majority-owned South African subsidiary to a local company. The United Kingdom parent group cited the shortage of skilled insurance staff due to the mounting emigration of professional white workers from the country, as well as the difficulty of finding foreign staff willing to relocate in South Africa (*Facts on File*, 1986, p. 828). Companies such as Eastman-Kodak (which in November 1986 announced the most complete withdrawal from South Africa to that date) initially indicated that they would still permit their products to be sold in South Africa, and when Kodak later attempted to prohibit distribution of its products South African retailers indicated that Kodak products would be available from other sources (Battersby, 1987). Other companies have been less complete in their disinvestment. In fact, it is the contention of this chapter that some companies which have disinvested have done so as part of a broader restructuring which responds to changed economic circumstances, and that this restructuring has been conveniently achieved (and undoubtedly partly effected by) the disinvestment issue. Thus, such is the importance of the computer in contemporary South Africa, and such the profits to be made, despite the deteriorating conditions of doing business, several companies have apparently chosen to restructure operations through locally owned subsidiaries in the interest of greater flexibility. The *de-construction* of the large conglomerate, and the increased flexibility of operations achieved thereby, is precisely the recommendation given by Alvin Toffler to AT&T in the 1960s in his commissioned report on the company. In the case of South Africa we see clear recognition of the advantages of such a 'devolution' of operations. Thus, Koenderman (1978, p. 56) suggested that '[o]ne way out of the dilemma facing multinationals in South Africa is to find a local company to go into partnership with, thus lowering the foreign companies profile here'. Many companies now have switched to operations through a local affiliate (Table 8.2), and as a result have obtained much favorable publicity in the United States and Europe for their disinvestment stand.

Thus, we are seeing a switching of emphasis between the three main forms of transnational corporation penetration of the South African economy: from direct investments and loans to the provision of technological assistance and management skills (Seidman and Makgetla, 1983). 'The spread of partnerships indicate that some major corporations are becoming more important for their provision of technology and management skills than as capital investors' (Gurney, 1980, p. 4). Such partnerships are not a new practice. ICI, for example, has shared control of the chemical company AE&CI with Anglo-American Corporation since 1926. The number and rate of increase in such affiliations and partnerships is, however, new. Chrysler provides a good example of the forging of such a relationship prior to the effects of the disinvestment campaign in the USA. In 1976 Chrysler sold a majority of its shares in its South African subsidiary to a subsidiary of Anglo-American Corporation. Chrysler retained a 25 percent

Table 8.2   Approval for foreign investors in decentralization and deconcentration points.

	No. of applications	Capital investment R'000	Employment opportunities
April 1, 1982–March 31, 1985			
Taiwan	63	94,738	16,293
Israel	23	44,864	5,020
United Kingdom	19	86,483	3,461
Hong Kong	10	10,849	2,580
United States	9	23,568	3,236
West Germany	9	20,884	698
Italy	9	10,585	724
Zimbabwe	7	3,034	344
Belgium	3	5,486	133
Australia	3	3,687	180
Philippines	1	5,300	1,410
France	1	1,528	22
Peru	1	1,255	72
Switzerland	1	283	12
	159	312,093	34,184

Source: Annual Reports, *Board for the Decentralization of Industry.*

equity in the new company, Sigma, and provides management and technical skills and knowledge for the company. 'On the one hand, Chrysler can now claim that it has no control over the operations of its South African associate and its name no longer appears on motor vehicles and advertisements all over South Africa. On the other hand, Sigma still depends on Chrysler for management, research and development and some components' (Gurney, 1980, p. 4).

NARMIC (1984, p. 8) gives the more recent example of Wang laboratories which in 1978 'withdrew' from South Africa, leaving its sales to be managed by a locally owned company: General Business Systems (GBS). Yet despite the fact that Wang claimed to have no direct subsidiary in South Africa, under a distribution agreement between GBS and the American parent company sales of Wang equipment there increased by 45 percent every year until 1979 and fourfold since 1982. Thus '[f]ar from hurting us, I feel the withdrawal has probably strengthened us in the South African marketplace' (Representative of Wang's South African distributor. Quoted in Purvis, 1979, p. 194). Under pressure from the United States that agreement was terminated in 1985, and a subsequent agreement was also terminated apparently under similar pressure (*New York Times*, 27 July 1987).

The case of IBM is particularly instructive. In 1987 IBM announced it intended to sell its South African holdings, but would continue to license and distribute its products in the country. According to a management letter leaked to *The Financial Times*, IBM operations will continue as normal with the creation of a locally owned company to handle IBM's business. The letter also claims that this new company will face fewer restrictions on its South African operations than is presently the case: '. . . the new company will be able to respond with greater flexibility than a wholly-owned IBM subsidiary. In the current international climate such flexibility will clearly be to our customer's advantage' (*Divestment Disc*, 1987). Simon Engineering – a British firm – announced in June 1987 that it would sell its South African subsidiary Simon Holdings to a consortium of local managers and directors. In making this move it was, according to the company, 'meeting both the strategic requirements of the present company and the aspiration of local management' (*Guardian*, 2 June 1987, p. 25).

As Massing (1987, p. 26) has recently pointed out:

For all the fanfare [over the pullout of General Motors, IBM and Coca-Cola], however, South Africans will still be able to buy IBM PCs and cars inspected by Mr. Goodwrench. And disinvestment will not interrupt profit flows. The three corporations are not closing up shop, only selling out to local businessmen, and all plan to maintain a presence through licensing agreements, distribution contracts, and technology transfers.[3]

These examples illustrate a changing organizational structure for the participation of foreign companies in South Africa. In particular, they raise the distinct possibility that as South Africa remains under siege from internal and international opposition groups, foreign corporate investment (especially from the United States because of the directedness of opposition to particular companies in the USA) will begin to make greater use of these more flexible arrangements. This has a direct effect on the geography of production.

Corporate restructuring, flexible specialization, and the regional division of labor in South Africa

In the light of apparent forms of restructuring under the guise of disinvestment, we must ask what are the implications of such changes in the organization of operations for the future? South Africa certainly loses some of the international investment it would otherwise receive, and with the sanctions campaign will now receive certain technologies only with greater difficulty and probably at greater cost. The negative aspects of disinvestment and sanctions have been widely publicized (Crocker, 1987; Schultz, 1987). The extent to which this resultant restructuring is 'formative' in the sense of reorganizing relations in a new way, and one with serious implications for labor relations and power, for the power of the state, and for the ability of business to reorganize to permit 'business as usual', has not yet been addressed.

The flexibility of company organization, production and distribution permitted by the new technologies, and encouraged by the pressure on international capital, is already resulting in far-reaching changes in the industrial geography of South Africa. Indeed, as the militancy of black labor has increased throughout the 1980s, and as its cost (relative to international labor costs) has risen, capital deepening or what Lipietz (1986, p. 25) has referred to as the move to an intensive regime of accumulation occurs (where labor is subsumed to capital as work is reorganized and technology use is increased).[4] Thus:

> The new 'telematics' is indeed revolutionising production, distribution and consumption patterns not only in the Silicon Valley but even in industries, like clothing, which some still regard as a relic of the past – a labor-intensive industry bypassed by modern technical developments, one best consigned to third world countries. Micro-electronics is contributing to the transformation of design, production and distribution – and the ways these are linked together – in 'old' industries as well as in new. And, unlike the earlier wave of innovation, it is affecting the world of work in small firms as well as large, in the tertiary as well as the primary and secondary sectors.   (Mahon, 1987, p. 6).

Moreover, Mahon (1987, p. 6) has pointed out that while the traditional fear of job loss due to capital deepening may be very real, more recently 'a perhaps more invidious threat has begun to claim attention: that the new technology could lead to the elimination of semiskilled blue- and white-collar jobs – and middle-level incomes – to yield a two-tier society in which a few enjoy stimulating jobs and high incomes while the vast majority scramble for the remaining jobs – which demand little skill and offer poor pay and working conditions.'

This 'polarization scenario' raises loud alarms for observers of the contemporary South African scene. Government policy of the 1970s and the reform movement of the 1980s have been predicated on the rapid development of a semiskilled urban relatively prosperous black middle class. Any structural change which prevents even this token release for the few presents a severe challenge to reformist claims and policies. This is perhaps reflected in the almost visible desperation in the scale of industrial decentralization incentives offered since 1982 for industries to locate in the development and deconcentration points away from the major industrial centers. The upgrading of the industrial decentralization policies is clearly part of a broader restructuring, but what this issue suggests is that it may

also be a policy aimed explicitly at controlling the labor market and
expectations in the urban areas, as well as all its more obvious effects in the
areas of deconcentration.

More specifically the potential of flexible production arrangements and
ad hoc investment strategies to restructure political and social relations
under apartheid must be considered. Such a restructuring may then represent an 'informalizing' of the central distinction adopted by the South
African government from the 1979 Riekert Report between 'insiders'
(those with rights in urban areas) and 'outsiders' (those with no rights in
urban areas). Recent changes in policy have resulted in a moving away
from the territorial dualism between 'white South Africa' and 'black
homelands' established under Verwoerd and extended under subsequent
governments. In its place has been adopted the notion of functional regions,
based on the actual operation of labor market areas. This is the basis of the
post 1985 'orderly urbanization' policy (*White Paper*, 1985). But, if
Mahon's polarization scenario holds for South Africa, the extent to which
Riekert's insider/outsider distinction has actually been abandoned (as
current literature is beginning to suggest, for example, Cobbett *et al.*, 1986)
must be questioned. Could it be the case that the distinction has been
'informalized' and hence hidden by the recent shift away from legislative
controls and towards 'market forces' brought about in part by the changes
in patterns in investment and production we have been considering.

Conclusion

The image of South Africa held by the general public is of a nation under
increasing internal and international pressure to abandon its apartheid
policies. Certainly the past decade has brought ever increasing difficulties
for the effective economic development of South Africa. Internal struggles
and international publicity, as well as the declining profitability of foreign
investments in South Africa, have severely curtailed the flow of foreign
capital and technology into the country. South Africans have been
repeatedly surprised and shocked by the withdrawal of companies such as
Barclays, Eastman-Kodak, General Motors and Ford, companies whose
products have been a central part of many white South Africans' way of life
for many years. As companies disinvest and capital has been withdrawn or
not reinvested, the South African government and business community
have been forced to change, but have thus far chosen to do so by
maintaining (and in some cases tightening) their control over the fundamental structures of power. In this regard the industrial and regional
restructuring currently occurring under apartheid can be seen as part of an
ongoing, long-term policy shift towards informal market-based controls
and the manipulation of space in order to achieve the desired goals of
stability under a regime in which minority (i.e. white) privilege is main-

tained through the entrenchment of group rights. This involves the entrenchment of the divisions between 'outsiders' (black residents of rural areas) and 'insiders' (black workers and families in the 'white' areas). Such a restructuring involves a reorganization of production and location, and it brings with it major changes in group power. Black organized labor is potentially undermined by the shift of new investments and plants to areas in which labor legislation has been removed (including the right to unionize and engage in collective bargaining and strikes, health and safety protections and minimum wage levels). Black labor in the new industrial development sites is controlled by strengthened state bureaucracies in the homelands and by the absence of hard-won legislative rights and protections that exist for urban dwellers.

Locational shifts in patterns of investment, production and employment depend directly upon the changing technologies which are used to organize production and to link plants and companies. Indirectly both existing and new industry depends upon the exercise of state control in the coercion of workers and their families. Profitability and hence flows of foreign capital also depend upon the degree to which the South African government is able to control the flow of information overseas and to control information and coerce action at home. Thus, below the surface appearance of the recent history of disinvestment lies an indication of another trend. Companies disinvesting are doing so because of declining rates of profitability, increasing pressure from consumers and shareholders (including in some cases boycotts), and the often immense public relations advantages to announcing withdrawal.

Several corporations have been very creative with their strategies for disinvestment and withdrawal. Selling to local subsidiaries or other South African companies, setting up trusts for workers with buy-back options written into the agreements, or the closure of factories and the selling of old plant and machinery on an ad hoc basis.

In this sense South African industry may be entering a phase of increasing organizational and fiscal flexibility, one which is more complex and multifaceted than the previous centralized operations of transnational corporations in South Africa. While this situation is causing immense financial burdens on the South African economy, it is also resulting in new opportunities for flexibly organized corporate structures or for renewed open investment for companies from those countries, such as Taiwan, Korea and Israel, where the pressure of public opinion is lower or less important to the general operations of the company. Companies from these countries in particular seem willing to take advantage of the large financial incentives available for setting up and conducting business in areas such as the Ciskei, an area recently described as 'a new Hong Kong' for business (Marsden, 1985, p. 20).

The future of South Africa will be linked to the achievements and realignments made by the various competing interest groups involved in the current struggle for power, including white politicians, well-entrenched state bureaucracies, opposition groups such as the United Democratic Front and the African National Congress, representatives of large and small capital in agriculture, mining and manufacture, trade union leaders and the increasingly entrenched elites and state bureaucracies of the homelands. The negotiation of changes in the current political economy of South Africa will be very much affected by the type of change which occur in the type, organization and location of industrial activity, and the way in which such changes are brought about and legitimized. The ability to reorganize production in regions (or countries such as Lesotho or Swaziland) offering the greatest potential benefits, to relocate easily and still maintain effective linkages, and the ability to continually reassess the current balance between conditions in the labor force and the substitution of technology as broader economic considerations require, have significant spatial implications. The spatial restructuring of production does not, however, occur independently of the broader restructuring of political relations. In the battle for profits space is implicated. But in the manipulation of space for profit, control over and the manipulation of informational environments and the creation of particular ideological settings, achieves solidly political ends.

Notes

The research on which this chapter was based was funded by a National Science Foundation grant, SES–8620016. The views expressed in the chapter are not necessarily those of the National Science Foundation. I would like to thank Jeff Woods for his comments on the manuscript.
1   In 1963 the UN agreed to an embargo on military sales to South Africa. In 1977 the embargo became mandatory, and in 1978 the United Nations extended the ban to cover all sales – not only the sales of arms – to South Africa's police and military. Despite the embargo US corporations continued to sell, lease and maintain computers and other technologies which support government, police and military operations in South Africa until the 1985 bill (NARMIC, 1984, p. 1). In 1979, for example, $60 million worth of US computers and other electronic equipment reached South Africa despite the embargo (Anderson, 1981, E51). This avoidance also involved US government agencies. For example, in October 1981 the US gave approval for the sale of a Sperry Univac computer to Atlas Aircraft, a state-owned weapons maker (*The Boston Globe*, 28 February 1982), whereas a few weeks later a request from the Mennonite Central Committee for permission to send pencils and rulers to school children in Kampuchea was turned down on the grounds that such material could be construed as development aid (NARMIC, 1981, p. 1).
      The reluctance of the Reagan Administration to impose even limited sanctions on South Africa is well illustrated in the legislative history of the sanctions Bill. On 10 June 1986 the US House of Representatives voted 25 to 13 for economic sanctions which sharply limited business activities, denied US landing rights to South African Airways and banned imports of South African uranium, steel and coal. In response to the continued reticence of the Reagan administration to accept the sanctions Bill, and in response to the mostly symbolic sanctions imposed by the President in 1985, on 18 June, by a voice

vote, the House passed a trade embargo on South Africa requiring all US companies and citizens to divest holdings in South Africa. On 14 August 1986 the US Senate voted 84 to 14 to impose sanctions on South Africa, which included halting commercial air service between the USA and South Africa, barring new US loans and investments in South Africa, banning the import-ation of South African coal, uranium, agricultural products, iron and steel, Krugerrand gold coins and other goods, and banning the export of crude or refined petroleum products and computers to South Africa (*Facts on File*, 15 August 1986, p. 594).

2   The ways in which the urban design and planning of townships represents the planned architecture of visibility, access and control requires further investi-gation.

3   Not all corporate disinvestment fits within this category. On 30 December 1986 Exxon Corporation announced that it would sell its South African affiliates to an independent trust that would channel future profits to employees and continue to finance social programs for Blacks. It is true, however, that this move came after Exxon failed to find local buyers for its operations (*Facts on File*, 1986, p. 972).

4   Real wages for black workers have actually declined in the late 1970s and early 1980s (Saul and Gelb, 1981, p. xx), a factor which has substantially contributed to the increased militancy of workers. They are, in effect, fighting to *retain* their position. However, relative to other sectors of labor and to international wage rates, the cost of black labor in South Africa has increased. The competitiveness of South African 'cheap labor' has thus, over the past decade, declined relative to other regions. This is undoubtedly a major cause of the restructuring of labor relations that the new regional industrial policy seeks to effect.

References

Adam, G. (1987), 'Behind Barclay's pull-out', *Africa Report*, May–June 32(3) pp. 23–5.

Anderson, J. (1981), 'Embargo on South Africa called farce', *The Washington Post*, 30 May, E51.

Battersby, J. D. (1987), 'US goods in SA: little impact of divestiture', *The New York Times*, 27 July.

Bell, T. (1984), 'The state, the market and the inter-regional distribution of industry in South Africa', Second Carnegie Inquiry into Poverty and Develop-ment in South Africa, Conference paper no. 243, Cape Town.

Claasen, P. E. (1980), 'Computers in local government', *Municipal Administration and Engineering*, 7 June, p. 7.

Cobbett, W., Glaser, D., Hindson, D., and Swilling, M. (1986), 'South Africa's regional political economy: a critical analysis of reform strategy in the 1980s', *South Africa Review*, 3 (Johannesburg: Ravan), pp. 137–68.

Crocker, C. A. (1987), 'A democratic future: the challenge for South Africans', address by the Assistant Secretary of State to the CUNY Conference on South Africa in Transition, 1 October, mimeo.

Department of Constitutional Development and Planning (1985), *White Paper on Urbanization* (Pretoria: Government Printer).

Department of Plural Affairs (1977), *Annual Report* (Pretoria: Government Printer).

Divestment Disc (1987), 'Philadelphia: buying back software'.

Gurney, C. (1980), *Recent Trends in the Policies of Transnational Corporations* (United Nations: Center Against Apartheid, Notes and Documents).

Harpers (1986), Harpers Index, *Harpers*, February, 272(1629), p. 13.

Hindson, D. C. (1985), 'Orderly urbanization and influx control: from territorial apartheid to regional spacial ordering in South Africa', *Cahiers d'Etudes africaines*, 99, 25(3), pp. 401–32.

Hirsch, A. (1986), 'Investment incentives and distorted development: industrial decentralization in the Ciskei', *Geoforum*, 17(2), pp. 187–200.

IDAF (International Defense and Aid Fund), *News Notes*, (Cambridge, Mass.: United States Committee of the International Defense and Aid Fund for Southern Africa).

Innes, D. (1984), *Anglo American and the Rise of Modern South Africa* (New York: Monthly Review).

Kaplan, D. E. (1983), 'The internationalisation of South African capital: South African direct foreign investment in the contemporary period', *Southern African Studies: Retrospect and Prospect* (Edinburgh: Centre of African Studies), pp. 193–248.

Koenderman, T. (1978), 'Computers take a byte', *Management* (South Africa), December, p. 56.

Lipietz, A. (1986), 'New tendencies in the international division of labor: regimes of accumulation and modes of regulation, in Scott, A. J. and Storper, M. (eds), *Production, Work, Territory: The Geographical Anatomy of Industrial Capitalism* (Boston: Allen & Unwin), pp. 16–40.

M., A. (1979), 'Information scandal revelations continue', *Southern Africa*, 12 May (4), pp. 10–12.

Mahon, R. (1987), 'From Fordism to ?: new technology, labour markets and unions', *Economic and Industrial Democracy*, 8, pp. 5–60.

Marsden, E. (1985), 'Black Homeland emerging as a new Hong Kong', *The Sunday Times*, 2 June, p. 20.

Massing, M. (1987), 'The business of fighting apartheid', *Atlantic*, February, pp. 26–32.

Moorsom, R. (1986), *The Scope for Sanctions: Economic Measures Against South Africa* (London: Catholic Institute for International Relations).

NARMIC (1982, Revised 1984), *Automating Apartheid: U.S. Computer Exports to S.A. and its Arms Embargo* (Philadelphia: American Friends Service Committee).

Pickles, J. (1987a), 'Industrial policy and the entrenchment of grand apartheid', Presented at the annual meeting of the Association of American Geographers, Portland, Oreg., April.

——(1987b), 'Recent changes in industrial policy in South Africa', Regional Research Institute, West Virginia University, Morgantown, mimeo.

Purvis, G. (1979), 'South Africa on upswing despite embargoes', *Datamation*, June, p. 194.

Rogerson, C. M. (1986), 'Third World multinationals and South Africa's decentralization programme', *South African Geographical Journal*, 68(2), pp. 132–41.

Saul, J. S., and Gelb, S. (1981), *The Crisis in South Africa: Class Defense, Class Revolution* (New York and London: Monthly Review Press).

Seidman, A., and Makgetla, N. (1983), 'US transnational corporate investment in strategic sectors of the South African military industrial complex', *Antipode*, 15(2), pp. 8–17.

Shultz, G. (1987), 'The democratic future of South Africa', speech to the Business Council for International Understanding, New York, 29 September, mimeo.

Smith, N. (1986), 'On the necessity of uneven development', *International Journal of Urban and Regional Research*, 10(1), March, pp. 87–103.

Smith, T. H. (1987), 'The corporate connection', *Africa Report*, May–June, 32(3), pp. 19–22.

Soja, E. W. (1987), 'The postmodernization of geography: a review', *Annals of the Associaton of American Geographers*, 77(2), pp. 289–323.

South Africa Reserve Bank. *Quarterly Bulletin*, Pretoria.

South African Defense Force (1979), *Paratus*, Pretoria, October.

Stohr, W. B. (1987), 'Regional economic development and the world economic crisis', *International Social Science Journal*, May, 112, pp. 187–97.

Toffler, A. (1980), *The Third Wave* (New York: Morrow).

——(1983), *Previews and Premises* (New York: Morrow).

——(1985), *The Adaptive Corporation* (New York: McGraw-Hill).

Wellings, P., and Black, A. (1986), 'Industrial decentralization under apartheid: the relocation of industry to the South African periphery, *World Development*, 14(1), January, pp. 1–38.

Zille, H. (1983), 'Restructuring the industrial decentralisation strategy', *South Africa Review*: One, pp. 58–71.

9 Telecommunications and international transactions in information services

JOHN V. LANGDALE

Introduction

International telecommunications is a key infrastructure underpinning the emergence of the global economy. This reflects in part the rapid changes in telecommunications technologies, which have led to a large increase in international traffic and a lowering of the cost of distance. These international information flows are most obvious in the transmission of television pictures by communications satellites: information on international political events, natural disasters and sporting activities is transmitted to our homes virtually instantaneously. While these mass media information flows are a highly visible illustration of the global information economy, they represent a relatively minor component of the total information flow. Much larger flows occur in the business and governmental areas, since rapid information flows are vital for the effective operation of transnational corporations (TNCs) and governments. The global information highways are primarily emerging to meet the needs of these large organizations.

New telecommunications technologies have had very important positive impacts in terms of facilitating international trade and social contacts. They also facilitate the operation of international markets and thus reduce the cost of distance. It would be impossible to conduct a substantial percentage of international trade today without modern means of telecommunications.

While the overall effect of telecommunications is positive, there are negative impacts as well. One is that there is greater instability in the international economy. For example, the impact of major political crises on foreign exchange markets is virtually immediate and may lead to the value of currencies fluctuating widely; such changes may cause significant harm to an economy. Financial markets may overreact to political and economic information. For example, a dealing room manager of a bank was quoted as saying: 'Information's lovely to have, but if everyone else has got it at the same time, each reaction is exaggerated. When the Shah left Iran, Reuters news kept pinging more news, and you assumed it was bad without even looking. Every time you heard that ping you sold dollars. In that sense you're clearly adding to volatility' (Winder, 1983, p. 223).

Telecommunications has increased the level of interdependency in the global information economy. Most research has focused on the national level (OECD, 1981; Porat, 1977) and has shown that there has been a consistent trend towards more people being employed in information occupations and that information industries are rising as a percentage of GDP. While these studies are important in demonstrating the extent of national changes, they do not focus on the rapidly growing interdependencies in the international economy.

The level of competition between companies and countries in international trade and overall industrial competitiveness is being accelerated by the shift into the information economy. For example, the competitive rivalry between the USA, Japan and to a lesser extent Western Europe has been particularly intense in high-technology electronics and other information equipment industries. Countries that are able to dominate these industries are likely to gain significant competitive advantages in a wide range of other industries. Similarly, countries with efficient telecommunications services industries will gain substantial competitive advantages in such telecommunications-intensive industries as banking, finance and computer services.

While there is an important shift towards an international information economy, the extent of this trend should not be exaggerated. An argument commonly advanced is that we are moving inexorably towards a global economy. While this argument is superficially appealing and it is true that there are important global processes, the actual situation is more complex. The nature of the international information economy is affected by forces operating at local, regional, national and international levels. International forces are on the ascendency at present; however, this trend is not irreversible nor is it necessarily going to dominate forces operating at other geographical scales. The terms globalization and internationalization of production now being extensively used in academic and policy circles need to be examined critically.

Reasons for growth of international telecommunications

Demand

Rapid growth in the importance of telecommunications in the international information economy partly reflects the need of large organizations to transfer information quickly and cheaply. This demand for information transfer is particularly important for large corporations and government users. TNCs are increasingly using telecommunications to link together various plants and offices located in different countries. The demand for information transfer reflects the increasing complexity of firms' international production: information on production scheduling, inventory control and marketing strategies must be transferred. TNCs also face a high level of uncertainty in their operating environments. Large amounts of information are transferred in an attempt to reduce this uncertainty by monitoring competitors' strategies, changes in various governments' policies and regulations as well as the possible impact of general economic and social changes on the firm's operations.

There is little public information on the geography of these international telecommunications flows in large organizations, but they are likely to follow the organizational hierarchies of firms and government departments. Headquarters of most large TNCs are based in major industrialized countries with branch offices scattered throughout the world. Corporate telecommunications flows reflect these organizational hierarchies, although the shift in some industries towards complex global production networks (e.g. banking and finance, automobiles and electronics) may lead to a less centralized pattern of information flows.

A substantial growth in international telecommunications demand has also resulted from the growing internationalization of government activity. Most countries are members of an increasing number of international political, economic and other purpose (e.g. scientific and health) intergovernmental organizations. Furthermore, there is a wide range of bilateral agreements. These international governmental activities generate significant international telecommunications flows. Most international government agencies are located in large cities in industrialized countries. For example, UN agencies are located in New York City and Geneva and the OECD and UNESCO are located in Paris. Consequently, it could be expected that there would be substantial telecommunications flows between these world cities and national capitals.

Technological change

Many commentators have argued that rapid change in information technology is the dominant factor in understanding the role of telecommunications in the international information economy. While technological change is very important, its role must be seen in the context of changes in demand and social conditions as well as in terms of political and regulatory factors.

Some information technologies have had an indirect impact on international telecommunications. For example, rapid adoption of electronic office technologies (e.g. computer, facsimile and electronic mail) in large organizations is stimulating growth of international telecommunications, since it is much easier for individuals to communicate using this equipment. Many organizations are attempting to standardize their purchases of information equipment so as to ensure easy communications within the entire organization: there is a rapid development of Local Area Networks (LANs) and Wide Area Networks (WANs) designed to achieve this aim. In addition, there are moves by equipment manufacturers and major users to adopt Open Systems Interconnection (OSI) standards; these will further enhance domestic and international communications.

Other technological changes such as new long-distance transmission technologies (e.g. fiber-optic submarine cables and communication satellites) are having a direct impact by lowering the cost and dramatically expanding the capacity of international telecommunications systems. The most important geographical impact of these technological changes is that the cost of distance has been dramatically reduced. These cost reductions are of particular significance for countries such as Australia and New Zealand, which are located long distances from their main trading and cultural partners. Australia is the sixth largest user of INTELSAT, the global cooperative which provides international communication satellite services. Furthermore, the Overseas Telecommunications Commission (Australia), which has a monopoly over Australia's international telecommunications, is the world's third largest investor in international submarine cables.

Communications satellites and fiber-optic cables have been competitive for a considerable period. At present, there is fierce competition between the two media, spurred on by rapid technological change in both. INTELSAT, the international satellite communications carrier cooperatively owned by over a hundred countries, has argued that even on the heavily trafficked North Atlantic route, where fiber-optic cables are likely to be most competitive, INTELSAT VI will be significantly cheaper than the first fiber-optic cable TAT-8 (CUMMINGS et al., 1985). However, an important long-term advantage for cable is that fiber-optic technology is advancing faster than that of communications satellites. Thus the distance over which cables are more competitive than satellites is likely to increase substantially by the 1990s (Figure 9.1). However, communications satellites will remain dominant for very long distance transmission and thinly trafficked routes, since costs are invariant with distance. They are also dominant for point-to-multipoint (television broadcasting) communications.

New transmission technologies (particularly international communications satellites) have also been very important for Third World countries. While these countries' usage of international telecommunications is relatively small in total and concentrated in major cities, improved international services have reduced the difficulties for business and government users to access international information sources. A problem for many Third World countries is that there is better telecommunications access from their major cities to New York and London than there is from these cities to rural villages only several hundred kilometers away. Unfortunately, national telecommunications networks are being expanded slowly and there is a danger that existing differentials in living standards between rural and urban areas may be increased. However, the cost of earth stations needed to access communications satellites is being sharply reduced; in the

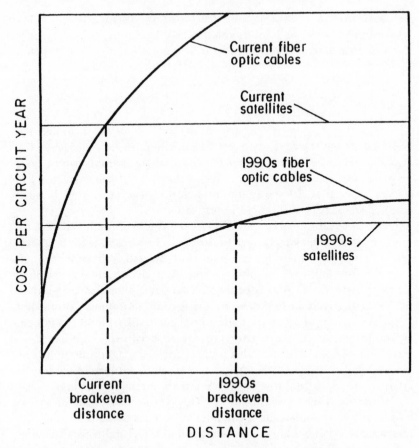

Figure 9.1 Relationship between telecommunications cost and distance for
fiber-optic cables and communications satellites
Source: US Office of Technology Assessment, 1985, p. 214

future it will be feasible for even isolated villages to be connected to
national and international telecommunications networks.

It is commonly argued that new telecommunications and information
technologies are reducing the distance component of telecommunications
costs, thus making distance irrelevant for many locational decisions. This
appears to be too simplistic a view. Not all locations on national and
international networks will have the same quality and range of services
available. It seems likely that this locational 'unevenness' in access to
telecommunications services will rise, given the pressures towards competition at national and international levels. Furthermore, telecommunications charges are only one component of the 'true' cost of distance.
Face-to-face meetings are very important in conducting international
business. Thus the cost of air travel, the frequency of flights as well as the
expense of hotel accommodation are a number of the factors which have to

be considered in evaluating the 'true' cost of distance. Even though telecommunications and information technologies are becoming more significant for international business, it is important not to overemphasize their role.

Social

The importance of social factors on the growth of international tele-communications is frequently overlooked, with most attention focused on business users. Unfortunately, telecommunications carriers do not disaggregate traffic statistics by type of user, so that it is difficult to compare the growth rate of social versus business traffic. It appears that business-oriented traffic is growing significantly faster.

One reason for the social use of international telecommunications is that migrants to a country tend to call relatives and friends in their country of origin. This is an important component of Australia's international telephone traffic. This may be inferred from Australia's outbound telephone traffic to countries such as Greece and Yugoslavia which rank eighth and twelfth respectively in 1985 (Overseas Telecommunications Commission, 1985); Australia's international trade with these countries is relatively small.

Another factor accounting for growth of social usage is international travel and tourism. It is likely that tourists generate a significant number of international telephone calls. More importantly, businesses serving the international travel and tourism market are becoming increasingly reliant on international telecommunications. This is particularly true of airlines, which use telecommunications and information technologies for passenger reservations, hotel bookings as well as for flight, fuel and crew scheduling.

It is likely that the social usage of international telecommunications is highest in industrialized countries. Third World countries' international usage is primarily business in orientation, apart from those countries with a significant number of foreign tourists (e.g. Caribbean countries).

Government regulation

One of the most contentious issues in telecommunications is the extent to which governments should regulate the industry. The US Government, supported recently by the UK and Japan, has argued that the level of government regulation should be greatly reduced. These countries have also argued that competition should be allowed in both domestic and international telecommunications. While there has been a trend towards deregulation, many countries have opposed these trends and have supported the continuation of their respective government monopolies by their Postal, Telegraph and Telephone (PTT) authorities in domestic and international telecommunications, as well as the continuation of INTELSAT's role as a monopoly international communications satellite carrier.

However, the major industrialized countries' goals are likely to be realized, given the size of their economies and their dominance of international telecommunications traffic. Some competition will be introduced in international telecommunications in the near future.

It is not possible to fully canvass the geographical implications of introducing competition in international telecommunications. Competition will be first introduced on the heavily trafficked North Atlantic route and rates on this route will be significantly reduced. In contrast, thinly trafficked routes are unlikely to attract much competition; rates on these routes will be relatively higher. It appears likely that North America and Western Europe will become much 'closer' in terms of telecommunications costs and in terms of the range of available services. In contrast, peripherally located countries will be relatively more distant from their major trading partners, given that they are predominantly located on medium and thinly trafficked routes. However, the most disadvantaged countries are likely to be those in the Third World which lie on thinly trafficked routes. While some Newly Industrializing Countries (e.g. Singapore and Hong Kong) are likely to be successful in operating as regional hubs for traffic in a deregulated and competitive international environment, it is likely that poorer countries (e.g. most of Africa) are likely to face higher telecommunications costs as compared with those available on the North Atlantic.

Telecommunications and international transactions in services

The shift of industrialized countries into service and information sectors has increased the volume of domestic and international transactions in these industries. There is increasing research and policy interest on international trade in services. Traditionally, most attention in international trade has been directed towards agricultural and mineral commodities and manufactured goods. However, it is becoming increasingly clear that trade in services is becoming more important. One estimate of world trade in the output of service industries, excluding foreign investment earnings, was over US $350,000 million in 1980 (US 1984, p. 8). While this figure is small compared to an estimated US $1,650,000 million for merchandise trade, actual trade in services is likely to be much higher than the recorded figures because of data measurement difficulties.

There are major difficulties in measuring trade in services. An important problem is that many international service transactions are not recorded in the trade statistics because they are intraorganizational in character and are not market-based. For example, the head office of a TNC may provide free information on international market trends to its worldwide subsidiaries. This service would not be included in international trade statistics because it is not a commercial transaction. Given that large TNCs as well as public organizations dominate international trade, it is likely that a significant percentage of intraorganizational transactions are not considered in official statistics. Geographers are interested in understanding the reasons for the complex web of *all* international transactions taking place between countries and regions, not just those that take place in the market.

Services which rely on telecommunications and electronic information systems are of particular interest because they are being increasingly traded internationally. This chapter focuses on international transactions in electronic information services (EISs). These services are provided by organizations involved in the collection, processing and/or transmission of electronic information. While the distinction between EIS and non-electronic information service firms is difficult to make, it is clear that today more firms are relying on electronic information systems. However, information transfer by non-electronic means (print and face-to-face communication) is still very important and likely to remain so in the future (Langdale, 1985b, pp. 2–3).

EISs are divided into primary and secondary categories, the distinction being based on whether the information handling is a primary or secondary activity for the organization. The distinction is based on research by Porat (1977) and the OECD (1981, pp. 34–40). *Primary EISs* include organizations whose main function is the handling of electronic forms of information, such as banking, finance, computer services, electronic media (television and radio) and business and financial information. Other information service firms (accountants, advertising and legal firms) have traditionally relied on non-electronic forms of information, but are utilizing electronic information systems more commonly, especially in the case of large firms.

Secondary EISs are provided by organizations primarily engaged in industries, such as manufacturing, mining or other services (e.g. health and transport). This category included EISs which are provided on an intraorganizational basis. Unfortunately, it is very difficult to obtain information on these services because most organizations do not separate these activities from their main manufacturing or service activities. Firms in high-technology manufacturing areas, such as electronics, computers, automobiles and aerospace, have sophisticated intrafirm electronic information systems linking production plants and offices throughout the world (Langdale, 1988).

International telecommunications

International telecommunications infrastructure

The international telecommunications infrastructure is important in understanding the changing geography of international telecommunications, since the introduction of new technologies is dramatically reducing the cost of distance and increasing the range of available services to users. However, not all countries and regions are receiving equal benefits from these changes. Major industrialized countries and the heavily trafficked routes connecting them are the first to receive new telecommunications technologies. The speed of introduction of new technology is increasing and while technology is diffusing to remote regions and countries over time, it is likely that these areas are relatively falling behind in terms of access to information.

International telecommunications are primarily carried by communication satellite and submarine cable, although a very small volume of traffic is still carried by high-frequency radio. Intercontinental telecommunications first commenced in the nineteenth century with submarine telegraph cables, the first of which connected North America and Western Europe in 1866. The first intercontinental telephone cable (TAT-1 with a capacity of forty-eight voice channels) was laid across the Atlantic in 1956. The first commercial communications satellite (Early Bird with a capacity of 240 voice circuits) was launched in 1965; this satellite provided services between North America and Western Europe and for the first time allowed television signals to be transmitted over intercontinental distances. The international communication satellite cooperative INTELSAT was formed in 1964; it has had a monopoly over international communications satellite transmission, although this position is increasingly under challenge. In 1965 INTELSAT served either directly or indirectly fifteen countries; by 1985 it served 166 countries (INTELSAT, 1986).

Submarine cables and communications satellites are competitive transmission media, although most carriers use both and would argue that it is important to maintain diversity of transmission media. This diversity is important if there is a fault or breakage of equipment. Furthermore, each medium has certain advantages over the other. For example, communications satellites are very efficient at distributing information in a point-to-multipoint manner (e.g. television broadcasting signals). Information sent on fiber-optic cables has the advantage of being more secure; satellite-transmitted information on the other hand is relatively easily intercepted. In addition, satellites suffer from a signal delay of approximately 0.25 of a second; this can be a problem for some types of applications and is a particular problem if a telephone circuit uses two satellite systems.

GEOGRAPHICAL EXPANSION OF FIBER–OPTIC CABLE SYSTEMS

The introduction of fiber-optic cable systems is likely to reverse the current dominance of communication satellites, at least on the heavily trafficked routes. It is unlikely that fiber-optic cables will completely displace satellites; they will be most competitive on the heavily trafficked routes such as the North Atlantic and North Pacific. However, fiber-optic cable costs are falling and the capacity of cable systems is rising dramatically, so that it is becoming more feasible to introduce them on lower-density routes.

In common with earlier international telecommunications innovations (e.g. submarine telegraph and telephone cable and communications satellites) the first intercontinental fiber-optic cables were built to carry the heavy traffic on the North Atlantic. The first cable (TAT-8 with a capacity of 8,000 voice equivalent circuits) was completed in 1988; the second (TAT-9 with a capacity of 15,000 voice-equivalent circuits) is due to be completed in 1991. While TAT-9 will have over twice the capacity of TAT-8, it will cost only 20 percent more to construct. In addition, two privately owned cables, PTAT-I (completed in 1989) and PTAT-2 (to be completed in 1992) will dramatically expand capacity. The viability of some of these private cables is rather doubtful, given the massive overcapacity likely to emerge on the North Atlantic route by the 1990s. One estimate is that by 1995 there will be a demand for 82,000 voice-equivalent circuits on the North Atlantic; there could be 650,000 circuits in place if the proposed cable and satellite projects actually take place (US Office of Technology Assessment, 1985, pp. 186, 211–13).

A North Pacific public fiber-optic cable was completed in 1988. The cable was built from California to Hawaii (HAW-4) and Guam (TPC-3). From Guam the cable branches with a northern cable connecting with Japan and a southern arm linking the Philippines and Hong Kong. The speed with which this cable was introduced following the North Atlantic one reflects in part Japan's rise as the second largest industrial power. However, at present Japan's international telecommunications traffic flows are relatively moderate, although this is expected to change in the future. There are a number of other factors which account for the rapid diffusion of this technology. One is that plans for a private fiber-optic cable (PPAC-I) have been announced. US fiber-optic cable manufacturers are eager to dominate the building of the early cables, so as to gain economies of scale in production. In addition, AT&T (and most other carriers) prefer cable to satellites, partly because they have a part ownership of the cable (whereas their influence or ownership of INTELSAT is far less direct), and in AT&T's case rate of return regulation encourages ownership of cables.

There are plans for a number of other fiber-optic cables in the Pacific Region. One will link Japan, South Korea, Hong Kong and Singapore, thus providing north–south connections with the east–west cables from the USA. In addition, the Overseas Telecommunications Commission (Australia) is planning a cable from Sydney to New Zealand with a later extension to Hawaii. A further cable may be built to East and Southeast Asia in the 1990s.

While fiber-optic cables were first introduced on heavily trafficked routes, the technology is rapidly diffusing to intermediate trafficked ones. Clearly, the very large capacity of cables at a relatively low cost is attractive to carriers. However, the pace and geographical extent of the diffusion of this innovation needs to be seen in the context of competition between public and private carriers, competition between telecommunications equipment manufacturers of major industrialized countries (particularly those of the USA and Japan), national security issues and the desire of various governments to dominate a key technological development.

International telecommunications traffic

There has been little research on the geographical distribution of international telecommunications traffic. This partly reflects the limited availability of information; many telecommunications carriers do not divulge information on international telephone and other traffic flows, regarding it as confidential. However, a rank ordering of major telephone traffic destinations for a number of countries is published (AT&T, 1983). This information source suffers from the problem that there are vast differences in the volume of outbound traffic for different countries, as for example between the USA and small Pacific Island nations. The largest outbound traffic flow for each country in the survey illustrates the dominance of major industrialized countries (Figure 9.2). With only a few exceptions, the USA dominates North, Central and South America. Traffic flows of the African countries in the survey largely reflect their colonial past. Ex-French colonies in North and Central Africa are still predominantly linked to France. Similarly, ex-British colonies of East, West and South Africa are still closely linked to the UK. The USA's influence is strong in the Middle East: it dominates the major flows from Egypt, Israel, Saudi Arabia and surprisingly Iran.

Within Europe, West Germany exerts a very strong dominance, especially for the countries of Central, Eastern and Southern Europe. Aside from the strength of the West German economy, there are also large numbers of migrant workers from Turkey and Yugoslavia living in Germany. A number of other European countries dominate smaller regional groupings: Spain and Portugal are linked to France; Ireland, Gibraltar and Monaco to the UK; and Norway, Finland and Denmark to Sweden.

Australia has a strong position in the South Pacific, although New Zealand also has a regional role. No clear pattern emerges in the Southeast and East Asian Regions, although this reflects the limited number of responding countries.

URBAN AND REGIONAL PATTERNS OF INTERNATIONAL TELECOMMUNICATIONS

A more throrough approach to the geographical pattern of international telecommunications flows would examine the linkages from cities and regions in one country to those in other countries. Unfortunately, because of data limitations, this approach has not been followed, although a number of studies have examined the urban and regional pattern of traffic at one end of the flows. While evidence is limited, studies in different

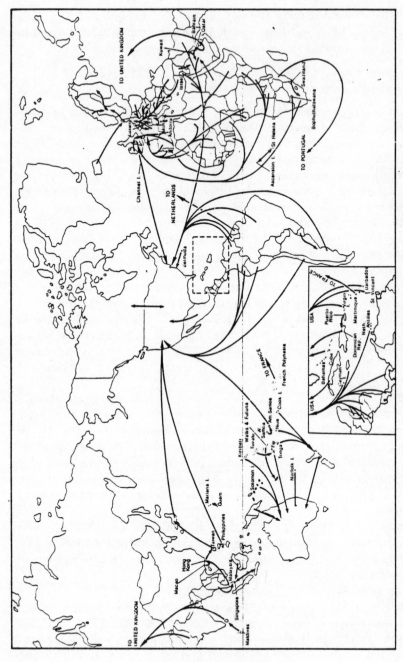

Figure 9.2 Destination of largest outbound telephone traffic flows from various countries

Source: AT&T, 1983

countries (France, Australia and the USA) have found that international telecommunications traffic is highly concentrated in major metropolitan areas.

Paris acts as the international hub for France's telephone and telex traffic. While the Paris Region (actually the Ile de France Region) has 19 percent of France's population in 1980, the region contributes 53 percent of outbound international telephone and 66 percent of international telex traffic (Bakis and Walford, 1987). The region's importance is particularly high on outbound telephone traffic to South America, with 71 percent of the total. Paris's importance reflects the concentration of regional and national offices of TNCs as well as its important role as a center for international intergovernmental organizations (e.g. OECD and UNESCO).

Australia's international telecommunications traffic is dominated by two major cities: Sydney and Melbourne. For incoming international telephone traffic in June 1983, Sydney (02 area code) attracted 38.3 percent and Melbourne (03 area code) 23.5 percent of the national total. In contrast, Sydney had 21.7 percent and Melbourne 18.6 percent of the national population. Sydney's dominance is even higher for business–oriented international telecommunications services. For outgoing international telex traffic in March 1981 over 50 percent of traffic originated from the Central Coastal Region of New South Wales (which includes Sydney, Newcastle and Wollongong), compared with 30 percent for Melbourne and 14 percent for all other capital cities (Langdale, 1982, p. 79).

While the US economic system is more decentralized than that of France and Australia, New York City dominates US international telecommunications traffic. One study in 1964 found that New York State accounted for 33 percent of revenue for outbound US telephone traffic. New York's dominance was even greater for traffic to the UK (43 percent) and Puerto Rico (48 percent). In contrast, California accounted for 34 percent of total revenue for outbound US telephone traffic destined for Japan with New York State second with 25 percent (Stanford Research Institute, 1965, B-2-9). A later study by Moss (1984) examined international telecommunications traffic from major US cities. New York City accounted for approximately 20 percent of the US total in 1982, while the greater New York City Region had just under 29 percent. In contrast, Los Angeles, the next highest ranking city, had only 8 percent.

The preeminent role of major metropolitan cities in international telecommunications partly reflects their role as head office locations for major national corporations and TNCs, which have telecommunications-intensive secondary EISs. The cities also have concentrations of primary EIS firms (e.g., banking, finance and computer services): these also have extensive demands for international telecommunications. The relative importance of these cities in terms of international telecommunications depends partly on their success in maintaining their dominance as head office locations. In some countries (the USA and the UK) traditionally dominant cites (New York and London) are losing head offices. It could be expected that their international telecommunications role would decline over time.

Internationalization of electronic information services

There has been growing interest on the part of geographers in the internationalization of production in manufacturing and service industries (Daniels, 1985; Dicken, 1986). The shift towards global production networks in some manufacturing (e.g. automobiles and electronics) and service (e.g. airlines and construction services) industries has fostered the internationalization of primary EIS firms (e.g. banking and finance and computer services).

The geographical pattern of international transactions in EISs reflects the location of head offices of major private and public organizations as well as the geographical distribution of their activities in various countries. The pattern is centered on world cities, such as New York, Tokyo and London, which are the head office locations of a large number of major TNCs in manufacturing, mining, service and EISs (Friedmann, 1986). There are large communications flows between these head offices and the organizations' regional and branch offices located in other countries (Figure 9.3).

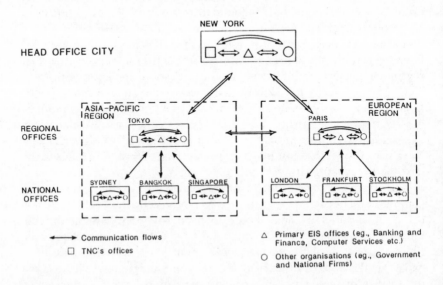

Figure 9.3 Hypothetical communications flows of New York-based transnational corporations

Complementing these vertical international organizational communications linkages are within-city communication flows at each level of the urban hierarchy; these occur between EIS firms and their clients in mining, manufacturing and service sectors. Thus at the world city level at the top of the international urban hierarchy, Cohen (1981) has examined the linkages within New York City between head offices of large TNCs and banking and finance, computer service, advertising, etc. firms. Similarly, a study of communications flows between industries located in Central London found strong linkages amongst them (Goddard, 1975).

It would be expected that in smaller cities further down the international urban hierarchy lower levels of specialization of primary EIS firms would be likely, given that the firms they serve are national corporations, government departments and branch offices of TNCs. Communications flows both within the city and to national and international locations would also be smaller for these lower-ranking cities.

Primary EISs

There have been significant moves towards an internationalization of production of primary EISs. Thus TNCs in banking, finance, computer services, advertising, accounting, etc. have expanded their network of branch offices in various countries and have also moved towards a higher level of integration of their global operations. A major reason for this development is that many of the industries' clients (which are themselves TNCs in manufacturing, mining and other services) have internationalized their own operations and expect their service firms to have branch offices in all the countries in which they operate.

There are significant variations in the nature of primary EIS firms' international operations. Some such as banking and finance have rapidly moved towards a tight integration of global operations. International telecommunications has played a critical role in the emergence of electronic funds transfer systems in the banking and finance industry (Langdale, 1985a). In contrast, other information services (e.g. law and accounting) operate in a far more decentralized manner, partly because of the diversity of national regulations and business practices, but also because much information is exchanged on a face-to-face basis. Yet even in these industries there has been growth in the importance of electronic information systems and an expansion in the coordination of international operations. For example, legal firms are making heavier usage of legal databases. In addition, many exchange information on a global basis between their offices or with clients; speed of information transfer is extremely important in legal business involving mergers and takeovers, or with complex international financing arrangements where numerous firms are involved.

In general, it is the largest of the firms in each EIS that have the most extensive international operations. For example, a significant percentage of international banking is conducted by the major banks. By the late 1970s the largest twelve banks in the USA accounted for 67 percent of all foreign lending by US banks. The degree of concentration is even higher in other countries, where the top three or four banks in Europe, for example, dominate foreign business (United Nations, 1981, p. 5). In the advertising industry the twenty-five largest US transnational advertising agencies accounted for 96 percent of the overseas business of the US advertising industry in 1970 increasing to 99 percent by 1981 (Aydin, Terpstra and Yaprak, 1984, p. 54). The top ten agencies alone accounted for 75 percent of the total international business by US agencies.

Internationalization of banking and finance

Banking and finance are of particular interest because of the rapid internationalization of production in the industry. This is reflected in the rapid expansion of large transnational banks (TNBs), the massive growth in the volume of international capital movements in the post World War II period and the growing reliance on electronic funds transfer (EFT). EFT is defined as the electronic transmission of payments or other money transmittals among financial institutions, individuals and businesses. It encompasses a wide range of payment systems and services directed towards the substitution of paper or coin movements by electronic means.

The rapid growth of the Eurodollar market is closely linked to these EFT developments. The market is 'offshore' in the sense that it minimizes national government regulations on banking and financial transactions. The Eurodollar market occupies no fixed location but relies on an international telecommunications network linking financial centers throughout the world. While participants in the market may be located anywhere, most activity takes place in the major financial cities: New York, London and Tokyo.

There has also been a rapid expansion in the importance of TNBs. A United Nations (1981, p. 3) report on TNBs defined them as deposit-taking banks with branches or majority-owned subsidiaries in five or more different countries. The key distinguishing characteristic of TNBs is the globally integrated network of banking facilities that respond to a common strategy. This global integration is achieved by using sophisticated international information systems. International telecommunications and computer systems provide the key infrastructure upon which the international financial markets rely (Langdale, 1985a).

TWENTY-FOUR HOUR GLOBAL TRADING

There has been a rapid expansion in the importance of twenty-four hour global trading in various financial markets. This reflects the growing internationalization of the banking and finance industry and the ease with which international EFT transfers can take place. The term twenty-four hour trading conjures up a vision of sleepless financiers working around the clock. While some large banks have staff rostered on such a schedule, there are a variety of other possible strategies. One is that financial institutions in one location are linked to others in different time zones. For example, futures exchanges in different countries are forming a mutual offset system: thus, a contract bought in Chicago could be sold in Singapore without additional commission fees. A number of futures exchanges are planning on establishing similar links. The advantage of the system is that it is possible for a contract to be bought or sold over a longer period of time in any day; this reduces uncertainty since financial markets are showing major fluctuations on an hourly basis.

The foreign exchange market also illustrates aspects of the shift towards twenty-four hour trading. Foreign exchange trading tends to be most active during the time period when the East Coast of the USA and the Western European trading day overlap. However, foreign exchange trading in the Asian Region has expanded rapidly in recent years. The turnover on the London foreign exchange market in 1986 averaged US $90 billion a day, compared with $50 billion in New York and $48 billion in Tokyo (Graham, 1986).

There has been some development towards a full twenty-four hour trading in foreign exchange. A number of financial institutions (e.g. TNBs, merchant and investment banks and large TNCs) which operate on a global basis, pass their trading positions in various currencies to the next office in a westerly direction, thus some foreign exchange trading follows the sun. A bank operating in New York City could pass on its foreign exchange position in some currencies to a regional office in Tokyo, Singapore or Hong Kong to cover the Asian trading day. At the close of trading, this office then passes its trading position to London for the European day and it finally comes back to New York for the cycle to begin again.

INTERNATIONAL ELECTRONIC FUNDS TRANSFER

International EFT is the vital infrastructure of global banking. EFT can be divided into two major markets: the wholesale market is oriented towards corporate and government customers; the retail market covers individuals and small business. While there have been a number of developments in international EFT in the retail market (e.g. individuals using their credit cards for purchases on overseas trips), the vast bulk of international EFT transactions take place for corporate and government customers.

The wholesale banking market makes use of a variety of international EFT networks. Public switched services (telephone, telex and data) are heavily used. The SWIFT (Society for Worldwide Interbank Financial Telecommunications) network provides cheaper, more reliable and more secure means for interbank transactions than does the public switched network. A third method is to use leased networks: these are lines leased from telecommunications carriers for the exclusive use of users. Large TNBs are very heavy users of international leased lines (Langdale, 1985a).

The SWIFT network was established to provide banks with a means of handling interbank transactions. It was originally European based, although its membership has expanded rapidly and by the middle of 1985 it had fifty-four member countries throughout the world. However, traffic is still dominated by Western Europe: in mid 1985 Western Europe accounted for 71.1 percent of the total number of messages, North America for 19.9 percent, Central and South America 1.2 percent and the Asia–Pacific Region (chiefly Japan, Hong Kong, Singapore, Australia and New Zealand) for 7.2 percent (Haney, 1985). The USA has the largest number of transactions on SWIFT. Within Western Europe, West Germany is the largest followed by the UK, Switzerland and Italy. Japanese banks are relatively small users of the network, despite Japan's number two international economic ranking.

These developments are having a major impact on the nature of distance. Capital is being transferred virtually instantaneously between the world's financial centers. While most activity is concentrated on the three large world financial cities (New York, London and Tokyo), more countries are being integrated into this global network as new telecommunications technologies are introduced throughout the world and even into formerly remote regions in Third World countries.

Secondary EISs

Some TNCs in manufacturing, mining and services which have extensive international operations also have sophisticated international information systems and make heavy use of intracorporate or secondary EISs to coordinate their international operations. However, it should not be assumed that all TNCs are moving towards this type of development; many have little need for substantial international telecommunications flows. Firms in relatively mature industries or those with fairly predictable supplies, markets and operating conditions do not need to transmit large amounts of information. In general, high technology manufacturing firms in information (electronics, computer and telecommunications) equipment, automobiles and aerospace make extensive use of information systems. Attention here is focused on the automobile industry: this industry is particularly interesting as a case study, given the massive investments in information technology and the rapidly increasing importance of international telecommunications.

CASE STUDY: THE AUTOMOBILE INDUSTRY

The automobile industry is increasingly dominated by large TNCs which operate on a global basis. Dicken (1986, p. 299) outlines two strategic options for these TNCs. One is the world car production strategy. This approach is based upon a worldwide integrated network of production, warehousing and marketing plants and offices. The alternative option is to manufacture automobiles in the TNC's home country for both domestic and export markets. Ford and General Motors have tended to pursue the world car strategy, while the Japanese manufacturers have exported from their domestic base. However, the recent rapid strengthening of the Japanese yen against other currencies is likely to encourage Japanese automobile manufacturers to follow the world car strategy or at least shift to a greater level of global production.

Production of a world car relies heavily on international information systems. It is necessary to integrate the individual computer systems at various sites specializing in design, manufacturing, assembly, warehousing and marketing. International telecommunications is being increasingly used to allow design work to be undertaken internationally. For example, Ford engineers in Western Europe are able to use the computer-aided design facilities in Detroit during the European working day while US usage is low. In addition, videoconferences are being extensively used by Ford engineers in the UK and West Germany to coordinate joint design work on new cars. A global information system is required to support the organizational and manufacturing complexity of the world car.

The above international information systems are within the one organization. However, automobile TNCs are increasingly insisting that firms which supply them with components have standardized or compatible computer systems. In the USA and Western Europe car manufacturers and component suppliers have established a common telecommunications system, which allows quotations, orders, shipping advices and invoices to take place electronically. These interorganizational and intraorganizational international information networks are leading to complex geographical patterns of information flows within and between countries.

Conclusions

This chapter has focused on changes occurring at an international level; a more thorough analysis would examine the interrelationships between forces operating at international, national/regional and intraurban scales. While international forces are having a greater impact in most countries, they in turn are modified by the operation of national and regional forces. The importance of the interrelationships between these different geographical scales of analysis needs to be stressed, since there is a significant body of opinion, especially amongst some policy makers, which argues that nation states are increasingly powerless against international forces of technological change and the globalization of industry.

The lack of understanding of the processes influencing the emergence of the international information economy partly reflects a lack of data on these industries. It is ironic that we are shifting into an information economy about which we know very little. The question needs to be posed: Where is the information on the international information economy? This is particularly true for geographical information: there is very little information on the origin/destination of traffic, or a geographical breakdown of the types of users of international services. Of even greater concern is the lack of conceptual frameworks in this field. Academics, policy makers and businessmen are used to thinking in terms of products, whether they be agricultural or mineral commodities or manufactured goods. Discussion of information service industries has been hampered by inadequate conceptual frameworks. It is difficult to place a value on information. Thus it is difficult to calculate the size of production and national and international trade in information service industries. It is not possible to consider the implications of these issues in this chapter.

Geographers have scarcely begun to consider the issues in this area. While many in the telecommunications and information industries literature have argued that distance is less important as a barrier to interaction, it is likely that the actual impact of new information technologies is likely to be quite complex. Some new technologies may reduce the cost of distance for some regions and countries, but increase it for others. To what extent are new electronic information technologies actually replacing face-to-face contact? There are many regulatory barriers to international telecommunications flows, which may offset the impact of reductions in the cost of distance resulting from technological change. There are also important organizational barriers to international information flows. For example, subsidiaries of large TNCs are not necessarily allowed to access information from other subsidiaries. We know very little about the geographical impact of these barriers to international information flow.

Frequently it has been assumed that the most profitable organizations are those which rapidly adopt information technology and develop global information networks. It is also argued that distance will be less important as a barrier to communication in these global corporations, since electronic information is transferred virtually instantaneously on a worldwide basis. However, global information networks do not necessarily lead to good decision making. For example, despite the fact that many TNBs have sophisticated international information systems, many made loans to Third World countries which have had little chance of paying them back. Similarly, in the automobile industry, rapid adoption of international information systems to support the production of world cars has not as yet given US firms a competitive advantage over Japanese automobile manufacturers.

It is clear that geographers need to undertake more research on the change taking place in the telecommunications and information services areas. While this paper has focused on the economic implications of these changes, there are critical cultural and political implications as well. Global information flows in the mass media industries (particularly in television broadcasting) are having a significant impact on cultural change. The trend towards the internationalization of media industries combined with efforts to liberalize trade in services, will increase international pressures on cultural change, particularly in Third World countries. Similarly, there are important political geographical implications as well. For example, concepts of national soverieignty are changing as a result of the adoption of new telecommunications and information technologies (Langdale, 1985b, 1987).

References

AT&T (1983), *The World's Telephones* (Morris Plains, NJ).

Aydin, N., Terpstra, V., and Yaprak, A. (1984), 'The American challenge in international advertising', *Journal of Advertising*, 13, 4, pp. 49–57.

Bakis, Henri, and Walford, Dominique (1987), 'Variables de télécommunications et variables socio-economiques: une analysis de données et son interpretation', in *Netcom*, no. 1, Paris: Notes-Etudes-Travaux du Groupe d'Etude Géographie de la Communication et des Télécommunications, Comité National Français de Géographie, pp. 40–75.

Cohen, Robert B. (1981), 'The new international division of labour, multinational corporations and urban hierarchy, in Dear, M., and Scott A. J. (eds), *Urbanization and Urban Planning in Capitalist Society* (London: Methuen), pp. 287–315.

Cummings, J. Michael *et al.* (1985), 'Satellite versus fiber optic cables', in Wedemeyer, Dan J. (ed.), *Pacific Telecommunications Conference Proceedings* (Honolulu), pp. 422–6.

Daniels, P. W. (1985), 'Service industries: some new directions', in Pacione, M. (ed.), *Progress in Industrial Geography* (London: Croom Helm), pp. 111–41.

Dickens, P. (1986), *Global Shift: Industrial Change in a Turbulent World* (London: Harper & Row).

Friedmann, John (1986), 'The world city hypothesis', *Development and Change*, 17, pp. 69–83.

Goddard, John B. (1975), *Office Location in Urban and Regional Development* (Oxford: Oxford University Press).

Graham, George (1986), 'London emerges as top foreign exchange centre', *Financial Times*, 20 August, p. 1.

Haney, Siobhan (1985), 'Conversion will increase system capacity', *Financial Times* Survey on Computers in Banking and Finance, 21 October, p. 8.

INTELSAT (1986), *INTELSAT Report 1985–86* (Washington, DC).

Langdale, John V. (1982), 'Telecommunications in Sydney: towards an information economy', in Cardew, R., Langdale, J., and Rich, D. (eds), *Why cities change: urban development and economic change in Sydney* (Sydney: Allen & Unwin), pp. 77–94.

——(1985a), 'Electronic funds transfer and the internationalisation of the banking and finance industry', *Geoforum*, 16, 1, pp. 1–13.

——(1985b), *Transborder Data Flow and International Trade in Electronic Information Services: An Australian Perspective* (Canberra: Australian Government Publishing Service).

——(1987), 'International information flows and national sovereignty', in Barr, T., and Garrow, C. (eds), *Choices, Challenges and Change: Australia's Information Society* (Melbourne: Oxford University Press), pp. 137–45.

——(1988), 'Telecommunications and electronic information services: Australian perspectives', in Hayter, R., and Wilde, P. (eds), *Industrial Transformation and Challenges in Australia and Canada: International Comparisons*, forthcoming.

Moss, M. L. (1984), 'New York isn't just New York any more', *InterMedia*, 12, 4/5, pp. 10–14.

OECD (1981), *Information Activities, Electronics and Telecommunications Technologies: Impact on Employment, Growth and Trade*, vol. 1 (Paris: OECD Information, Computer, Communications Policy, no. 6).

Overseas Telecommunications Commission (1985), *Annual Report*, Sydney.

Porat, Marc (1977), *The Information Economy* (Washington, DC: Office of Telecommunications Special Publication 77-12, US Department of Commerce).

Stanford Research Institute (1965), *Study of International Telecommunications Policies, Technology and Economics*, Vol. II, Menlo Park, Calif.: Stanford Research Institute.

United Nations (1981), *Transnational Banks: Operations, Strategies and their Effects in Developing Countries* (New York: Centre on Transnational Corporations).

US (1984), *US National Study on Trade in Services. A Submission by the United States Government to the General Agreement on Tariffs and Trade* (Washington, DC: US Government Printing Office).

US Office of Technology Assessment (1985), *International Cooperation and Competition in Civilian Space Activities* Washington, DC.

Winder, R. (1983), 'From information to communication', *Euromoney*, May, p.223.

PART III

Communications, technology and regional development

10 *The role of information technology in the planning and development of Singapore*

KENNETH E. COREY

After having been in this country for little more than a year, I have come to appreciate that one of the advantages the average educated Asian has over the average educated American is that he knows more about you than you do about him. I speak your language. You don't speak mine. I have read your great books, which I have come to love as if they were my own.

. . . The knowledge the Japanese, South Koreans, Taiwanese and Singaporeans have about your language, customs and practices gives them [an] advantage. Your well-wishers would urge you to restore the balance.

> Ex-Singapore President C. V. Devan Nair,
> speaking at Indiana University (1987)

The intention of this chapter is to provide information that can be used to restore the balance noted above by former President Nair. There is much to learn from development cases in the Pacific basin. The following examination of Singapore is offered as documentation on how one metropolis has been planning itself for and into the information age. It is written in the belief that it contains lessons and suggestions for other cities and places, especially for the officials of metropolitan areas that are searching for examples of policies and programs that could be used as models in formulating their own information–age development planning strategies.

The method used here is to draw on documents, words and plans from the Singapore government to tell the story of this contemporary development case. These extracts are supplemented by reference to the relevant literature. Enough of the operational details of Singapore's information technology planning and implementation are offered here to suggest potential adaptation elsewhere.

Singapore

Singapore is both a country and a metropolis. This metropolitan area has developed on a small island off the southern tip of the Malaysian peninsula. Today the population of Singapore is nearly 2.6 million people.

For the student of modern urban development Singapore is an extra-ordinary laboratory that is rich with learnings. Even though the city-state of Singapore is unique, its long-term, futuristic approach to development has produced remarkable results (Yeung, 1987). Were other metropolitan areas able to devise their own analogous planning insights and development commitments, then it is likely that they would be in a better position to meet the unknowns and challenges of the information age.

Singapore's reliance on intervention and planning

The role of the state is fundamental to Singapore's development (Dicken, 1987; Grice and Drakakis-Smith, 1985). Singapore became independent in 1965. Since then it has been governed by the People's Action Party under the leadership of Prime Minister Lee Kuan Yew. The government of Singapore has taken up the task of development in an activist manner that has used intervention and planning in pursuit of the ultimate goal of realizing a high quality of life for Singaporeans.

Nearly 200 years ago Singapore began as a port and it developed a predominantly entrepôt economy. In 1963, just two years before independence, a program of industrialization was conceived. While manufacturing has been and will continue to be of major importance to Singapore's economy, by the late 1960s and the early 1970s the government also had begun to consider complementary development policies that importantly involved scientific and technical knowledge. In 1972, the government had formulated its Development Plan for the 70s which included planned pro-grams for an important future sector of high-technology and brain services (Hon, 1972).

1979 saw government intervention in support of its continued objective of restructuring the economy toward higher productivity and higher-skilled industries and services. This strategic position was sustained by the Singapore government in 1981, when it announced its Economic Develop-ment Plan for the 1980s; the principal objective of which was 'to develop Singapore into a modern industrial economy based on science, technology, skills and knowledge' (Goh, 1981).

The 1980 through 1984 period was assessed by the government as having 'been years of exceptional prosperity for Singapore and Singaporeans. Singapore's gross national product (GNP) grew at an average rate of 8.5 per cent per annum in real terms' during these five years (Tan, 1985). However, in 1985, after twenty years of double-digit average annual economic growth, the Singapore economy, as measured by the gross domestic product (GDP), declined by 1.8 percent.

Related to the worldwide recession, this downturn catalyzed Singapore policy makers and planners into an immediate reassessment of their approach to economic development. This resulted in the early 1986 publication of the *New Directions* report (Economic Committee, 1986). Basically, this evaluation reaffirmed the earlier goals of restructuring the economy and for the future recommended continued high priority for develpment policies based on high technology, research and development, information technology and the services sector – among other policies and sectors. By the end of 1987, the government's short-term measures for stimulating the economy had produced two consecutive years of economic growth; 1986 growth was 1.8 percent and 1987 growth was an unexpectedly high 8.6 percent (*The Wall Street Journal*, 1988, p. 10).

This brief summary of economic policies and results is suggestive of Singapore's style of development. It is characterized by active government planning (Krause, Koh and Lee, 1987) that is premised on long-term goals that have remained relatively constant and on interventions in the short run that are designed to make necessary mid-course corrections that bring development back toward the ultimate imperatives again. In short, Singapore's government-dominated development planning has been elitist, pragmatic, flexible, experimental, far-sighted, anticipatory and innovative (Corey, 1987). One of the more innovative policies is the use of information technology in the planning and development of Singapore for the information age.

Singapore and the information age

As a result of the industrialization initiatives of the 1960s, by the early 1970s Singapore was well on its way to becoming an information economy. But what is this type of economy?

The *information economy* consists of functions and activities that are:

> . . . supported by the establishment of industries and markets wherein the necessary technical infrastructure are produced and information as a commodity is sold respectively. As these activities expand in scope and volume, there is a concomitant increase in the share of GNP arising from the value adding which originates from the production and distribution of information goods and services. (Jussawalla and Cheah, 1983, p. 162).

Using Singapore's input–output table for 1973, Jussawalla and Cheah (1983) concluded that indeed 'the results of our empirical study of Singapore's information sector . . . suggest that Singapore is proceeding towards an information economy' (p. 168). Their study began the important process of accounting for the production generated by the information sector of the Singapore economy. Jussawalla and Cheah found that the information sector accounted for 24 percent of Singapore's gross domestic product in 1973. Also, they concluded that

> the public sector appeared to be the most information-intensive with approximately 2/3 of all government services being information-based. Public bureaucracies in administration, planning and development agencies accounted for 49 per cent of total value added. (p. 169).

Singapore's workforce has become dominated by white-collar occupations. In 1921, 27.2 percent of employed persons in Singapore were in white-collar occupations; by 1980, that percentage had risen to 41.5 percent; currently this is the largest occupational category in Singapore's employment structure; blue-collar occupations form the next largest classification at 40.4 percent (Kuo and Chen, 1985).

The 1980 Singapore workforce of 1,077,090 has 34 percent (366,912) of its workers employed in information occupations. In 1921, only 18.6 percent of its workforce was engaged in information occupations (Kuo and Chen, 1985).

In an age when more future employment can be expected to be in white-collar and information occupations, it is instructive to examine some of the policies, programs and actions that have been formulated and utilized by the city-state of Singapore in its goal of becoming an information age metropolis.

The national information technology plan

A national computerization effort began in 1981. It was part of Singapore's Economic Development Plan for the 1980s (Goh, 1981). The government's National Computer Board was established then to encourage Singapore's computerization.

With this initiative and the previous history of higher-technology and higher-level services policy experiments and program planning, Singapore by 1985 had positioned itself to be able to formulate a national development plan based on information technology. For Singapore, the operational definition of information technology (IT) 'embraces the use of computer, telecommunication and office systems technologies for the collection, processing, storing, packaging and dissemination of information' (National IT Plan Working Committee, 1985).

Based on the potential of information technology, the national plan that was constructed consisted of seven building blocks. These include:

(a) information technology manpower; (b) information technology culture; (c) information communication infrastructure; (d) information technology applications; (e) information technology industry; (f) climate for creativity and entrepreneurship in information technology; and (g) coordination and collaboration of the information technology plan.

Information technology manpower

One of the most fundamental of the building blocks in Singapore's national information technology plan is to achieve a workforce that can fully exploit the current and future potential of information technology – both in innovative and economic terms. Singapore has evolved its own philosophy for its manpower policies. These reject Europe's industrial democracy approaches; Singapore has developed a response to its manpower needs that accommodates to its unique ethnic, cultural, economic and locational attributes (Wilkinson, 1986). Through its Economic Development Board, the government of Singapore seeks to attract high value-added industry; this is possible only with a skilled workforce which is computer literate, comfortable in automated workplaces and capable of technological creativity. As a consequence, information technology manpower development is emphasized at all levels, including: 'assembly workers, technicians and technologists, and manufacturing engineering' (Christiansen, 1987).

Manpower development activities are being conducted in an environment of nearly full employment in Singapore (i.e. 3.6 percent unemployment rate in September 1987), and in an environment of continued worker shortage in the industries and services driven by information technologies (Christiansen, 1987; Ministry of Trade and Industry, 1987). To respond to these needs, the Singapore National Computer Board coordinates computer training and education programs; many of these programs are operated by other public-sector and public-funded organizations.

CIVIL SERVICE COMPUTERIZATION PROGRAM

One of the earlier and most important activities of Singapore's information technology initiatives is the Civil Service Computerization Program. This consists of using government agencies as computer pioneers to test the technological and organizational implications of computerizing offices and office staffs. Approximately $50 million (all dollar figures are given in US dollars) have been invested in computerizing the public sector (Khan, 1987, p. 12). Seventeen government ministries and units are having their computer needs supported by the National Computer Board. The lessons of the Civil Service Computerization Program are being disseminated into Singapore's private sector so as to further develop the local economy.

OFFICE SYSTEMS PROGRAM

The National Computer Board introduced a pilot office automation effort in 1985. Its purpose was to demonstrate the use of the technology and to show how it would increase productivity. The National Computer Board and the Ministry of Finance were the principal actors in this program.

USER EDUCATION PROGRAM

By means of courses for civil servants, the National Computer Board is introducing computer approaches and usage throughout the government agencies involved in the Civil Service Computerization Program.

CIVIL SERVICE INFORMATION NETWORK

This program will link the computer systems of the government ministries and public agencies. It is a collaborative effort between the National Computer Board, Telecoms (i.e. Telecommunication Authority of Singapore) and the Management Services Department. The end result is the realization of the required infrastructure for electronic information flows across government units.

These and other programs have surpassed original expectations. 'The original target to have 5,800 computer professionals by 1990 has already been exceeded, three years early' (Khan, 1987, p. 12).

Information technology culture

In order to provide a climate that is supportive and receptive to information technology and to encourage widespread understanding of its benefits, the Singapore government has developed many educational and promotional campaigns. Information technology permeates the daily life of Singapore. For example, October 1987 was designated 'technology month'. At the checkout counter of the supermarket and at the ubiquitous automatic teller machine, the Singaporean is constantly reminded of the country's objectives for cashless shopping and cashless banking. Especially if a Singaporean is a public servant, then he or she has probably been using computers in the office for some time; this is part of the strategy of using the government as the pilot laboratory for computerizing the world of work. Local newspapers routinely publish articles and graphics that educate the citizenry about information technology. An international information technology exhibition has been held three times in recent years; 'Singapore Informatics' exhibits attract not only computer technicians, vendors and government and business professionals, they also attract curious citizens and interested school-children and their parents. The Ministry of Education is implementing information technology education programs throughout Singapore's schools; it also seeks to develop the 'whole person' and to provide for more creativity (Study Team, 1987).

Another complementary exhibition, designed to inform the populace and to aid in creating the climate for familiarity with information technology, is the Civil Service Productivity Exhibition. It is a means by which the public can be shown the progress being made by the public sector in its drive for improving productivity via information technology. Progress on computerization projects are demonstrated in the areas of financial and economic services, training and education, infrastructure, housing and environment, defense and security, health and community services, communications and information services and the modern office. Twenty-five ministries and government units were able to demonstrate their progress in computerization at the 1986 exhibition. These are just some of the examples that have been used to overcome resistance to change and to cultivate a widespread appreciation of information technology as one of the most important resources for Singapore as it plans itself into the information age.

Information communication infrastructure

One of the other fundamental building blocks of the information technology plan is the development of the actual plant and equipment of the technology itself.

> This infrastructure is the electronic highway of the Information Age:
> it provides the communication grid of the Singapore information
> society and the various information-based business and industries.
> (National Computer Board, 1986b, p. 2).

These telecommunications facilities enable the Telecommunication Authority of Singapore to offer these services to private and corporate customers: (a) international telephone service, including international direct dialing; (b) regional telephone services, including subscriber trunk dialing to Malaysia; (c) private leased circuits, including high-speed international leased circuits for multinational corporations with locations in Singapore; (d) data services, including 'telepac', a remote computer access service with data links to thirty-five overseas destinations; (e) public message services including facsimile (Fax) machines for sale or lease; telefax, a counter service at Telecoms outlets and post offices for the general public; telex, with 6.9 lines per 1,000 population; telemail, which is telex for small businesses; and telebox, which is an electronic mailbox service; (f) an optical fiber network service has been completed that links all of Singapore's twenty-six telephone exchanges enabling data, text and video services; (g) various radio services, including widely used radio pagers, radio memo pagers that can store and display 1,200 characters, landmobile telephone service − i.e. car phones for rent and sale, maritime radio communication services, including radiotelex, radiotelephone and radiotelegram, and maritime satellite communication services, including inter-

national telephone traffic and international telex traffic; (h) national telephone service with a telephone network that is 100 percent push-button
and Telecoms is proud that 98.4 percent of applications for telephones are
met by the date needed by customers; telephone service can be provided
within six working days. Telephone services include both business and
residential direct exchange lines; at forty-three telephones per 100 population, Singapore, after Japan, has the second-highest telephone density in
Asia. Some other telephone services include the 'phone plus service', which
consists of abbreviated dialing, absentee message, automatic redial, call
transfer, call waiting and three-way calling.

The Telecommunication Authority of Singapore also delivers the more
conventional information services of the post office. These complement the
telecommunications services in an efficient and low-cost manner.

For the near future, Singapore is taking the steps to provide Integrated
Services Digital Network (ISDN) linkages to business by the early 1990s.
'ISDN will essentially be an enhanced telephone network which provides
end-to-end digital connectivity. This will allow voice, data, video and text
communications simultaneously from one network' (Telecommunication
Authority of Singapore, 1987, pp. 18–19). Because separate networks
become redundant, one of the major values of ISDN is its one-network
efficiency, with resultant cost savings. Teleview is another program for the
future; teleview is an interactive system for making electronic transactions
that are based on computer-stored information displayed on monitors and
home television sets in combination with the use of the telephone network
for off-air communication. A design and testing period will be necessary
before the full development of Teleview. Other future telecommunications
infrastructure and services are under development, including a new cellular
radiotelephone system and a consultancy company to provide technical
assistance and telecommunications planning services overseas. (The above
material on Singapore's information communication infrastructure was
abstracted from the 'Operating Review' section of *Telecoms Annual Report
1986/1987*, 1987. Also see Chia, 1986.)

Information technology applications

The government of Singapore continues to promote the widespread and
active use of information technology throughout the economy. This is
expected to result in greater efficiency in business and a higher quality of
life for Singaporeans. Some of the applications mentioned include civil
service computerization, widespread availability of automatic teller
machines for consumer banking and, among others, the use of credit cards
in retail sales so as to move Singapore toward becoming a 'cashless society'.
The *National IT Plan* (National IT Plan Working Committee, 1985)
includes the identification of a wide range of information technology
applications for Singapore; in the primary activities of the manufacturing

sector, information technology can be applied to: (a) 'inbound logistics', as in the case of receiving and storing the resources and commodities that go into the manufacturing process, an automated warehouse is an example of such an application; (b) operations, as in the case of a factory production line, e.g. flexible manufacturing; (c) 'outbound logistics', as in handling the finished manufactured product, e.g. automated order processing; (d) marketing and sales, as in the case of telemarketing and remote terminals for retail employees; and (e) service, as in the case of repairing a product. The application of information technology to activities in support of manufacturing includes: (a) firm infrastructure, as in the case of management inside the firm, e.g. planning models; (b) human resource management, as in the case of personnel matters, e.g. automated personnel scheduling; (c) technology development, including research and development; computer-aided design is such an example; and (d) procurement, as in the case of input purchasing, e.g. on-line procurement of parts.

Information technology applications in the service sector, according to the *National IT Plan*, include among the 'primary activities' the following: (a) 'inbound logistics', such as receiving and storing information, e.g. receiving survey returns; (b) operations such as the production of reports, for example, data editing, data capture, data processing, analysis and report generation; (c) 'outbound logistics', as in processing the completed report, e.g. storing the report's information in a database; and (d) marketing and sales, as in the case of distributing report information to a client, for example, 'on-line retrieval by clients and follow-up on-line query', and (e) service, such as updating information provided earlier and transmitting it to clients. The application of information technology to activities in support of service-sector functions includes: (a) firm infrastructure, such as in management, e.g. payroll; (b) human resource management, such as personnel matters, e.g. staff training in the application of information technology; (c) technology development, including research, as in electronic market research; and (d) procurement, 'input purchase, e.g. liaising with clients, third-party information providers locally and overseas' (National IT Plan Working Committee, 1985, p. 39).

Information technology industry

The national information technology plan intends that information technology industry will be an increasingly important force in Singapore's economy and its development. The underlying concept is to develop a robust information technology industry and in the process to increase Singapore's industrial productivity and thereby its international competitiveness (National Productivity Board, 1986). The principal sectors of Singapore's information technology industry planning include: (a) computer hardware manufacturing, (b) computer services and (c) telecommunication services.

The manufacturing of high technology products in Singapore has been led by multinational corporations.

Over the last 20 years, about 600 international manufacturing companies have set up business in Singapore. Most of them are engaged in technologies the government identified as desirable, such as computers, microelectronics, biotechnology and communications. (Cheng, 1987).

The information technology strategy of Singapore recognizes that multinational corporations will continue to be essential in the transfer of high technology into Singapore. However, increasingly there will be emphasis on ensuring that Singapore's local industry in information technology will be fully developed. In the manufacturing sector, the following industrial functions have been and will be nurtured and developed in Singapore; computer integrated manufacturing, computer aided engineering, computer aided design, automated product systems and computer numerical control as in machine tool fabrication. These kinds of information technology innovations in manufacturing may be most useful in industries such as consumer electronics and integrated circuit assembly (National IT Plan Working Committee, 1985, p. 39).

The service sector of the Singapore economy will be stimulated by various initiatives and incentives, and 'will concentrate on services which have the potential to be internationally tradable, which make maximal use of our limited brainpower resources, and which generate domestic spin-offs for economic activities' (Singapore Economic Development Board, 1986, p. 53). The Economic Committee recommended the promotion of these services subsectors: sea transport, air transport, land transport, telecommunications, warehousing and distribution, computer services, laboratory and testing services, 'agrotechnology', publishing, and the professional services of: accounting and auditing, legal services, advertising and public relations, management and business consulting, hotel management services, medical services, educational services, and cultural and entertainment services (Economic Committee, 1986, pp. 177–80).

The third sector of Singapore's information technology industrial priorities is the telecommunication services industry. Telecommunications are seen as the major facilitator of both information technology manufacturing and information technology services. The Telecommunication Authority of Singapore is charged with implementing Telecoms infrastructure; its principal activities are listed above (Chia, 1986).

Climate for creativity and entrepreneurship in information technology

The national information technology plan seeks to have Singapore and Singaporeans become increasingly more innovative and developmental in their information technology activities. Toward that end several public-sector research and development institutions have been created. The Institute of Systems Science is researching bilingual systems, intelligent public information systems and office automation (National Computer Board, 1986a, p. 16). The National University of Singapore is conducting applied research in artificial intelligence, local area networks, software engineering, microprocessor applications and semiconductors. Information technology research also is underway at the Grumman International-Nanyang Technological Institute CAD/CAM (Computer aided design/computer aided manufacturing) Research Center and at the Information Technology Institute of the National Computer Board. As noted above, the Telecommunication Authority of Singapore also conducts its own research and development.

International corporations receive incentives to establish research and development centers in Singapore. The Economic Development Board of Singapore offers overseas companies a variety of tax incentives to stimulate industrial research and development. Several incentives are particularly supportive of information technology development activities. For example, there is the venture capital incentive; it is for eligible companies and individuals in approved new technology projects. Such companies must be (a) 50 percent or more owned by Singaporeans and (b) incorporated and resident in Singapore for tax purposes; individuals must be citizens or permanent residents of Singapore. The tax concession is for losses incurred from the sale of shares for up to 100 percent of equity invested; such losses can be offset against the other taxable income of the investor (Singapore Economic Development Board, n.d.). Other Singapore investment incentives include tax breaks on: international consultancy services; pioneer (i.e. new industries) and post–pioneer status; expansion for large investments in manufacturing machinery and equipment; investment allowance for specified allowances in research and development activities, among others; headquarters operations; warehousing and technical and engineering services; an approved foreign loan scheme; and on approved royalties.

In addition to these incentives, a Science Park has been developed to serve as a locational focal point for research and development activities in Singapore. Both private-sector and public-sector firms may locate in the Science Park, the site of which is near the National University of Singapore; the Park, a 125-hectare estate, permits early start-up of high technology research and incubator activities. The National Computer Board headquarters also is located on the site. The research interests of the National Computer Board's Information Technology Institute focus on: (a) software technology so as to increase the productivity of computer professionals; (b) computer and communications technology, so as to improve the efficiency of information workers by means of the integrated use of computers and office equipment; and (c) knowledge systems that are concerned with the development of application software to some problems usually addressed only by humans (Information Technology Institute, n.d.).

Information technology professionals in Singapore are encouraged to be innovative by having access to venture capital financing, incubator facilities, a legal context that provides for copyright protection of software, specialized technical assistance, and a climate for creativity that comes from the synergy that results from the close proximity of others also engaged in related developmental work in information technology (National Computer Board, n.d., p. 10).

Coordination and collaboration of the information technology plan

The actors involved in implementing the above six elements of the *National IT Plan* (1985) require linkage and active attention if they are to realize their objectives. Principal actors are: (a) information technology manpower development through training programs has involved many actors, including the National University of Singapore, the Japan–Singapore Institute of Software Technology, and Ngee Ann Polytechnic, among other organizations; (b) the Ministry of Education, the National Trade Union Council, People's Association, the Ministry of Communications and Information and others play important roles in promoting information technology culture; (c) information communication infrastructure development has been the primary responsibility of the Telecommunication Authority of Singapore; over the years, a modern foundation of facilities has been laid; over the next several years Telecoms will continue to invest heavily (i.e. $1.5 billion) in enhancing its networks and developing new services (National Computer Board, 1986b, p. 3); (d) information technology applications are the business of many private-sector and public-sector organizations, but the National Computer Board has the explicit mandate to stimulate and coordinate the implementation of the computerization of the Singapore economy and society; (e) information technology industry creation ultimately is the responsibility of the government of Singapore, particularly the Economic Development Board, but in partnership with multinational corporations from overseas and local companies and indigenous innovators and workers; (f) the climate for creativity and entrepreneurship in information technology is intended to be achieved through effective innovation nurtured by Singapore's research institutions such as the Information Technology Institute, the Institute of Systems Science and the National University of Singapore, among others; and (g) in addition to its charge to promote national computerization, the National Computer Board has the responsibility also for developing Singapore's computer service industry. The implementation of these functions is shared with the many ministries, boards and organizations noted earlier, with the principal coordinating body composed of the National Computer

Board, Telecoms, the Economic Development Board and the educational institutions, with the professional information technology societies and selected private-sector firms in support.

Singapore's information technology planning in sum

From the preceding, it should be concluded that it is possible to scan a city's external and internal economic and technological environments so as to identify needs and niches that may be used in steering future development. In just over twenty years Singaporeans have been able to plan and implement themselves from a preindustrial economy and society to one that is importantly industrial and to one that is importantly 'postindustrial' or information-driven.

This has been possible in large part because of an often renewed contract between the elected leaders and the voting electorate. The government and its support bureaucracy have taken up their mandates by formulating policies and engaging in planning that are intended to bring Singaporeans a high quality of life by means, in part, of preparing for and exploiting the potential of information technology. By laying a modern and effective foundation for its development strategies and by being meticulous, tenacious and flexible in its short-term and long-term planning, the government of Singapore has brought its city-state the third highest level of prosperity in Asia – after only Japan and Brunei.

As Singapore's aging leadership is replaced and as Singapore's citizenry comes to grips with its growing education and affluence, the country will be faced with different and new needs. Lessons from some of the planning behavior of the previous generation have not been lost on the younger generation as it faces the future. Singapore can be expected to continue to be a rich source of learning for students of planning and development (Krause, Koh and Lee, 1987).

Lessons from other countries

Singapore is unique in that so much of its achievement and development may be attributed to government leadership, planning and intervention in an 'open economy' (Singapore Economic Development Board, 1987, p. 9). The United States is unique in that much of its achievement and development has been realized as a result of *laissez faire* governments, with relatively more reliance on market forces and the disaggregated initiatives and responses of the private sector. Both are systems based on capitalism. Given these contrasts and common grounds, it would seem that a great deal can be learned from each other. For example, while some of the United States consider the formulation of Federal information policies (Hernon and McClure, 1986), it is of immense value to be able to trace the approach of other countries with longer experience in a similar policy sector.

This chapter began with former President of Singapore Nair challenging Americans to become more competitive by getting to know about Asian-Pacific countries which, in the last two decades, have been hailed as models of planned economic development. By analyzing part of the future development strategy of the tiniest of these newly industrialized countries, it has been the intent here to demonstrate that others have been able to scan the future and have planned so as to exploit their comparative advantages. As you read this and the lessons contained in this volume from other countries and places, you would do well to grapple with their potential applicability – with appropriate modification – to your own city, state and nation (Corey, 1987; Masser and Williams, 1986).

Note

This research was made possible through a Fulbright research grant and the support of K. S. Sandhu and the staff of the Institute of Southeast Asian Studies. For the preparation of this chapter special thanks are due to Chia Choon Wei, Loh Chee Meng, Ong Choon Hwa and Martin Anderson.

References

Chia, C. W. (1986), 'Economic aspects of the information revolution: the Singapore experience', in Jussawalla, M., Wedemeyer, D., and Menon, V. (ed.), *The Passing of Remoteness? Information Revolution in the Asia-Pacific* (Singapore: Institute of Southeast Asian Studies), pp. 42–50.

Cheng, G. (1987), 'The Singapore way: acquiring technology through MNCs', *The Straits Times*, 3 October, p. 16.

Christiansen, D. (1987), 'An electronic nation', *IEEE (Institute of Electrical and Electronics Engineers) Spectrum*, 24, 12, p. 30.

Corey, K. (1987), 'Planning the information-age metropolis: the case of Singapore', in Guelke, L., and Preston, R. (eds), *Abstract thoughts: concrete solutions, essays in honour of Peter Nash* (Waterloo, Ontario: Department of Geography, University of Waterloo), pp. 49–72.

Dickens, P. (1987), 'A tale of two NICs: Hong Kong and Singapore at the crossroads', *Geoforum*, 18, pp. 151–64.

Economic Committee (1986), *The Singapore economy: new directions* (Singapore: Ministry of Trade and Industry).

Goh, C. T. (1981), 'Towards higher achievement', Budget speech, 6 March 1981 (Singapore: Information Division, Ministry of Culture).

Grice, K., and Drakakis-Smith, D. (1985), 'The role of the state in shaping development: two decades of growth in Singapore', *Transactions Institute of British Geographers*, 10, pp. 347–59.

Hernon, P., and McClure, C. (1986), *Federal Information Policies in the 1980's: Conflicts and Issues* (Norwood, NJ: Ablex).

Hon, S. S. (1972), *Singapore: Economic Pattern in the Seventies* (Singapore: Ministry of Culture).

Information Technology Institute (n.d.), *Information Technology Institute* (Singapore: National Computer Board).

Jussawalla, M., and Cheah, C. W. (1983), 'Towards an information economy: the case of Singapore', *Information economics and policy*, 1, pp. 161–76.

Khan, T. L. (1987), Going for IT, *Mirror*, 23, 18, pp. 12–13.

Krause, L., Koh Ai Tee, and Lee (Tsao) Yuan (1987), *The Singapore Economy Reconsidered* (Singapore: Institute of Southeast Asian Studies).

Kuo, E. C. Y., and Chen, H. T. (1985), *Towards an Information Society, Changing Occupational Structure in Singapore* (Singapore: Select Books for the Department of Sociology, National University of Singapore).

Masser, I., and Williams, R. (eds) (1986), *Learning from other Countries* (Norwich: Geo Books).

Ministry of Trade and Industry (1987), *Economic Survey of Singapore, Third Quarter 1987* (Singapore: Ministry of Trade and Industry).

Nair, C. V. D. (1987), 'Notable & quotable', *The Wall Street Journal*, 29 May, p. 26.

National Computer Board (1986a), *National Computer Board Yearbook FY85/86* (Singapore: National Computer Board).

——(1986b), *National IT Plan* (Singapore: National Computer Board).

——(n.d.), *The Singapore Information Technology Center* (Singapore: National Computer Board).

National IT Plan Working Committee (1985), *National IT Plan: A Strategic Framework* (Singapore: National Computer Board).

National Productivity Board (1986), *The Report of the Committee on Productivity in the Manufacturing Sector (1985)* (Singapore: National Productivity Board).

Singapore Economic Development Board (1986), *Economic Development Board Annual Report 1985/86* (Singapore: Singapore Economic Development Board).

——(1987), *Economic Development Board Yearbook 1986/1987* (Singapore: Singapore Economic Development Board).

——(n.d.), *Singapore, the Business Center of Asia* (Singapore: Singapore Economic Development Board).

Study Team (1987), *Towards Excellence in Schools: A Report to the Minister for Education* (Singapore: Ministry of Education).

Tan, T. K. Y. (1985), *Budget Statement 1985* (Singapore: Information Division, Ministry of Communications and Information).

Telecommunication Authority of Singapore (1987), 'Operating review', *Telecoms annual report 1986/1987*.

The Wall Street Journal (1988), 'Singapore's growth accelerated in 1987 but is seen slowing', *The Wall Street Journal*, 4 January, p. 10.

Wilkinson, B. (1986), 'Human resources in Singapore's second industrial revolution', *Industrial Relations Journal*, 17, 2, pp. 99–114.

Yeung, Y. M. (1987), 'Cities that work: Hong Kong and Singapore', in Fuchs, R., Jones, G., and Pernia, E. (eds), *Urbanization and Urban Policies in Pacific Asia* (Boulder, Colo.: Westview Press), pp. 257–74.

11 *Telecommunications in the Pacific region: the PEACESAT experiment*

NANCY DAVIS LEWIS & LORI VAN
DUSEN MUKAIDA

The Pacific ocean covers approximately one-third of the earth's surface. The small nations and territories of the Pacific Basin exhibit the extremes of insularity, isolation and small size, and face the challenge of bridging great distances between islands within a single nation, within the region and with the rest of the globe. As the microstates strive to become an integral part of the global economy and at the same time gain greater control over their political, economic and social destinies, the demands for accessing communication and information have grown and are gaining increasing momentum.

This chapter, after briefly reviewing the Pacific Basin and the development of its telecommunications systems, presents a case study of the ATS-1 public service satellite communication system implemented through the PEACESAT (Pan Pacific Education and Communication Experiments by Satellite) Project. The case study examined the success of the PEACESAT Project in breaking down political, social, economic and ideological barriers to accessing communication and information. The Pacific experiment can be viewed as a model of communication and information access for developing countries.

An overview of the Pacific Basin

Thousands of square miles of ocean separate the 21 nations and territories of the Pacific Basin from one another and from the more industrialized Pacific rim nations. They vary in size, population, resource base and level of development and are among the smallest and most isolated of all the globe's territories. The region maintains its prominence due to its strategic military importance, continued role in global shipping, fisheries resources and deep-sea mining potential. The Pacific Basin has also been used for the dumping of toxic wastes, nuclear testing and other defense related activities. It has long been perceived of and promoted as a tropical paradise for tourism development (Shand, R. T., 1980).

Figure 11.1 Map of Pacific

Table 11.1 Demographic and telecommunication profile of the Pacific Basin.

State	1985 population	% urban population	Population growth % 1980–1985	Land area (sq. km.)/ Sea area (sq. km.)	Newspapers/ magazines	No. of radio stations	No. of telephones installed	Television service*
American Samoa	35,500	43	1.8	197 216,000	2	1	4,500	Y
Belau (Palau)	13,800	62	0.7	509 629,000	1	1	681	Y
Cook Islands	17,600	56	−0.3	240 2,000,000	1	2	2,400	N
Federated States of Micronesia	1,300	22	3.5	704 3,367,000	3	5	844	Y
Fiji	705,000	37	2.0	18,222 1,290,000	11	2	51,071	N
French Polynesia	172,800	59	3.1	4,000 4,000,000	7	6	NR	Y
Guam	114,700	96	1.6	541 218,000	6	5	23,354	Y
Kiribati	64,000	36	2.0	719 3,500,000	1	1	865	N
Marshall Islands	35,700	60	2.9	171 2,000,000	1	1	440	Y
Naura	8,400	100	1.2	26 320,000	1	1	1,194	N
New Caledonia	151,300	61	1.7	19,000 1,740,000	4	2	28,217	Y
Niue	2,900	21	−2.6	259 168,000	1	1	385	N

Northern Marianas	19,900	79	3.3	478 / 777,000	3	2	2,325	Y
Papua New Guinea	3,320,700	13	2.1	462,840 / 3,120,000	5	1	50,050	N
Solomon Islands	272,500	10	3.5	27,556 / 1,340,000	2	1	4,110	N
Tokelau Islands	1,600	0	0	12 / 288,000	1	1	Radiotelephony 1 line	N
Tonga	94,000	26	0.3	668 / 700,000	4	1	3,629	Y
Tuvalu	8,600	30	2.8	26 / 900,000	1	1	120	N
Vanuatu	135,600	18	3.3	11,800 / 680,000	2	1	2,376	N
Wallis & Futuna	13,700	7	4.9	274 / 300,000	2	1	Radiotelephony 1 line	N
Western Samoa	160,000	21	0.6	2,860 / 120,000	5	1	7,499	Y

*In most cases, television service consists of rebroadcast US programming during a specific period in the day. Both Western Samoa and Tonga can receive television from American Samoa.

Sources: SPC, 1984; SPC, 1987   Taylor, Lewis and Levy, 1987   1987 Pacific World Directory, 1986.

The microstates, offset to the west, span the Pacific between the tropics, 23.5° north and south extending from approximately 130° west to 130° east. Three culture regions were identified by eighteenth- and nineteenth-century explorers: Melanesia to the west; Micronesia to the east of Melanesia, largely north of the equator; and Polynesia, a vast triangle anchored by Hawaii, New Zealand and Easter Island (Figure 11.1). The islands of Melanesia are larger continental islands with a greater variety of natural resources, a longer settlement history, much greater linguistic diversity and larger populations. This subregion is, however, the least developed. The islands of Micronesia are complex remnants of continental masses and volcanic and coraline formations. Some of the microstates are composed of tens to hundreds of coral atolls. All are relatively small and resource-limited. The islands of Polynesia span the greatest distance and were the last to be populated. They are a mix of volcanic and coraline formations, and range in size from moderate to small, the larger volcanic islands having greater terrestrial resource potential than the coral atolls. The cultures and languages of this subregion share the greatest similarity.

Papua New Guinea, the region's giant, has a total land area (465,243 km²) over five times that of all other countries combined and 61 percent of the regional population. (Table 11.1). Tokelau, a territory of New Zealand, is comprised of three coral atolls and has a population of only 1,700. Seven microstates have fewer than 15,000 inhabitants and only Papua New Guinea and Fiji over 500,000 (South Pacific Commission, 1986).

The political status of the island groups varies. Fifteen of the twenty-one are either independent or internally self-governing. The last three decades have been characterized by the move towards independence. With the exception of Tonga, all of the fifteen have gained their autonomous status since the early 1960s, beginning with Western Samoa in 1962 and continuing through the status negotiations for the former American insular jurisdictions of Micronesia, (TTPI), now being concluded. Guam and American Samoa remain territories of the USA; French Polynesia, New Caledonia and Wallis and Fatuna are overseas territories of France. The major players in the region are the United States, France, Australia and New Zealand. The USSR and the People's Republic of China have both become increasingly interested in the island Pacific. Japan exerts significant economic influence and this is expected to increase. The territories under French control have experienced considerable, often violent, unrest in recent years in response to independence and nuclear testing issues. Fiji, an independent state since 1970, experienced two political coups in 1987. Closely tied to the West in the post World War II period, this allegiance is weakening in some of the nations. There is no general Pacific policy, but intergovernmental organizations (e.g. the Pacific Forum and the Association of Pacific Island Legislatures in the North Pacific) have grown in importance and have played a role in the formulation of island governmental policies with respect to the region and the world at large.

Generalizations about the region must be made with caution. The microstates vary greatly in terms of resource base and level of economic development but, with the exception of phosphate-rich Nauru, all rely more or less heavily on some kind of foreign economic assistance (Table 11.2). The continental fishlands of Melanesia have deposits of nickel, gold, copper and other ores. The economies of most Pacific states, however, depend largely on subsistence agriculture and the production of tropical crops: copra, sugar cane, oil palm, etc. Tourism and fisheries, most often in joint venture with outside interests, are the region's major industries.

The microstates also exhibit varying stages of demographic and epidemiological transitions. Birthrates and dependency ratios remain high throughout most of the region. Infant mortality rates range from 13 per 1,000 in Niue to over 90 per 1,000 in Kiribati and Vanuatu (Table 11.2). Not surprisingly, most health status indicators are correlated with levels of development although there are important exceptions (Taylor, Lewis and Levy, 1987). Primary education is available for much of the region's population but access to secondary and advanced education varies greatly from state to state (Table 11.2). There are major tertiary institutions in Papua New Guinea, Fiji and Guam and plans are underway for universities in the French territories. Numerous students attend institutions in New Zealand, Australia and the United States. Overseas education is only one of the factors contributing to permanent out-migration. Emigration is a major demographic factor particularly for some of the smaller island microstates and remittances are commonly an important component of the economy in these locations. High rates of migration to the main island and urban areas (urban being a relative term in the Pacific context) are also characteristic of most microstates in the Pacific, mirroring similar patterns throughout the developing world.

Telecommunications system development in the Pacific Basin

While telecommunications is taken for granted as a key factor in . . . industrialized countries . . . as an engine of growth, in most developing countries the telecommunication system is not adequate to sustain essential services. In many areas there is no system at all. Neither in the name of common humanity nor on grounds of common interest is such a disparity acceptable. (International Telecommunication Union, 1984, p. 3).

Two of the defining characteristics of islands are isolation and small size. These reach their extreme in the Pacific. Great distances and small popu-

Table 11.2 Political, social and economic profile of the Pacific Basin.

State	Capital	Political status	Development assistance per capita (A $)	Estimated gross domestic product (A $000)	Per capita income (A $)	Long distance cost from state to Hawaii*	Infant mortality rate/1000	% of those 5–19 *NOT* in school
American Samoa	Pago Pago (Tutuila)	Unorganized/Unincorporated territory of US	1,091	111,500	4,280	1.10 per mn.	18	9
Belau (Palau)	Koror	Republic in association with US	1,147	14,600	1,097	12.00 fst. 3 mn. 4.00 add'l mn.	28	7
Cook Islands	Avarua (Rarotonga)	Self-governing country in free association with NZ	520	17,400	983	8.40 fst. 3 mn. 2.80 add'l mn.	29	NR
Federated States of Micronesia	Kolonia (Pohnpei)	Self-governing country in free association with US	666	106,500	1,249	9.00 fst. 3 mn. 3.00 add't mn.	45	4
Fiji	Suva (Viti Levu)	Independent country since 1970	47	1,372,000	1,794	2.17 per min.	33	30
French Polynesia	Papeete (Tahiti)	French Overseas Territory	944	1,141,844	6,733	12.00 fst. 3 mn. 9.00 add'l mn.	57	6
Guam	Agana	US organized, unincorporated territory	791	161,000	4,223	3.75 fst. 3 mn. 1.25 add'l mn.	13	10
Kiribati	Bairiki (Tarawa)	Independent country since 1979	287	23,000	384	9.00 fst. 3 mn. 3.00 add'l mn.	93	37
Marshall Islands	Majuro	Self-governing Republic in free association with US	498	12,225	1,619	9.00 fst. 3 mn. 3.00 add'l mn.	45	4
Nauru	Yaren	Independent country since 1968	0	125,000	NR	9.00 fst. 3 mn. 3.000 add'l mn.	31	NR
New Caledonia	Noumea	French Overseas Territory	1,234	837,800	5,879	12.00 fst. 3 mn. 4.00 add'l mn.	27	1
Niue	Alofi	Self-governing country in free association with NZ	970	3,300	1,406	2.55 evry 3 mn.	11	NR

Northern Marianas	Saipan	US Commonwealth	804	165,000	4,046	4.00 fst. 3 mn. 1.21 add'l mn.	26	8
Papua New Guinea	Port Moresby	Independent Country since 1975	95	2,257,243	279	2.50 per mn.	77	67
Solomon Islands	Honiara (Guadalcanal)	Independent Country since 1978	137	128,500	547	9.00 fst. 3 mn. 3.00 add'l mn.	53	45
Tokelau Islands	Tokelau (administered by Western Samoa)	Non-self-governing NZ territory	1,063	530	susbistence economy	9.56 fst. 3 mn. 3.85 add'l mn.	37	NR
Tonga	Nuku'alofa (Tongatapu)	Independent Kingdom Constitutional Monarchy	134	51,800	526	11.55 fst. 3 mn. 3.15 add'l mn.	41	19
Tuvalu	Funafuti	Independent country since 1978	573	3,732	254	8.00 fst. 3 mn. 2.66 add'l mn.	43	42
Vanuatu	Port Vila (Efate)	Independent Republic since 1980	326	76,700	639	9.00 fst. 3 mn. 3.25 add'l mn.	94	42
Wallis and Futuna	Matu Utu (Uvea)	French Overseas Territory	696	168	649	12.00 fst. 3 mn. 4.00 add'l mn.	49	16
Western Samoa	Apia (Upolu)	Independent Country since 1962	136	43,500	590	9.00 fst. 3 mn. 3.65 add'l mn.	33	16

*US currency   Australian $ = .72 US, 1985.
Sources: SPC 1984; SPC 1987   Taylor, Lewis and Levy, 1987.

lations have long hampered transportation and communication in the Pacific. Ties with metropolitan powers have often resulted in better linkages to the outside world than with the rest of the region or even between rural areas and outer islands and the capital within a single microstate (Couper, 1973; Kissling, 1984). Linkage is frustrated further by historic colonial patterns which inhibit Islanders in the North Pacific from communicating with the Islanders in the South Pacific.

All but the smallest microstates have international airports. Domestic interisland flights serve high traffic routes. Other interisland routes are serviced by sea, often following an infrequent and irregular schedule. Shipping, along with communication and air transport, has been identified as a key to development in the future. Only by providing efficiently run and cost-competitive ports of the standard required by international shipping lines will islands be able to retain direct services. Attempts are being made to upgrade domestic, interregional and international shipping structure as well as road networks on individual islands.

In the colonial era, telecommunications development in the region was haphazard. Prior to and immediately after World War II, telecommunications between island territories and metropolitan powers and between island capitals and outer islands was via high frequency or 'shortwave' radio. By the mid 1960s, many main island centers had reticulated cable networks serving the urban areas; both Fiji and Papua New Guinea are nodes on oceanic cable networks. During the 1970s, the major Pacific microstates gained access to the INTELSAT system with Standard A or smaller Standard B earth stations. In many Pacific microstates, the telecommunications traffic is split in two with the more profitable international traffic conducted by corporations and government/private ventures on a semi-commercial basis.

Communication with rural areas is still underdeveloped and underfinanced. The SPEC 'Rural Telecommunications Study of the South Pacific', (SPEC, 1986), addressed this and now supports programs to make rural telephony the highest development priority. It is realized that improved telecommunications is an important component of rural and outer island development yet Pacific island countries, given their current economic status, must deal with conflicting developmental objectives and competing costs. It is difficult to develop and provide essential services (health care, power, water, sanitation and transportation) to rural areas, and provide telephony beyond urban centers at the same time (Dator, Jones, and Moir, 1986).

Most Pacific island microstates have access to a substantial segment of modern telecommunications technologies: radio, telephone, data communication, telegram, facsimile and videocassette recorders. With the exception of limited radio communications to outer-island locations, access to these is not uncommonly limited to those in urban centers. The major exception to the above generalizations concerning urban concentration is the VCR, which has reached many extremely remote rural locations. This penetration, in conjunction with television available in a few Pacific microstates, is viewed by many with great concern as a pervasive element of culture change and a stimulus to further migration. A major concern is the limited ability of Islanders to provide their own programming. At the present time, a number of microstates are debating the issue of licensing television, available via satellite.

While the Pacific Island urban centers as a whole are served by almost the full range of current telecommunications technologies, access to, maintenance of, and training in the practical applications of such technology has been extremely limited. Much of the island telecommunications technology is already obsolete, and the Pacific microstates lack the infrastructure to support and maintain various elements of their telecommunications systems, resulting in unreliable, poor quality and inappropriate telecommunications applications. Recognizing this, the South Pacific Bureau for Economic Cooperation has made training an important component of its telecommunications efforts (SPEC, 1986).

Inadequate training and consultant support for hardware and software technology maintenance and use are major barriers to effective technology transfer in the Pacific islands. The Pacific microstates are experiencing a significant brain-drain in mass out-migrations of skilled and professional Islanders. Those who are trained to facilitate effective technology transfer are trained at risk.

Effective technology transfer also relies heavily upon how technically, environmentally and culturally appropriate the supplied technology is in facilitating the needs of the user. What the Pacific microstates deem as most appropriate or cost-effective in telecommunications development systems, however, is ultimately dependent upon what the telecommunications market will bear, which services are available and desired, and, most probably, the influence of foreign donor countries. In the past, the telecommunications industry has often failed to make appropriate distinctions between industrialized countries and developing country consumers in its commercial applications, particularly for microstates. Although efforts are being made to develop large-scale international commercial telecommunications networks in the Pacific Basin by the telecommunications sector, these technologies, however appropriate, include a price tag which is beyond the reach of many users in island nations (US Senate Committee on Commerce, Science, and Transportation, 1987).

Faced with limited resource bases, small and dispersed populations, geographic isolation and the necessity for large commitments of foreign economic assistance, Pacific Island telecommunications development goals remain elusive. What is certain is that the Pacific microstates each will either independently or within a framework of regional cooperation continue to strive to reach those goals.

The PEACESAT project

One very important alternative communication bridge linking the Pacific microstates to each other and the world worthy of study is PEACESAT (the Pan Pacific Education and Communication Experiments by Satellite) Project. PEACESAT Project objectives over the past sixteen years have been: (1) to provide access to communication and information in the areas of health, education, research and community service through voice, data and slow-scan video services via low cost technology; and (2) to educate users in the Pacific Basin in practical and viable telecommunication methods and applications. As a multinetworked, regional, public service satellite communication system designed for development, PEACESAT must continuously break through political, social, economic and ideological barriers to communication and information access in the Pacific Basin in order to achieve its objectives.

PEACESAT Project history

In 1966 the US National Aeronautics and Space Administration (NASA) launched the ATS-1, the first in a series of six Applications Technology Satellites, to conduct weather information experiments. When the weather experiments were completed in 1969, NASA offered free access to the satellite to any non-profit, non-commercial group desiring to benefit from experimental communication programs by satellite. In response, University of Hawaii professor John Bystrom founded the PEACESAT Project utilizing the NASA ATS-1 satellite. Bystrom proposed to demonstrate 'the benefits of currently available telecommunications technology when applied specifically to the needs of sparsely populated, less industrialized areas'. (Bystrom, 1971, p. 1).

The PEACESAT project began with ground terminals at the University of Hawaii campuses at Manoa, Maui and Hilo. PEACESAT became international during its first year of operation, 1971, with the addition of a terminal initiated at Wellington Polytechnic in New Zealand. These terminals formed the nucleus of the PEACESAT Network, the first of seven communication networks to develop in the Pacific Basin through ATS-1. The PEACESAT Network rapidly grew with the addition of ground terminals in the Cook Islands, Papua New Guinea, New Caledonia, American Samoa, Tonga and Fiji (Figure 11.2).

In the early 1970s, the University of the South Pacific in Fiji was established as the regional university for eleven member countries in the South Pacific. Through an agreement with the PEACESAT Project and NASA in 1974, time on the ATS-1 satellite was made available to the University of the South Pacific for its distance education program extending educational opportunities to students throughout the region. Consequently, the USP Network under the PEACESAT Project was established with ground termi-

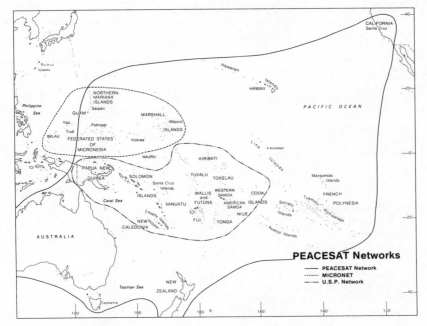

Figure 11.2 ATS-1 PEACESAT Project Networks

nals in each of the participating island countries, including Fiji, Tonga, Papua New Guinea, New Caledonia, Cook Islands, Niue, Solomon Islands, Kiribati, Vanuatu, Western Samoa, American Samoa and Tuvalu (Figure 11.2).

A subsequent network established to utilize the ATS-1 satellite was the Department of Interior Satellite Project (DISP) Network. DISP Net linked the various district centers of Micronesia and also permitted telephone service to the United States via a phone patching device located at the University of Hawaii terminal. After commercial COMSAT stations were established in Micronesia in 1983 and telephone patching services were no longer required, the time allocated to the DISP Network was made available to the University of Guam for the development of a distance education network in Micronesia. This network became known as MICRONET and includes Guam, Kosrae, Pohnpei, Truk, Yap, Palau, the Marshall Islands and the Northern Marianas (Figure 11.2).

To conserve fuel on the satellite in 1983, a decision was made by NASA to move ATS-1 from 149° west longitude (or approximately over Kiritimati, Kiribati) to a more westward position of permanent geostationary orbit. With the satellite at 162° east longitude (or approximately over Honiara, Solomon Islands), it cast a new footprint excluding all of the United States except the west coast, and including all of Australia as well as points in Southeast Asia, such as Singapore, Kuala Lumpur, Bangkok, Hong Kong

Table 11.3   ATS-1 PEACESAT project programming.

Sample general programming	Average number of times per year
DATA TRANSMISSION EXCHANGES	
Computers	184
Computer Assisted Exchanges	430
EDUCATION	
Adult Education	39
Broadcast for Development	18
Curriculum Development	25
Education for Self-Government	22
Head Start	21
High School Lifestyles	33
Hotel Operation	12
Improving Career Decision-Making	40
International Association for School Librarians	5
Law of the Sea	10
Library Morning Exchanges	42
Pacific Language Conference	8
Public Transport/Communications	16
Small Business Development	21
Social Welfare Education	11
Tertiary Education for Credit	280
Theological Education	8
ISSUES	
Island Relations	52
Community Development Exchange	46
Consortium for Pacific Art and Culture	18
International Physicians for Prevention of Nuclear War	21
Pacific Island Migration	7
Pacific Island Development Program	30
PEACESAT Consortium Council	9
Physicians for Social Responsibility	42
Preservation of Oral Tradition	31
South Pacific Commission News	49
Veterans' Affairs	46
Women for Peace	8
World Health Organization Seminars	10

and the People's Republic of China. This new position gave rise to the Kangaroo and Asia Networks.

KANGAROONET consists of more than fifty ground terminals throughout Australia, many of which were active participants in the exchanges conducted by the PEACESAT Network. ASIANET consists of terminals in Japan and Thailand, and proposed terminals in China, Singapore and Kuala Lumpur. Through the use of the ATS-1 over the years, OCEANNET and COMPUTERNET were also established. OCEANNET grew out of the need for

Table 11.3 (continued) ATS-1 PEACESAT project programming.

Sample general programming	Average number of times per year
MEDICAL	
Breast Feeding	35
Equipment for the Handicapped	7
Epidemiology	12
Health Manpower	25
Handicapped Children/Totally Disabled	11
Natural Family Planning	15
Nutrition Education	42
Professional Home Economists	8
Red Cross	10
Rehabilitation, Research and Training	80
Student Nurses Exchange	45
SCIENCE	
Asian Pacific Regional Information Network	12
Building Engineering	5
Disaster Preparedness	12
Fisheries Management	26
Natural Science Ecology	9
Renewable Energy	10
Sea Grant	16
Soil Science	8
South Pacific Regional Agriculture Development Program	49
Tropical Fruits	13
Turf Culture	13
Volcanology	9
COMMUNITY SERVICE PROGRAMS	
American Field Service	9
Boys Brigade	12
Girl Guides	13
Global Ministries	22
Media Women	4
Pacific Youth	15
Pacific Lifestyles	33
Rotary International	6
Soroptomists	5
YWCA	14

ship-to-shore communications by research vessels in the Pacific with small satellite terminals aboard. COMPUTERNET is a a promising new network involving ground terminals from all networks interested in increasing their data communications potential.

Since 1971, the PEACESAT Project umbrella has served seven overlapping satellite communication networks: PEACESAT NET; USP NET; MICRONET; KANGAROONET; ASIANET; OCEANNET; and COMPUTERNET. Over sixteen years these seven networks have involved twenty-nine participating countries, over a hundred ground terminals and hundreds of thousands of users in the Pacific Rim and Basin. With over 8,000 airtime hours per year, these networks have facilitated the shared exchange of more than 300 distinct programs covering the areas of health, education, telecommunications, research and community services (Figure 11.2, Table 11.3).

The technological significance of the PEACESAT Project

One key to the PEACESAT Project's success is its technology. The ATS-1, NASA's first geostationary satellite, possesses two characteristics supportive of development communications: (1) the ATS-1 has an eight-element phased array VHF antenna, which allows for transmission in an elliptical beam creating a footprint covering 42 percent of the earth's surface; and, (2) it is a high-powered satellite requiring relatively low power on the ground (NASA, 1984).

The eight-element phased array VHF antenna allows users anywhere within the footprint of the satellite to access the PEACESAT Project. The satellite's high power means that expensive large-dish ground terminals are not required. The satellite's low power requirement on the ground makes it possible for the PEACESAT Project to establish new ground terminals anywhere within the footprint for a very low cost, approximately US $1,200. Some users of ATS-1 have gained access with as little as $640 worth of off-the-shelf equipment. The design and cost of the satellite ground terminal equipment facilitate minimal staff training, regardless of cultural or educational background, and simplified maintenance of component parts. ATS-1's narrowband VNF frequency has served PEACESAT primarily for two-way voice communication, but telex, facsimile, slow-scan video and data transmissions are also possible between many of the ground stations.

A key to the PEACESAT Project, then, is its technology; it is affordable, appropriate, easily maintained in the tropical environment, simple to use and provides public access to communication and information to any location within its footprint. This access to the satellite, by itself, facilitates the education of users in practical and viable telecommunication methods and applications.

The PEACESAT Project – A management model based upon need

Another key to the PEACESAT Project's success is its management model. All PEACESAT Project ground terminals are autonomous. The PEACESAT Project characteristically has had no central funding. These facts place the burden of participation in communication and information access on the users of the ground terminal rather than on any administrative body or management board.

Ground terminal applicants in remote areas of the Pacific Basin desiring PEACESAT Project Network membership must be willing to equip, install, maintain, staff and manage their terminal. Additionally, it is the responsibility of the ground terminal applicants to request approval from their respective licensing authority to access programming on the ATS-1, although PEACESAT does not require this. Assistance in setting up a terminal may be sought by the PEACESAT Project on behalf of an applicant, in the event the terminal applicant requests it, and where it is evident that the lack of a PEACESAT Project ground terminal in the applicant's locus would be a disservice.

The director and the coordinator of the PEACESAT Project Networks are housed and supported by the University of Hawaii. Both the director and the coordinator access the PEACESAT networks through the Hawaii PEACESAT ground terminal located on the university campus. The director represents the needs of the users to state, national and international agencies or organizations, such as the University of Hawaii, the US Congress and the South Pacific Forum, and the coordinator responds to the needs of the terminals during network time over the satellite. Both director and coordinator confer with NASA regarding project status and satellite station keeping. NASA maintains the ATS-1 and monitors the status of the experimental satellite networks. In the history of PEACESAT, the United States government has neither censored program content nor rejected a terminal applicant submitted by PEACESAT for access to the satellite.

Before a terminal applicant may access the satellite, the PEACESAT Project requests that representatives of the terminal commit that: (1) no commercial activity will take place on the satellite; (2) no personal or intimate traffic will take place over the air; (3) they will respond to medical and environmental emergencies on the satellite as a top priority overriding all other air time needs; (4) they will always conduct their terminal in a professional manner, will monitor their participants and instruct them to do so also; (5) they understand that the PEACESAT Project is not political – that all terminals desire the free flow of ideas and the professional discussion of the same (they do not select ideas for others or advocate one over another), and they will not censor content other than in the above restrictions; and (6) that they will operate their terminal on the satellite in the accepted air-time PEACESAT protocol, will actively educate their terminal constituency about PEACESAT and its applications and will encourage its use.

All program scheduling is done through Network Management, which takes place the first day of each week for one to two hours. Representing its constituency, the original terminal introduces the program and asks for other terminals interested in the program to provide participants. Two or more terminals in any one network or networks must respond with commitments of participation in order for the proposed program to be scheduled.

Terminals can not be forced to participate, and so no one terminal can be

dominated by other terminals. If programming is inappropriate or inadequate, participation in that particular program drops off and the program is
canceled. If a program is canceled twice consecutvely, efficient use of air
time requires that the program be dropped unless a request to continue it is
voiced and agreed upon during management sessions on the air.

In part because of these commitments, and because the hurdles of cost
and technology are low, sixteen years of PEACESAT Project experience has
indicated that a terminal's viability is directly related to the level of need for
access to communication and information by the applicants or their constituency. The management structure within the PEACESAT Project serves
only a facilitating role. If there is a need, there is programming. The model,
then, is one of self-management.

Conclusion

PEACESAT has successfully broken down political, social, economic and
ideological barriers by opening up a common access to communication and
information to all Pacific microstates, where political, social, economic,
health and education leaders have met over the satellite to discuss issues of
common concern, to work through problems together and to share
solutions. Environmental and medical emergencies have resulted in peoples
in the North and South Pacific and from island state to island state banding
together, regardless of differences, to battle epidemics such as dengue fever
in the 1970s and the cholera outbreaks in 1979–82, and disasters, e.g.
tsunamis and typhoons, which have been particularly severe in the 1980s.
Boys Brigade, Girl Guides, Pacific Youth, Rotarians, Physicians for Social
Responsibility and many other groups have been brought together in one
voice through rich cross-cultural exchanges. The universities of the South
Pacific, Papua New Guinea, Guam, Australia, New Zealand and Hawaii
were all linked to each other and extension students throughout the Pacific,
bringing the benefits of international tertiary and continuing education to
thousands of people in remote island communities, who might otherwise
never have received advanced education. PEACESAT provides experiences in
low-cost telecommunications in a very non-threatening way to people
whose development is dependent upon being able to link with other
peoples in other countries who have similar problems, aspirations, conditions and futures (Table 11.3). A Pacific Island user commented, 'Perhaps
the most significant foreign aid the United States has ever given the Pacific
Basin is indeed the free use of ATS-1.' (Mukaida, 1985, p. 93).

In August of 1985, after fourteen years of service to PEACESAT, ATS-1
was turned off when it began to drift east and potentially interfere with
other satellite transmissions. The US Congress immediately ordered studies
to see if PEACESAT was still critical in the Pacific. A NASA survey showed that
PEACESAT's success can not yet be duplicated by commercial satellite pro-

viders; simply, the Pacific microstates cannot support the financial burden of expensive ground terminals (NASA, 1985). Although much of the Pacific was left in silence, PEACESAT survives, servicing one-tenth of the Pacific Basin (microstates east of the international dateline) on a NASA sister satellite, the ATS-3, which orbits over North and South America. The US Congress is currently considering a $3.4 million appropriation for the restoration of PEACESAT for a two-year period on a replacement satellite. This bill has successfully passed through major committees. In case the US Congress fails to replace the ATS-1, Japan has offered to provide a replacement satellite system.

PEACESAT is not a competitor or replacement for commercial telecommunications systems development in the Pacific Basin. It is a public supplement serving a need for communication and information access for as long as a need exists, or as long as PEACESAT exists. But foremost, it does not remove or take the place of the commercial responsibility to respond to demand appropriately and efficiently.

The realities of telecommunications system development in the Pacific Basin, with or without a public service satellite system such as PEACESAT, are that the most appropriate and cost-effective infrastructure is to provide modern telecommunication services beyond the urban centers to vast, extremely remote areas and to other Pacific island urban and rural communities and the development of high-powered satellite networks with other technologies employed where appropriate, e.g. microwave or fiber-optic networks, linking low-cost, easily maintained ground terminals. (Dator, Jones, and Moir, 1986; Mukaida, 1985). No commercial carrier provides this type of service and access.

INTELSAT, the International Telecommunications Satellite Organization, which is comprised of 110 member nations, enjoys a monopoly on commercial satellite services in the whole of the Pacific Basin. Access to INTELSAT requires an investment of $40,000 plus a ground terminal and monthly satellite access charges. It is beyond the ability of most island states to provide such service to outer island locations. INTELSAT has been accused of global cooperation for profit wherein member countries charge the end user fees much higher than actual operating costs (Martinez, 1986).

AUSSAT-3, the Australian satellite, and Papua New Guinea's PACSTAR, are intended in the near future to be Pacific subregional satellites only capable of servicing the South Pacific. Recognizing the need for competitive, commercial satellite services in November 1985, President Reagan signed a Presidential determination (PD-No. 85-2) declaring that alternative satellite systems were required in the national interest. The FCC implemented this deregulating position and began accepting applications for separate satellite systems. To date, the FCC has received and approved more than eight applications. Approved applicants for 'separate systems' face tremendous economic and regulatory hurdles before they can play a role in the Pacific Basin and it is not conclusive that the element of competition will bring about more appropriate and compatible technologies for commercial service in the region.

The PEACESAT Project's sixteen-year history and the heavy use and adoption of the system by Pacific islanders attest to its success as a public service satellite for development. PEACESAT's technology and self-management model are keys to this success. The continued need for the project is evident in government studies and network terminal requests to gain access to a replacement satellite. PEACESAT's future development, though, is constantly dependent upon the availability of funding.

References

Bystrom, J. (1971), 'The PEACESAT Project', Original Proposal to the National Aeronautics and Space Administration, 1.

Couper, A. (1973), 'Islanders at Sea: Change and the Maritime Economies of the Pacific', in *The Pacific in Transition: Geographical Perspectives on Adaptation and Change*, Brookfield, H. C. (ed.) (London: Edward Arnold), pp. 229–47.

Dator, J., Jones, C., and Moir, B. (1986), Pacific International Center for High Technology Research (PICHTR), A Project by PICHTR for GTE Corporation and Hawaiian Telephone Company. 'A Study of Preferred Futures for Telecommunications in Six Pacific Island Countries'.

Independent Commission for World-wide Telecommunications Development (1984), *The Missing Link* (Executive Summary) (Geneva: International Telecommunication Union), p. 3.

Kissling, C. (ed.) (1984), *Transport and Communications for Pacific Microstates: Issues in Organization and Management* (Suva, Fiji: Institute of Pacific Studies).

Martinez, L. (1985), *Communication Satellites: Power and Politics in Space* (Dedham, Mass.: Artech House, Inc.).

Mukaida, L. (1985), 'The PEACESAT Experiment' (Washington, DC: NASA publication).

——(1986), *The PEACESAT Video Documentary* (Washington, DC: NASA publication).

National Aeronautics and Space Administration (NASA) (1984), 'The ATS-1 & ATS-3 Experimenters' Guide', Revision 5, (Washington, DC: Communications Division, NASA Headquarters).

The SPEC Rural Telecommunications Study of the South Pacific, South Pacific Bureau for Economic Cooperation (SPEC), (1986), 'South Pacific Telecommunications Development Programme Planning Report', p. 9.

SPC (1984), *South Pacific Economies Statistical Summary* (Noumea: South Pacific Commission).

——(1987), *South Pacific Economies Statistical Summary* (Noumea: South Pacific Commission).

Shand, R. T. (ed.) (1980), *The Island States of the Pacific and Indian Ocean: Anatomy of Development* (Australian National University Development Study Center), Monograph # 23.

Taylor, R., Lewis, N., and Levy, S. (1987), *Pacific Island Mortality: A Review Circa 1980* (Noumea: South Pacific Commission).

Uludong, Francisco T. (1986), *Pacific World Directory 1987*. Saipan, Pacific Information Bank.

US Senate Committee on Commerce, Science and Transportation, S. 828, (1987), 1st Session of the 100th Congress, Senate Report No. 100–92 (Washington, DC: US Government Printing Office).

Yamaguchi, N. (1985), 'Telecommunications and development in the Island Pacific: prospects for a regional satellite based telecommunications network', MA thesis, University of Hawaii.

Young, E., and Hurd, J., Public Service Satellite Consortium (1981, 1982), *Satellite Communications for the Pacific Islands*, 2 vols. Prepared by the Public Service Satellite Consortium for the US National Aeronautics and Space Administration: Washington, DC.

12 The role of telecommunications in assisting peripherally located countries: the case of Israel

AHARON KELLERMAN

It has already become common wisdom to identify three global economic cores, namely North America, Europe (either Western Europe only, or to some degree, Eastern Europe as well), and the emerging Pacific rim, centered on Japan and including some NICs (Newly Industrialized Countries), as well as Australia and New Zealand. The rest of the world is, thus, economic and geographical periphery to these three cores. The periphery consists mostly of LDCs (less-developed countries). Telecommunications may assist in connections among the three cores and between the cores and the peripheries. It may also become part of the development process of LDCs.

The first part of this chapter will be devoted to a general discussion of the roles of international communications, with a particular emphasis on its possible impacts on peripherally located countries. The detailed discussion of an example which will follow will focus on Israel, which is *not* a typical example of a peripherally located LDC. On the contrary; the argument will be that, from several viewpoints, Israel is a core-extension that is geographically separated from the core. The analysis will therefore aim at an exposition of the use made of telecommunications to close this geographical gap.

Telecommunications and international relations and development

There are several contemporary examples of telecommunications assisting peripherally located countries. One is the development of international tourism in Nepal following the introduction of international telecommunications infrastructure (Saunders et al., 1983). Other on-line examples include programmers in Chile and Mexico working for a California company (Rohwer, 1985), and keypunch operators in the People's Republic of China working for American companies (*Time*, 1986). Geographers have usually focused more on the role of telecommunications in regional development rather than on its role at the national–international level (e.g. Kellerman, 1986a; Robinson, 1984; Salomon and Razin, 1985; Stern, 1983). The two levels might be interrelated. If the British case is paving the way (Hepworth *et al.*, 1987), then increased international telecommunications tends to concentrate in major urban centers. This leads to enhanced domestic peripheralization on one hand, and may lead to additional pressures to close the gap on the other. After all, further development of international telecommunications would mean not only better business and producer services, but more entertainment and data services for households as well.

International telecommunications may be related to several aspects of national economic development and international relations. These aspects may be classified as economic, political, social and geographical. Since the literature on the possible relationships between international telecommunications and economic development is relatively rich the discussion will begin with this aspect.

International communications and economic development

A major problem that is inherent in discussions of telecommunications is the joint treatment of telecommunications *channels* and the information moved through them. The two are commonly joined together as IT (Information Technologies). In transportation analyses a clear distinction is made between a vehicle and its load. It is, however, more difficult to make such a distinction as far as IT is concerned (Kellerman, 1984). However, the very existence of a good system of international telecommunications does not yet assure its use for R & D (research and development) purposes, or for on-line production of goods and services. It might also be used for simpler business calls, or it might be used mainly for social purposes.

In any case, it is common to find statements that claim an unequal global distribution of IT, so that LDCs, or the periphery, have less of it while the cores enjoy more and richer IT (Jussawalla, 1982; Lyon, 1986; Smith, 1980). For example, in the statement that 'the great telecommunication highways which distribute information within nations and between nations have displaced transportation highways as the core communication systems of a modern society – and of the modern world' (Smith, 1980, p. 111). On the other hand, information is a renewable resource the moving costs of which, using innovative telecommunications technologies, are low (Jussawalla, 1982). As such, some would argue for an 'advantage of backwardness', that would assist LDCs to move directly from an agricultural orientation to service economies (Singelmann, 1978). Telecommunications is used most intensively by the tertiary and quarternary sectors, less by the secondary one, and little by the primary one (Saunders *et al.*, 1983, p. 87). However, since telecommunications is just a vehicle, the question still remains open whether it will be used for routine information

geared to the development of industrial (secondary) production and for
routine services (finances, key-punching), or also for non-routine services
(research and development). If it is used for the first option, then international telecommunications is just another utility needed for industrial
mass-production, thus making core countries more specialized in research,
development and non-routine production. For the second option, many
other conditions have to be met, especially the existence of trained human
resources, which is the most important locational factor for high-tech
industries and services. This condition is usually not met in LDCs.

There is, therefore, a question whether telecommunications in general,
and international in particular, should be considered as development
factors, or looked upon as being more of development aids, or facilitators.
Some argue that 'telecommunications infrastructure may be viewed as an
input to a productive process, a "factor of production" like petroleum or
electricity' (Saunders et al., 1983, p. 73), or that telecommunications is 'one
of the world's most effective productivity raisers' (Rohwer, 1985, p. 5).
Similar views were stated also by Smith (1980, p. 117). Others presented
this same notion as a controversial one (Jussawalla and Lamberton, 1982,
p. 11), while geographers have taken a more cautious position (Kellerman,
1986a; Salomon and Razin, 1985). Interesting in this regard is Hardy's
study reported by the ITU (International Telecommunications Union,
1983). This study covered fifteen industrialized countries and thirty-seven
developing ones over fourteen years. It was found that economic development 'causes' or 'leads' telephone development *and vice versa*. However, in
the developed countries the contribution of telecommunications to
economic development is somewhat weaker than in LDCs. This is
explained by the larger marginal utility of additional phones when the
system is still small. Also, residential telephones lead economic development in LDCs. Quantitatively, an increase of 1 percent in the number of
telephones per 100 population between 1950 and 1955 in all fifty-two
countries studied yielded a rise of about 3 percent in per capital income
between 1955 and 1962. The poorer the country the larger the contribution
of telephone developments to the rise in incomes. The importance of
residential telephones and the importance of telephones for poorer countries is quite contrary to economic notions, but attests to the universal
contribution of telecommunications and to the two-way relationship it has
with economic development.

The political, social and geographical importance of international telecommunications

International telecommunications has several positive and negative effects
from a political viewpoint. Telecommunications may prevent the outbreak
of wars (Lyon, 1986, p. 580), (though this assumes the existence of direct
and indirect communications, which at least in the Arab-Israeli conflicts are

almost non-existent). At the other end of the spectrum telecommunications may assist in the promotion of international understanding. It may also provide for better journalistic coverage of LDCs (Galtung, 1982). On the other hand, however, information access may become a source of power for better equipped core countries (O'Brien and Helleiner, 1982). Free information flows may detract LDCs from the development of their own informatics in areas such as banking and education (Smith, 1980, p. 128). It may further involve inconsistency in flow regulations between countries at the two ends of a communications line (Sauvant, 1986; Snow, 1985).

Socially, international telecommunications provides a service that raises the quality of life especially if direct dialing is available. It permits better social and family contacts especially in immigrant countries. It further provides for better international understanding at the individual level and assists in the development of richer intercultural exchanges.

From a geographical perspective, international telecommunications reduces the friction of distance to a much larger extent than in domestic telecommunications. It may permit the evolution of international regional telecommunications junctions such as Singapore. It may also assist core-extensions within the periphery, as the following case of Israel will exemplify.

Israel and international telecommunications

In this section we shall review some of the general characteristics of Israeli international telecommunications before going into detailed analyses of some pertinent data. The domestic geography of the Israeli telephone system has been portrayed elsewhere (Salomon and Razin, 1986), so that the focus here will be mainly on international telecommunications.

Israel on the international telecommunications map

Israel has a population of about 4.5 million and it extends over some 21,000 km^2 (not including Judea, Samaria and Gaza). In terms of its GNP and size of the telephone system it is considered a developed country, so that in 1981 its population numbered 0.1 percent of world population, while its number of telephones and its GNP were 0.2 percent of world telephones and GNP (Saunders, 1983, p. 6). Plotting telephone stations against GNP in 1979, Israel occupied a position similar to other developed Mediterranean countries, namely Spain, Italy and Greece (and Hong Kong) at the lower level of core countries, and much above the East European countries and LDCs (Pelton, 1981). In 1985, Israel ranked 21 in the number of telephone lines per 100 population (25.9; 28.2 on 31 March 1986), lower than Singapore and higher than Spain. It also ranked 20 in the world list of telex ranking (Bezek, Israel Telecommunications Co., 1986). (Interestingly enough the telephone ranking is headed by Sweden and Switzerland with the US ranked 6, while the telex ranking puts international junctions such as Singapore, Switzerland and Hong Kong first.)

In its shortage of natural resources and in its reliance on trained human resources Israel presents some similarity to Japan, and in its size and location at the heart of the Middle East it is similar to Singapore. However, while Japan was able to create an economic spillover effect for the whole Pacific rim, and Singapore has become a regional telecommunications center, Israel is almost totally isolated from the region that surrounds it. With the exception of Cyprus and Egypt with which Israel maintains formal tele-communications, transportation and limited trade relations, there are almost no other direct contacts with other surrounding countries (some indirect trade exists through Samaria, Gaza and southern Lebanon).

In order to maintain a modern economy Israel has developed close economic ties with the European and North American cores. These have found their expression in the orientation of exports and imports to these markets and in the evolution of a two-way tourism between the two cores and Israel. The ties between the two cores and Israel go even deeper. Not only are people and goods moving between Israel and the cores but information does so as well. Israel was ranked 24 in the list of leading service exporters in 1984, and was ranked 25 in the equivalent list of service importers (US Congress, OTA, 1987). The country maintains a growing high-tech sector in both R & D and specialized production. Israel was, thus, one of the first countries to join the BITNET/EARN system that permits open telecommunications among universities worldwide (Kellerman, 1986b). Israel considers itself an integral part of Western culture in areas such as music and the arts. In addition, Israel maintains close relations with the Jewish Diaspora which is scattered mainly in major Western countries (and the USSR). All these are major factors in the two-way relationship between telecommunications on one hand and social and economic life on the other. Some analysis of this relationship will be provided in a later section. One additional aspect has to be mentioned, however. Because it is frequently in the focus of international news coverage, the world press and media maintain a large body of representatives in Israel that may use telecommunications intensively.

The evolution of international telecommunications channels in Israel

There are several landmarks in the evolution of the Israeli international telecommunications system. Until 1968 the system was based on poor radio connection with between ten and thirty channels in operation. In 1960 a telex service started, and in 1968 a maritime cable with 128 available channels started operations between Israel and France. In 1972 Israel first leased transponders in communications satellites. The number of available voice-grade channels via satellites increased from 288 in 1976 to 696 in 1986. Israel has used transponders in two satellites that permit connections with Europe and North America, and since 1980 it leased a transponder in a third satellite located over the Indian Ocean which permits communications with Asia as well. The expansion in the availability of international channels permitted the introduction of nationwide direct international dialing in 1972–3. In 1975 a second maritime cable was put into service, this time ending in Italy and having a capacity of 1,380 channels. Israel also leased channels in European continental cables and in transatlantic ones. In 1982 the Knesset (parliament) approved the separation of postal and telecommunications services and the creation of a government-owned telecommunications company. In 1984 Bezek-Israel Telecommunications Company started operations.

The voice-grade channels enjoyed continuous growth between 1976 and 1986 (Table 12.1). The average annual growth rate was 11.5 percent. Until 1984 the growth rate presented a biannual cyclical form, so that a large increase in one year was followed by a lower one in the next year. This trend may be related to a policy of careful use of international channels since, as we shall see in the next section, the growth in telecommunications traffic has not shown a biannual cyclical form. Two specific years present the highest growth rates, namely 1979 (24.1 percent) and 1977 (16.4 percent). This is probably related to the liberalization in policies with respect to foreign currency and international trade declared by the Likkud administration that came to power in 1977. Since 1982–84 the annual growth rate has fluctuated between 8.1 and 10.6 percent, thus reflecting a continuous growth in international telecommunications traffic. Though, as we shall see in the next section, this growth has not been equal since 1982.

Another trend is the stability and even decline in the use of older transmission systems when new ones are ushered in. Thus, the trend in the use of radio channels, which doubled in 1979, has remained steady while that for the use of cable 1 has even declined.

Telegraph channels devoted to telex and telegrams have not shown high growth rates as voice-grade channels have, with two exceptions. In 1977 there was a growth rate of 19.5 percent in telegraph channels compared to 16.4 percent in voice-grade channels. This is probably due to the 1977 political change. In 1983 there was a 10.7 percent increase in telegraph channels, which may be related to the Lebanon war, and an increased use of telegraph by journalists. As we shall see later, Middle Eastern wars had an effect on international telecommunications in Israel. Most of the growth in telegraph channels is attributable to telex, while the older form of telegrams has declined.

Transitions in international telecommunications traffic of Israel

Three major means of telecommunications will be discussed, namely telegrams, telephone and telex. Generally, the volume of the total two-way

Table 12.1 International telecommunications channels for Israel 1976–1986.

Year	Operated voice-grade channels		System						Telegraph channels			
	Total	% Annual growth	Cable 1	Cable 2	Satellite	Radio	Telephone	Telegraph[1]	Total	% Annual growth	Telex	Telegrams
1976	422	—	126	116	160	20	397	25	308	—	254	54
1977	499	16.4	124	188	161	18	463	28	368	19.5	344	52
1978	531	8.1	123	218	173	17	498	33	396	7.6	344	52
1979	659	24.1	124	312	188	35	592	67	405	2.3	352	53
1980	692	5.0	124	336	197	35	619	73	431	6.4	373	58
1981	795	14.9	125	404	232	34	715	80	448	3.9	389	59
1982	863	8.6	126	433	270	34	791	72	460	2.7	401	59
1983	953	10.4	125	504	290	34	875	78	509	10.7	449	60
1984	1,030	8.1	113	547	336	34	947	83	528	3.7	474	54
1985	1,118	8.5	113	592	380	34	1,027	91	530	0.4	480	50
1986	1,236	10.6	113	665	425	34	1,142	94	526	−0.8	480	46

[1] Includes 28 channels for Cyprus in 1979–80; 26 in 1981 and 13 in 1982–86.
Source: Bezek, Israel Telecommunications Company, *Statistical Yearbook 1985/86.*

international telephone traffic has increased tremendously during the last thirty-six years. In 1950/1, two years after the establishment of the State of Israel, this traffic amounted to 14,627 calls, compared to 62,164 in 1961/2 or a growth of 325 percent in just 10 or 11 years (State of Israel, Central Bureau of Statistics, 1954–1963). This growth was channeled through the rather limited and technologically inferior radio system. In 1961/2 the number of telephone-minute calls was 311,000 compared to 138,710,000 minute calls in 1985/6, or a growth of 445 times! (Bezek, 1986; State of Israel, Ministry of Postal Services, 1964). This even more tremendous growth has been channeled since 1968 through the more superior cable and satellite technologies.

Telex service, which started in 1960, also presented a high growth rate. In 1961/2 the number of international minute calls was 115,600, while in 1984/5 the equivalent figure was 16,500,000 or a growth of 142 times. In 1985/6, for the first time, a decline of 4.8 percent in telex calls was registered. Telegrams numbered in 1950/1 some 900,000, while in 1985/6 only 360,000 were sent or received, a decline of 60 percent! (Bezek, 1986; State of Israel, Central Bureau of Statistics, 1954, 1986).

The continuous growth of telephone use on the one hand, compared to the extensive decrease in telegrams and the beginning of decline in telex use on the other, call for a more detailed comparison among these three forms of international telecommunications.

Annual changes in Israeli international telecommunications

Annual percentage growth rates for the total traffic of telephone, telex and telegrams are presented in Figure 12.1. The growth of international tele-communications in Israel in general is determined by three major factors, namely technology, economy and politics. These three factors partially overlap in time. The early 1950s were marked by severe austerity measures in order to permit the absorption of the huge immigration waves which doubled the population of the young state, so that declines were registered in both telephone calls and the number of telegrams. However, while for telephone contacts this was the only period of absolute decline, it marked the beginning of a decline in telegrams that would resume in 1964/5 and continue since then. Fluctuations in the moderate growth rate and in the decline of telegrams over the whole period of thirty-six years match those in telephones and telex services, thus reflecting the more general factors of change. The Sinai War in 1956 is marked by growth in both telephones and telegrams, and after four more years the first of two periods of high growth rates appears in the early 1960s. These were years of most intensive expansion of the Israeli economy, typified by rapid industrialization and high percentage growth of the GNP. This was, however, also the time when telex service was initiated, while telephone service was still limited to radio, operator-assisted service. In the early 1960s, the number of telephone

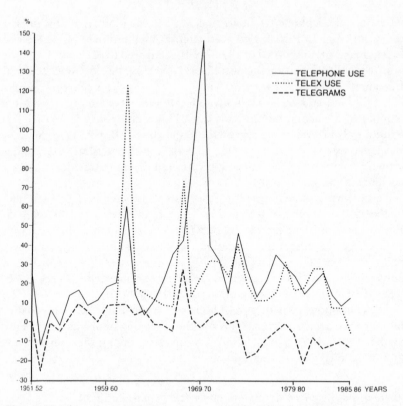

Figure 12.1 Annual percentage growth rates for the total international telecommunications traffic of Israel, March 1950–March 1986★

Source: Bezek (1986), pp. 27–31

★Computations of growth in telephone traffic are based on number of calls (1950/51–1960/61) and
number of minutes (1960/61–1985/86). Computations of growth in telex traffic are based on the
number of minutes, and computations of telegrams growth are based on the number of telegrams.

lines in the country was still small (67,900 in 1960/1 compared to 1,208,000 in
1985/6). Thus the growth rate of the telex service exceeded that of the telephone. The mid 1960s were typified by an economic recession, so that
growth rates for all forms of international telecommunications were low and
those of telegrams negative. The mid 1960s mark also the overtaking of the
telegrams by the telex as the preferred form of international telegraph service.

It was only in 1967 that telegrams grew substantially again, and even
more so did the telex service. This growth is due to the 1967 Six Days War,
which received extensive international press coverage. However, the war
also marked a renewed and more intensified economic growth, especially
with regard to exports and imports from Western countries. The late 1960s
also witnessed stronger bonds between Israel and the Jewish Diaspora and
an increase in tourism. In 1967, however, telephone service was still limited
so the expansion in international telecommunications traffic is more
evident in telex and telegrams. It was only a year after the war, in 1968, that

there came a breakthrough in the telephone service with the completion of the maritime cable between Israel and France. This resulted in a 146 percent growth in telephone use in 1968 alone. It further marked the beginning of the era when telephone service will be the dominant mode in the system. The telephone is more flexible than telex. It is available to both households and businesses, while the telex is more restricted to business. Later major technological changes such as the second international cable, satellite communications and direct international dialing did not have the same impact on annual growth rates as had the first cable, which permitted, for the first time, a reliable and extensive service at times of increased economic, social and political/journalistic international ties.

Since 1970 changes in the size of international traffic present cycles of peaks and troughs of about five years each. These cycles again represent a mixture of technological changes, economic cycles and major political events. The high growth rates in 1973/4 are obviously related to the 1973 Yom Kippur War. However, in 1972 Israel was first connected to satellite communications and direct international dialing was introduced. After several years of economic stagnation, 1977 was marked by the first change in political power in Israeli history. The new Likkud administration and its liberal monetary and international trade policies created more international traffic. Another peak occurred when an even more liberal policy, especially regarding imports, was declared by the Likkud administration in 1982/3. Interestingly enough, the Lebanon War had no impact in the form of increased traffic as had earlier wars. In 1985 a new increase in growth rates of telephone use is apparent, reflecting economic stability achieved after years of high inflation. The year 1985/6 presents also a first absolute decline in international telex. This is despite a continued growth in the number of telex subscribers (5,440 in 1985/6 compared to 5,005 in 1984/5). It might be attributed to the increase in computer telecommunications. The number of data transmission lines increased from 5,300 in 1984/5 to 6,170 in 1985/6. This relatively new form of telecommunications exceeded the older telex for the first time in 1984/5 in terms of the number of subscribers.

The general growth in international telecommunications has had another expression, namely a decline in outbound airmail service. This service reached a peak in 1980, when 55 million pieces of mail were sent from Israel. The figures for 1984 and 1985 are just 40 and 41 millions respectively. At the same time, however, inbound airmail service continued to expand, probably due to the expansion in professional and commercial literatures (56, 58 and 59 millions of pieces in 1980, 1984 and 1985 respectively) (State of Israel, Central Bureau of Statistics, 1986).

Inbound versus outbound international telephone traffic

An interesting trend is revealed when the volume of inbound and outbound telephone traffic is compared. The ratio of inbound/outbound calls was 4:6 in 1950/1 (5,898 incoming calls versus 8,729 outgoing ones). This ratio gradually changed towards a balanced distribution, which was reached in 1961/2 (31,235 incoming versus 30,929 outgoing calls). Since 1961/2, which marked the first massive expansion of international telephone use, the ratio has constantly changed in favor of inbound calls. This ratio reached 7:3 in 1985/6, or on the average, a ratio of more than two incoming calls for each outgoing one (97.5 millions of inbound minute calls versus 41.2 millions of outbound ones).

There might be several reasons for this constant ratio change. First the cost of international calling in Israel is higher than abroad. This makes it cheaper to use collect calls placed in Israel or to have the foreign party call in. If this happens, then revenues of the Israeli telephone service are higher in hard currency than when calls are equally distributed. These net revenues (from both telephone and telex calls and after deduction of payments to foreign companies) reached $61 million in 1985/6, which comprised some 40 percent of the total revenues of Bezek in that year (Bezek, 1986). While this policy might look attractive in terms of short-run national balances of payments, and in terms of revenues of the telephone company, this policy is detrimental from the viewpoint of economic development and social welfare. It increases the cost of international business in Israel at times when exports and imports of goods, services and information are growing and becoming more important (Kellerman, 1986c). It further reduces the number of social calls, while declared policy has always called for increased ties between Israel and the West in general and with the Jewish Diaspora in particular. Several years ago an international economist proposed to reduce the rates of international calls in Israel to those of domestic long-distance ones, since the cost of the infrastructure of the system was paid off already. This proposal was not adopted since the important role of telecommunications was not appreciated.

A second and related cause for the increased inbound/outbound gap is the tremendous expansion in international telecommunications worldwide and its reduced cost, which permits more social one-way calling in other countries. A third factor, which has a less continuous effect, are the wars, during which more social inbound calls than outbound are generated.

International telephone traffic of Israel by country

The telephone has emerged as the most dominant means for international telecommunications in Israel. It is therefore of interest to see the distribution of telephone traffic by country during the period 1950/1–1985/6 (Table 12.2).

There were nine countries that over the years comprised the lion's share of Israeli total telephone traffic, namely: US, UK, West Germany, France, Switzerland, Italy, Canada, the Netherlands and Belgium. These countries

Table 12.2 Ranking by country of Israeli total international telephone traffic 1950/1–1985/6.

	1950/51	1953/4	1955/6	1956/7	1961/2	1966/7	1973/4	1977/8	1980/1	1982/3	1983/4	1984/5	1985/6
1	US 22.8%	Switzerland 19.9%	US 19.0%	US 18.3%	US 22.2%	US 24.9%	US 34.1%	US 29.9%	US 34.8%	US 40.7%	US 45.0%	US 45.2%	US 42.2%
2	France 14.7%	US 16.1%	Switzerland 13.9%	Switzerland 10.4%	UK 13.8%	UK 13.8%	West Germany 12.9%	France 17.3%	UK 13.2%	UK 11.9%	UK 11.4%	UK 11.6%	UK 11.5%
3	UK 13.0%	UK 13.7%	UK 13.0%	UK 10.1%	West Germany 11.3%	France 11.5%	UK 11.8%	UK 13.2%	France 12.9%	France 11.6%	France 10.3%	France 9.4%	West Germany 9.3%
4	Switzerland 12.1%	France 11.6%	France 7.5%	France 7.6%	Switzerland 10.9%	Switzerland 11.0%	France 11.1%	West Germany 7.8%	West Germany 8.3%	West Germany 8.3%	West Germany 7.7%	West Germany 8.3%	France 9.1%
5	Italy 7.8%	Italy 6.4%	West Germany 6.1%	West Germany 5.8%	France 10.2%	West Germany 8.6%	Switzerland 10.2%	Switzerland 4.8%	Switzerland 3.9%	Switzerland 3.9%	Switzerland 3.7%	Switzerland 3.7%	Switzerland 3.8%
6	Belgium 5.8%	Belgium 5.5%	Italy 5.2%	Belgium 5.5%	Italy 6.7%	Italy 5.0%	Italy 6.1%	Italy 4.2%	Canada 3.8%	Italy 3.8%	Italy 3.4%	Italy 3.1%	Italy 3.7%
7	West Germany 4.2%	West Germany 4.6%	Belgium 4.2%	Italy 4.8%	Belgium 4.1%	Belgium 4.4%	Belgium n.a.	Canada 3.4%	Italy 3.0%	Canada 3.6%	Canada 3.3%	Canada 2.9%	Canada 3.3%
8	Netherlands 2.5%	Netherlands 3.0%	Netherlands 4.0%	Netherlands 3.7%	Netherlands 2.6%	Netherlands 2.7%	Canada 3.4%	Netherlands 3.0%	Netherlands 2.9%	Netherlands 2.4%	Netherlands 2.1%	Netherlands 2.1%	Netherlands 2.3%
9	Canada 0.9%	Canada 1.0%	Canada 1.7%	Canada 1.9%	Canada n.a.	Canada 1.4%	Netherlands 2.5%	Belgium 2.8%	Belgium 2.6%	Belgium 2.0%	Belgium 1.5%	Belgium 1.9%	Belgium 2.1%
Total:	83.8%	81.8%	74.6%	68.1%	81.8%	83.3%	92.1%	86.4%	85.4%	88.2%	88.8%	88.2%	87.3%

Sources: Bezek, Israel Telecommunications Company, *Statistical Yearbook 1985/86*; State of Israel, Central Bureau of Statistics, *Post of Israel Survey and Statistics 1953/54–1961/62; 1962/63–1968/69; 1970/71–1972/73; 1969/70–1981/82*.

are also major countries in the two Western cores. The inner ring of countries that surround Israel, which should have taken a large share of the telephone traffic, is absent, due to the Arab–Israeli conflict. However, even an outer ring of more distant countries such as Iran, Turkey, Cyprus and Ethiopia is missing. These are countries in which Israel has invested much effort in order to build friendly relations, so that a ring of friendly countries would surround the inner ring of hostile Arab countries. The data reveal, however, an almost constant increase in the share of much more distant Western Europe and North America. Obviously, the non-appearance of closer countries has partially to do with the lack of modern telecommunications in LDCs until the 1970s. More important, however, is the development of an economy in Israel that exports to and imports from the industrialized nations and the evolution of a society that is culturally oriented to the West. The existence of large Jewish communities in Western countries that can afford international telecommunications may be another factor. These factors will be formally examined in the next section.

Generally speaking, the share of the nine most intensively communicated nations has comprised over 80 percent of the total telephone traffic for most of the period, with values around 88 percent in recent years. This leads to the statement made earlier that Israel is more of a core extension than a classic case of a peripherally located country. The fact that Israel is now the only country in the world that has signed free-trade treaties with both the USA and the EEC provides it with a unique status with regard to the two Western cores. This status may potentially become a source of unique economic development and telephone traffic.

There were two exceptional periods in terms of the share of the nine countries in Israeli international telephone traffic. First, in the mid 1950s, when this share dropped to 68.1 percent in 1956/7. That this has nothing to do with the 1956 Sinai War is demonstrated by the low figure of 74.6 percent a year earlier, in 1955/6. (These low values were registered after the high values in the early 1950s, which are probably due to ties with Jewish communities, since the international economy was still limited.) The mid 1950s was an era of intensive effort to build relations with East-bloc countries and the new states in Africa and Asia. However, long-run trends in both politics and economics have strengthened the ties with the West. The year 1973/4 marks an exceptionally high value of traffic to the nine nations, 92.1 percent. This was related to the 1973 Yom Kippur War, when both military and moral support came from most of these countries and when international news coverage was very intense.

Major trends by country

The nine countries may be divided into two groups. The first five countries, USA, UK, West Germany, France and Switzerland, have taken a larger percentage of Israeli international telephone traffic over the years, with Germany joining this group since the mid 1950s. The other four countries, Italy, Canada, Netherlands and Belgium, have taken a much lower percentage. In recent years the Swiss percentage of less than 4 percent brings it closer to the second group. These two groups represent international realities of major economic powers and smaller ones. With two exceptions, however, they match the relative sizes of Jewish populations. The two exceptions are West Germany with a small Jewish community and much telephone traffic with Israel, and Canada with a much larger Jewish community and a small share in the traffic.

Several major trends may be identified with regard to specific countries. The most striking one relates to the USA. With the exception in 1953/4, the remotely located USA has led the list. Its share increased from a modest 16.1 percent in 1953/4 to 45.2 percent in 1984/5. Since the early 1980s the figure has continuously exceeded 40 percent. The major increases occurred since the 1967 Six Days War because of a combination of several effects. Political and economic relations have continuously improved since 1967, with American military and economic aid becoming a major factor in Israeli life. While the military aid reached its culmination in the 1973 war, the economic ties gained important status with the signing of the free trade zone agreement in the early 1980s. The economic and defense ties between the two countries have found expression in research and development activities and in the evolution of high-tech industries in Israel. The flow of information between the two countries has thus represented more than mere business and management contacts. Several companies have leased phone lines between the two countries for R & D purposes. The ties between the academic communities in the two countries have also been enhanced as a result of these developments. In addition, press coverage of Israel by US media has become extensive over the years, so that much use is being made up of various forms of telecommunications. However, there is also a social aspect to these increased international contacts. The Jewish community in the USA is the largest in the world and many of its members and institutes have strong bonds with Israel. There is also a growing Israeli community in the USA which is coupled with more extensive social ties between Israelis and Americans in general as a result of the more extensive business ties between the two countries. These ties can be channeled through the telephone especially since 1972 when satellites permitted direct telephone interaction between the two countries. Direct dialing in Israel and reduced rates especially in the USA have assisted this process as well. It is, therefore, again a blend of economic, political, social and technological factors that have caused the size, growth and change in the Israeli telecommunications system.

Another interesting trend is the change in the size of contact with Switzerland, which used to be high in the mid 1950s (12.1–19.9 percent), declined to an intermediate in the 1960s and early 1970s (10.2–11.0 percent), and stabilized at a relatively low level in the 1980s (3.7–3.9 percent). This may be related to the important roles played by Switzerland in the past as an intermediator between Israel and the Arabs and as a major financial outlet. Switzerland has continued over the years to be a popular resort destination for Israelis and its current ranking reflects its importance in the global economy despite its small size.

The ranking of West Germany has also undergone marked changes since the early 1950s. Ties with Germany then were limited, due to the Israeli hesitation to maintain close ties with Germany following the Holocaust. However, the German compensation monies that allowed for much of Israeli economic development in the 1950s and 1960s, and the increasing economic importance of Germany, have opened up relations between the two countries so that in 1985/6 Germany became third in importance following the USA and the UK. (Its rank as second in 1973/4 reflects the close ties during the 1973 war.)

While the UK maintained an almost constant share of 11 to 13 percent of the traffic, the share of France has fluctuated. The close political and military ties that developed between the mid 1950s and the mid 1960s were not translated into increased telephone ties until 1961/2 when the system grew generally due to economic development in Israel. For one year (in 1977/8), the percentage share of France increased to 17.3 percent though it is difficult to relate it to the change in the administration in Israel during that year. Since then, the French share has steadily declined to 9.1 percent, making the ties with Germany higher than those with France.

The transitions in the second group of leading countries have been less dramatic. Canada increased its share from a mere 0.9 percent in 1950/1 to 3.6 percent in 1982/3 following the growth in the traffic with the US. Italy's share declined and Netherlands has become more important than Belgium. Though Brussels serves as the headquarters for the EEC, with which Israel maintains special relations, the Netherlands has developed as a country which provides special services for Israeli imports (oil) and exports (e.g. flower market, potash storage).

Factors for Israeli international telephone traffic

Several aspects have been mentioned so far as factors of the Israeli international telecommunications system, especially economic, social, political and technological ones. The data on change in the system as a whole (Figure 12.1) displayed correspondence between economic cycles and growth in international telecommunications. In addition, the data show a high variation among countries that communicate frequently with Israel (Table 12.2). Given these observations, a more formal explanatory analysis was attempted, namely a multiple regression analysis. Three such analyses were performed using outbound calls, inbound calls and exports for 1985/6

Table 12.3   Variables in the multivariate regression analysis (all for 1985/86).

Country	Outbound calls from Israel in thou. min.	Inbound calls to Israel in thou. min.	Import to Israel in $ mil.	Export from Israel in $ mil.	Jewish pop. in thou.	Tourists to Israel in thou.	Call rate from Israel in NIS
Argentina	555	736	36.5	12.4	233.0	14.5	23.32
Australia	523	1,336	45.5	58.5	75.0	19.9	23.32
Austria	582	783	45.2	31.1	7.5	26.0	23.32
Belgium	1,401	1,481	900.0	230.0	32.5	19.7	16.88
Brazil	347	512	37.8	31.4	100.0	9.6	23.32
Canada	1,489	2,968	103.5	65.3	308.0	38.1	21.12
Cyprus	222	317	2.8	17.2	0.0	4.6	6.32
Denmark	303	871	36.4	19.6	6.9	16.6	16.88
Finland	151	186	44.3	22.7	1.0	10.7	16.88
France	4,320	7,968	303.2	262.7	530.0	140.7	16.88
Greece	421	537	16.6	53.0	5.0	9.9	11.52
Italy	2,098	2,940	411.0	249.1	32.0	44.4	16.88
Mexico	150	121	2.7	10.0	35.0	12.0	26.24
Netherlands	1,181	1,970	221.0	276.1	26.5	32.1	16.88
Norway	166	329	17.5	24.8	0.95	10.0	16.88
Rumania	473	100	20.3	10.0	30.0	3.0	18.36
South Africa	964	2,001	174.7	63.8	119.0	18.0	23.32
Spain	644	853	77.1	29.9	12.0	14.2	16.88
Sweden	376	900	74.3	34.9	15.0	24.9	16.88
Switzerland	2,207	2,864	545.9	133.3	19.0	36.2	16.88
UK	5,311	10,807	753.9	477.0	350.0	129.3	16.88
US	11,013	46,024	1,679.0	2,138.0	5,705.0	367.0	21.12
West Germany	4,238	8,329	898.3	329.4	33.5	145.8	16.88

Sources: See Table 12.2.

as dependent variables. The assumption, thus, was that the two-way telephone traffic and exports might be interrelated. Rather than using the short list of the nine most frequently called countries, an expanded list of twenty-three countries with the highest number of calls to and from Israel was used. The independent variables that were used reflected economic, social and cost aspects only, since for a one-year analysis, technological change and political events were irrelevant. These variables included exports and imports between the twenty-three countries and Israel, the Jewish population in each country, the number of tourists from each country visiting Israel in 1985/6, and the call rates from Israel to every country (this last variable is not country-specific since there exist multi-country rate categories only) (Table 12.3).

The intercorrelation among several of the independent variables is very high. Exports and imports are correlated by .86, and the Jewish population is highly correlated with exports ($r = .96$) and imports ($r = .74$). Exports and imports are also highly correlated with the number of tourists from each country ($r = .85$ and .75 respectively). This correlation matrix demonstrates the complex relationships that exist between Israel and these twenty-

Table 12.4   Multiple regression analyses of Israeli international telephone traffic 1985/86.

Dependent variable	Independent variable	R^2 change	Cumulative R^2
Outbound calls	Exports	0.80	0.80
	Jewish pop.	0.08	0.88
	Imports	0.02	0.90
	Tourists	0.02	0.92
	Call rates	0.003	0.92
Inbound calls	Exports	0.98	0.98
	Tourism	0.002	0.98
	Jewish pop.	0.01	0.99
	Imports	0.006	0.99
	Call rates	0.001	0.99
Exports	Inbound calls	0.98	0.98
	Outbound calls	0.004	0.98

three countries, most of which represent the two Western cores. It caused the R^2 change to be low after the first variable was entered into the regression procedure (call rates showed no relationship with other independent variables).

The results of the several runs are presented in Table 12.4. When outbound calls were used as the dependent variable, exports seemed to be the most important explanatory variable with $R^2 = .80$. The Jewish population was second with an added R^2 of 8 points. Imports and tourists added 2 points each, while call rates added very little. The total R^2 reached .92. It seems, therefore, that exports create outgoing telephone calls more than do imports and this to a high degree, and that the Jewish population in the various countries provides an additional explanation. When inbound calls served as the dependent variable, then the role of exports was much higher, reaching .98, so that very little room was left for the other independent variables. The creation of more inbound rather than outbound calls by exports attests again to the higher proportion of incoming versus outgoing calls, which is mainly due to the higher call rates in Israel than abroad.

When the analysis is reversed and exports serve as the independent variable, then obviously inbound calls explain again 98 percent of the variation among counties (inbound and outbound calls are interrelated by $r = .95$, so that the gap between the two persists for all countries). The relationship between international economic relations in the short run (exports and imports) on one hand, and telephone traffic on the other might be a one-way or two-way relation. There is no doubt, however, that in the case of Israel exports and telecommunications go together, and that the role of telephones in this area is even more important than with respect to tourism and social ties. Unfortunately, this important relationship does not find its expression in the rate schedules. A recent study focused on communications means in a sample of 17 percent of Israel's high-tech plants (Shefer and Frenkel, 1986). It was found that the telephone is widely used for marketing purposes to Europe (in 73.1 percent of the plants) and to the USA (76.9 percent) and less so to other parts of the world (38.5 percent). Telex is used even more than the telephone for international marketing, to Europe (84 percent), to the USA (78 percent) and to the rest of the world (50 percent). Management considers international telecommunications crucial to their prosperity.

The view from the American side

We have noted earlier the high proportion of the USA in Israeli international telephone traffic, amounting to 42.2 percent in 1985/6. That this is a high value may be seen by comparing it to the 19.5 percent that the UK reached at the top of the US ranking of outbound calls in 1985 (US, FCC, 1987) (excluding Canada and Mexico). For this reason, it is of interest to view the ranking of Israel in the USA and compare it to other countries (Table 12.5).

The ranking of countries by US outgoing calls is, obviously, of much interest beyond the focus on Israel here. Several general comments are therefore appropriate, before the analysis of Israel's status.

Changes in country ranking of US outbound calls 1961–85

It is easily possible to recognize the 'big five' on the American list, as was done for the Israeli counterpart. These five countries have been, since the late 1960s, UK, West Germany, Japan, France and Italy. Until then, Cuba generated a very large number of calls that made it first in the ranking of 1961. Other Caribbean countries have been excluded from the ranking, since their proximity to the USA has turned them into popular resort and investment destinations for Americans. The number of calls made to them from the USA is, thus, very high, and they are part of the US calling system as much as are Canada and Mexico (cf. the almost nonexistence of telephone ties between Israel and its neighboring countries). The inclusion of Cuba in the ranking, for obvious reasons, provides a criterion when comparisons are made to the more distant parts of the world. It further shows a loss of a big communications destination. Cuba was first in the ranking of 1961, second in 1965, fourth in 1967, tenth in 1970, and since then did not make it any more as one of the leading fifteen.

The top five countries that emerged in the 1960s were the five largest countries of the two cores, and if Canada is added, then one may conclude that the USA maintains the closest telephone ties with the major industrialized countries that have turned or will soon turn into service economies

Table 12.5 Ranking by country of US outbound international telephone calls (by number of calls) 1961–1985.

	1961	1965	1967	1970	1973	1976	1978	1981	1984	1985
1	Cuba 306,844	United Kingdom 421,833	United Kingdom 607,933	United Kingdom 1,246,119	United Kingdom 2,474,357	United Kingdom 4,764,630	United Kingdom 8,633,119	United Kingdom 17,850,208	United Kingdom 30,671,203	United Kingdom 38,128,930
2	United Kingdom 207,026	Cuba 331,959	West Germany 489,706	West Germany 855,477	West Germany 1,769,950	West Germany 3,146,421	West Germany 5,402,758	West Germany 10,363,631	West Germany 18,110,902	West Germany 21,220,761
3	West Germany 182,802	West Germany 307,115	Japan 278,218	Italy 413,485	Italy 998,401	Italy 1,640,950	Italy 2,520,337	France 4,703,769	Japan 9,892,400	Japan 12,917,084
4	France 83,586	France 158,103	Cuba 239,402	Japan 382,114	France 643,409	France 1,251,064	France 2,166,667	Italy 4,313,463	France 8,742,605	France 10,531,006
5	Italy 54,971	Japan 131,203	France 216,057	France 342,523	Japan 564,302	Japan 1,043,920	Japan 1,808,836	Japan 4,247,283	Italy 7,703,402	Italy 9,156,493
6	Japan 51,201	Italy 116,500	Italy 174,696	Philippines 224,298	Philippines 443,670	Venezuela 762,255	Venezuela 1,362,626	Venezuela 3,039,686	Colombia 6,190,486	Korea 6,726,847
7	Switzerland 43,317	Switzerland 83,654	Philippines 160,952	Vietnam 221,637	Israel 439,880	Greece 758,423	Greece 1,156,878	Columbia 2,503,523	Korea 5,187,962	Colombia 6,301,648
8	Netherland 24,117	Panama 67,212	Venezuela 146,380	Venezuela 217,439	Greece 437,137	Switzerland 640,917	Iran 1,101,636	Switzerland 2,407,462	Philippines 4,760,889	Taiwan 5,317,546
9	Venezuela 19,270	Philippines 60,314	Switzerland 115,701	Switzerland 198,741	Switzerland 361,014	Philippines 620,023	Switzerland 1,052,477	Israel 2,166,473	Israel 4,489,239	Philippines 5,271,349
10	Argentina 18,570	Netherlands 54,557	Panama 107,639	Cuba 187,668	Venezuela 338,708	Israel 604,668	Israel 1,022,014	Philippines 2,157,387	Switzerland 4,231,503	Israel 5,142,125
11	Panama 17,206	Spain 47,295	Netherlands 74,632	Israel 181,277	Spain 330,216	Brazil 537,514	Philippines 952,943	Korea 2,071,342	Taiwan 3,950,596	Switzerland 4,974,299
12	Brazil 16,691	Venezuela 39,366	Hong Kong 68,839	Australia 147,003	Cuba 281,464	Netherlands 535,611	Netherlands 945,906	Greece 1,923,033	Venezuela 3,678,107	Greece 4,105,532

Table 12.5 Ranking by country of US outbound international telephone calls (by number of calls) 1961–1985.

	1961	1965	1967	1970	1973	1976	1978	1981	1984	1985
13	Belgium 16,176	Australia 36,252	Spain 59,709	Netherlands 137,168	Netherlands 273,375	Korea 465,375	Brazil 838,990	Netherlands 1,819,457	Greece 3,555,894	Brazil 4,103,065
14	Sweden 16,137	Belgium 35,944	Belgium 55,897	Panama 133,559	Brazil 254,507	Spain 418,395	Korea 820,485	Brazil 1,721,971	Brazil 3,379,462	Hong Kong 3,642,739
15	Colombia 13,638	Sweden 35,593	Australia 53,935	Spain 132,281	Panama 233,722	Belgium 407,426	Colombia 777,316	Panama 1,470,156	Saudi Arabia 3,329,281	Venezuela 3,532,626
27	Israel 6,403	Israel 9,855	Israel 31,954							

Sources: 36–7.
Note: Calls from the coterminous US only. Not considered for ranking: Alaska, Hawaii, Canada, Mexico, Caribbean states and territories (excluding Cuba), Bermuda, Guam.

(Kellerman, 1985). Though the economies of these countries are largely interdependent, the high volume of calls to these countries involves a social dimension as well. There are special cultural ties between the USA and the UK. There are large Italian and Japanese ethnic groups in the USA and the Vatican serves as a religious center for many Catholic Americans. With the exclusion of Japan and the lower role of Italy, this list is similar to the Israeli list of top countries, thus pointing again to the core-oriented telecommunications traffic of Israel.

Total US calls to the top five countries amounted to 38.8 percent of the outbound US international calls in 1985 (excluding Canada and Mexico). The UK and West Germany have not changed their positions as first and second countries over the years, with Britain greatly exceeding Germany. This attests to the special role of London in world markets and to the special cultural bonds between the two nations. Japan has emerged from sixth in the early 1960s to third in the 1980s, thus demonstrating the increasing economic relations with the USA. Italy and France have competed over the years on the fourth and fifth positions.

While the five top countries have not changed for many years now, there have been many changes among the following ten, the second-highest group of countries. Theoretically, this group could consist of two types of countries, namely smaller core countries and peripheral countries. It is, therefore, first interesting to note which countries have *not* made it into the list. There has not been included any African country, not even South Africa. India is not on the list. Also, no East-bloc country is included. The countries which appear on the list are usually smaller core countries, and some South American countries. Three countries have been constantly present, at least since the late 1960s, namely Venezuela, the Philippines (*not* including Guam) and Israel. Of the three, one (Venezuela) is a relatively close South American country with heavy US investments. The other (the Philippines), is a Pacific country that was on the list even before it turned into a NIC, since it has been governed by the USA in the past, and has maintained special relations with the USA. The third (Israel) is a core-extension country with special relations with the USA.

Over the years most smaller European core countries have left the list: Sweden (1967), Spain (1976), Belgium (1976), the Netherlands (1981). This is despite their unique functions and/or relations with the USA. It has happened mainly due to the evolution of the new core in the Pacific rim, the countries of which had developed strong bonds with the USA, even before they became NIC. Thus Taiwan entered the list in 1984 and ranked already eighth in 1985. Korea entered the list in 1976 and was sixth in 1985. Hong Kong was on the list in 1967 and appeared in 1985. Soon the same will happen with Australia. There are, however, two European exceptions to this rule. One is Switzerland, which constantly occupies a 8–11 rank (cf. similar recent stability in Israel!). Here again, the special role of Switzerland as a center of finance, international organizations and tourism can be seen.

The second country is Greece, which made it into the list in 1973 with the introduction of modern telecommunications. The contact here is mainly due to a large Greek minority in the USA and due to tourism. Greece was seventh and eighth between 1973 and 1978 but declined to 12–13 later, probably since social ties alone are not enough to sustain ongoing high growth in traffic.

There are two other notable changes on the ranking list. One is the change among South American countries. While Venezuela constantly made it to the list, Argentina has not since the mid 1960s. Panama (*excluding* the Canal Zone) was high in 1965 and has been out since the mid 1970s. Brazil was in during 1961 and then reappeared in 1973. Colombia was on the list in 1961 and reappeared in 1978. Since then it has been very high on the list (6–7 rank in 1981–85). This represents changing US interests in the South American continent and variable economic development. A similar explanation is pertinent for several countries in other parts of the world, such as Vietnam (7 in 1970), Iran (8 in 1978) and Saudi Arabia (15 in 1984).

Israel's ranking in US outbound calls 1961–85

Israel has emerged as among the highest destination countries for US calls. In the early 1960s about 0.3 percent of US outbound calls were made to Israel. This percentage increased to 1.5 percent in the mid 1970s and reached 2.1–2.2 percent in the mid 1980s (US, FCC, 1962–87). In 1961, a year of high growth in Israeli traffic (Figure 12.1), Israel's position reached 27 in US ranking. In 1965, a year of economic recession in Israel, this ranking fell to 34. The war in 1967 raised Israel's rank to 20, but it was only after the first Israeli maritime cable came into operation, and economic and political relations between the two countries became stronger, that Israel's rank reached 11 in 1970. It is, therefore, in the late 1960s that there was a clear breakthrough in terms of outgoing calls to Israel, because of the mix of social, economic, political and technological factors discussed above. Since the early 1970s Israel has occupied the relatively high 9–11 positions, with the exclusion of the war year of 1973 when it was ranked 7.

Israel's ranking is similar to countries such as Switzerland, the Philippines and Taiwan. The first is a small country and the two others enjoy special relations with the USA. Both aspects pertain to Israel as well. In other respects Israel is similar to the much larger Italy. This relates mainly to the Jewish population in the USA and its special attitude to Israel and its people. We have noted earlier other similarities between Israel and Italy regarding the correlation between economic development and telephone density. On another level the high volume of traffic to Israel stems from a factor similar to US–Japan relations, namely the strong bonds between the high-tech and R & D sectors in the two countries.

The deregulation of international calling in the USA may open up the Israeli market to companies other than AT&T, depending on agreements with the Israeli-regulated Bezek company. It would be of interest to see whether the possible reduction of rates would result in more (business or social?) calling from the USA.

Conclusion

At the beginning of this chapter several aspects and factors were presented with respect to international telecommunications in general and regarding peripherally located countries in particular. Though the Israeli case has been presented as a unique peripherally located core-extension, some of the conclusions may pertain to a wider group of countries. It seems that the social element in international telecommunications may partially influence the *size* of the traffic. It would account less, however, for annual change (except for extreme events such as war). This is apparent also from the Greek and Italian shares of the American traffic.

It is rather an economic aspect of business cycles and economic growth that strongly influence *annual* fluctuations, and differential traffic growth rates. This is not only true for variation across time, but for diversity across international space as well. Differing levels of economic ties by country are reflected in telecommunications traffic, as both the US and Israeli data demonstrate. The question whether economic development proceeds or follows international telecommunications in the long run cannot be answered from the data used here. However, exports and international telecommunications were found to be highly interrelated in Israel. Rate policies may cause an unbalanced two-way telecommunications traffic.

The political aspect has been shown to determine the group of countries or areas to which traffic is oriented, though this aspect cannot be isolated from economic aspects. Telecommunications can serve as a vehicle for international understanding as long as it is accompanied by other ties such as two-way tourism and business. Viewing the very existence of international telecommunications as providing for better relations among countries would be analogous to a car moving from one country to the other without people or goods in it.

An important lesson that can be learned from the Israeli experience is that distance is completely meaningless when it comes to international telecommunications. The share of the remotely located USA and Canada in Israeli traffic was larger in 1985/6 than that of the UK, West Germany, France, Switzerland, Belgium and the Netherlands combined. In additional to the political, social and economic conditions necessary for overcoming space, there needs to be a technologically reliable system. In Israel, the breakthrough in this regard came relatively early, in 1968, with the inauguration of the first maritime cable.

In summary, international telecommunications for peripherally located countries cannot be viewed as an independent issue that is determined by technology only. It cannot even be looked upon as merely an economic issue. In as much as communications among individuals is a complex phenomenon, it is not less so among countries as well.

As far as Israel is concerned, international telephone communications have undergone a sophistication process in recent years through the introduction of data tranmission lines and services, cellular phones and facsimile service. The major problems of the Israeli system are more domestic than international, namely the need to meet as fast as possible the demand for new lines, and the need to upgrade the system which is still partially based on old equipment. The telephone system is transformed now into digital equipment, and will soon have to cope with ISDN (Integrated Services Digital Network) which is planned for the mid 1990s (State of Israel, Ministry of Energy and Development, and Ministry of Communications, 1985).

The Arab–Israeli conflict takes its toll even in the area of current and future telecommunications. The Israeli infrastructure could have served neighboring countries, especially Jordan, which recently developed its own international infrastructure. Israel plans to build its own satellite (maybe even two), in order to use the slots allocated in the geosynchronous orbit, so that they will not be taken by additional ARABSATs. In addition, Israel plans a third maritime cable to Italy and Spain and its participation in the first transatlantic fiber–optic cable.

Note

I wish to thank Ms Daphna Reichenbaum for her assistance. This paper was written while the author served as Visiting Lecturer in the Department of Geography, University of Maryland, College Park.

References

Bezek, Israel Telecommunications Company (1986), *Statistical Yearbook 1985/86*. Jerusalem (Hebrew).
——(1987), *Rate Schedule for International Telephone Calls*, Jerusalem.
Galtung, J. (1982), 'The new international order: economics and communication', in Jussawalla, M., and Lamberton, D. M., (eds), *Communication Economics and Development* (New York: Pergamon), pp. 133–43.
Hepworth, M. E., Green, A. E., and Gillespie, A. E. (1987), 'The spatial division of information labour in Great Britain', *Environment and Planning A*, 19, pp. 793–806.
International Telecommunications Union (1983), *Telecommunications for Development*, Geneva.
Jussawalla, M. (1982), 'International trade theory and communications', in Jussawalla, M., and Lamberton, D. M. (eds), *Communication Economics and Development* (New York: Pergamon), pp. 82–99.
Jussawalla, M. and Lamberton, D. M. (1982), 'Communication economics and

development: an economics of information perspective', in Jussawalla, M., and Lamberton, D. M. (eds), *Communication Economics and Development* (New York: Pergamon), pp. 1–15.

Kellerman, A. (1984), 'Telecommunications and the geography of metropolitan areas', *Progress in Human Geography*, 8, pp. 222–46.

——(1985), 'The evolution of service economies: a geographical perspective', *The Professional Geographer*, 37, pp. 133–43.

——(1986a), 'Telecommunications as a tool for closing center-periphery gaps', *Economics Quarterly*, 128, pp. 547–54 (Hebrew).

——(1986b), 'The diffusion of BITNET: a communications system for universities', *Telecommunications Policy*, 10, pp. 88–92.

——(1986c), 'Characteristics and trends in the Israeli service economy', *The Service Industries Journal*, 6, pp. 205–26.

Lyon, D. (1986), 'From "post-industrialism" to "information society": a new social transformation?', *Sociology*, 20, pp. 577–88.

O'Brien, R. C., and Helleiner, G. K. (1982), 'The political economy of information in a changing international economic order', in Jussawalla, M., and Lamberton, D. M. (eds), *Communication Economics and Development* (New York: Pergamon), pp. 100–32.

Peder, Y. (ed.) (1986), *Israeli General Encyclopedia* (Jerusalem: Keter) (Hebrew).

Pelton, J. N. (1981), *Global Talk* (Netherlands: Sijthoff & Noordhoff).

Robinson, F. (1984), 'Regional implications of information technology', *Cities*, 1, pp. 356–61.

Rohwer, J. (1985), 'The world on the line', *The Economist*, 297, pp. 5–40.

Salomon, I., and Razin, E. (1985), 'Potential impacts of telecommunications on the economic activities in sparsely populated regions', in Y. Gradus (ed.), *Desert Development* (Netherlands: D. Reidel), pp. 218–32.

——(1986), *The Geography of the Israeli Telecommunications System: Patterns and Implications* (Jerusalem: The Jerusalem Institute for Israel Studies 19) (Hebrew with English abstract).

Saunders, R. J., Warford, J. J., and Wellenius, B. (1983), *Telecommunications and Economic Development* (Baltimore, Md.: The Johns Hopkins University Press, for the World Bank).

Sauvant, K. P. (1986), 'Trade in data services: the international context', *Telecommunications Policy*, 10, pp. 282–98.

Shefer, D., and Frenkel, A. (1986), *The Effects of Advanced Means of Communication on the Operation and Location of High Technology in Israel* (Haifa: Technion, The Samuel Neaman Institute for Advanced Studies in Science and Technology) (Hebrew with English Abstract).

Singelmann, J. (1978), *From Agriculture to Services: The Transformation of Industrial Employment* (Beverly Hills, Calif.: Sage).

Smith, A. (1980), *The Geopolitics of Information* (New York: Oxford University Press).

Snow, M. S. (1985), 'Regulation to deregulation: the telecommuncations sector and industrialization', *Telecommunications Policy*, 9, pp. 281–90.

State of Israel, Central Bureau of Statistics (1955–63), *Post of Israel Survey and Statistics 1953/54–1961/62*. Special Publications Series 30, 67, 83, 92, 118, 141 (Jerusalem) (Hebrew).

——(1983–86), *Statistical Yearbook 1983–1986* (Jerusalem).

State of Israel, Ministry of Postal Services (1964–70), *Survey and Statistics 1962/63–1968/69* (Jerusalem) (Hebrew).

State of Israel, Ministry of Communications (1971–74), *Report 1970/71–1972/73*. (Jerusalem) (Hebrew).

——(1971–83), *Statistical Yearbook 1969/70–1981/82* (Jerusalem) (Hebrew).

State of Israel, Ministry of Science and Development and Ministry of Communications (1985), The Telecommunications Steering Committee, *Summary Report* (Jerusalem) (Hebrew).

Stern, E. (1983), 'Communication and future spatial structure – some possible scenarios', *Town and Regional Planning*, 15, pp. 19–22.

Time (1986), 'Have data, will travel', 23 June, p. 36.

US Congress, Office of Technology Assessment (1987), *International Competition in Services*, OTA–ITE–328 (Washington, DC: US Government Printing Office).

US Federal Communications Commission (1962–85), *Statistics of Communications Common Carrier 1961–1984* (Washington, DC: US Government Printing Office).

US Federal Communications Commission (1987), *International Communications Traffic Data Report for 1985* (Washington, DC: FCC, Industry Analysis Division, Common Carrier Bureau).

13 *Towards a schematic model of communications media and development in Latin America*

ERICK HOWENSTINE

Communications development – the elusive field

To say that the study of communications media and regional development is undergoing a reformation may be exaggerating the initial acceptance of simplistic models of communications and change now several decades old. Perhaps it would be more honest to say that development communications is still in its infancy, though it is certainly a very big baby. Most simply defined as the transfer of information, communications is an overwhelming field of study. It has not received the attention it deserves, I suggest, from either scholars or policy makers. Gilling's observation that, 'the communications environment is taken for granted, much as fish might regard the water in which he swims' (1975) aptly pointed out the pervasive nature of communications when taken as a whole. One might as well try to study 'interaction'. Each subfield of communications in regional development (by media, by actor, by progress, or however you want to cut it) is still so complex that the basis principles continue to elude scholars.

But despite the rudimentary understanding of the overall effect of communications on development, much work has in fact been done. In this chapter I will discuss the progress made in the study of mass communications and structural change and the changing posture of Third World governments in terms of communications policies and will suggest a theoretical framework for analysis of communications media in regional development. Only the media of mass communications, particularly newspapers, popular magazines and broadcast media, will be considered. Although many of the issues raised pertain as well to other developing countries and industrial countries, this paper's regional focus is Latin America.

By reducing the friction of time and space, communications media effect national growth and development – this is hardly debated. On the other hand, it is also apparent that economic growth leads to greater affordability of, and therefore greater access to, media of communication. In fact, strong correlations between media availability and both economic growth and social well-being have long been recognized throughout the world (for a summary of such works, see Saunders *et al.*, 1983).

In these studies, the typical approach is to correlate United Nations data on per capita estimates of telephones, radio, televisions, newspaper circulation, or combinations of these, with various measures of social change or economic well-being, such as gross domestic product and/or infant mortality rate. The resulting correlations are usually high, which is of interest but is hardly surprising. After all, newspapers, televisions, telephones, and to a lesser extent, radios are concentrated in urban areas. This is particularly true in developing countries, where metropolitan areas generally offer the best health care, higher income and a higher standard of living by most measures. It is expected then, that development, loosely defined for the moment as increased economic transaction and improved living conditions, is associated with availability of mass media and the growth of urban centers. However, little progress has been made in measuring a direction of causation between media availability and development.

Furthermore, the *mechanisms* of these associations are not known. The traditional approach in economic analyses is to avoid the communications issue altogether with an assumption of omniscience, thereby making analyses possible with current techniques but at the same time making results much less valid. Geographers have also skirted the issue. Spatial analyses have long emphasized transportation costs and have virtually ignored communications; even though some forms of communications are often considered a substitute for transportation (Nicol, 1985a), rarely are structures of information flow incorporated into locational models. It has been suggested that improved *two-way* channels of communications are usually considered to have a decentralizing influence on urban structures (Nicol, 1984a, 1984b), but there is no theoretical framework to defend this implication. Nor, to my knowledge, has there been a rigorous theoretical investigation of the effects of one-way media on concentration of population.

A generalized scheme of the complex dynamic relationships between mass media and development is shown in Figure 13.1. As this figure indicates, governments and/or market forces cause qualitative changes through programming regulation, expansion (or contraction) of media distribution, restructuring of ownership and control and/or development of new media technologies. These changes may have a variety of effects. The list of five suggested in Figure 13.1 – social well-being, political structures, economic systems, individual psyches and migration – is neither exhaustive nor mutually exclusive. It should, however, be recognized that communications media may affect social, political, economic, individual and spatial structures. It is generally true that market forces encourage expansion, development of new technologies and agglomeration of media ownership. In some cases, unrestricted market forces have also been credited with successful international competition, e.g. the 'telenovas' programs

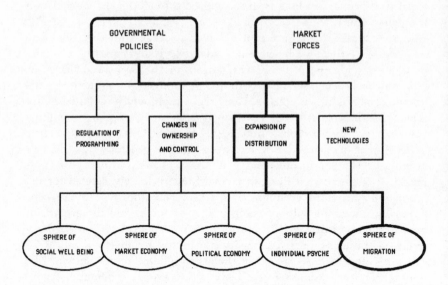

Figure 13.1   Mass media and national development

in Brazil and Mexico (Rogers and Antola, 1985). Government regulation, on the other hand, has more diverse effects. While it is often used to promote the same 'market goals' mentioned above, it commonly focuses on program regulation (especially educational and cultural programming), and shifting media control away from large corporations into more diverse private or public management.

Besides giving a general overview of the changes and effects of communication media, portions of this model can be shown in more detail, illustrating the dynamics involved. Later in this chapter I will expand the highlighted 'track' of Figure 13.1 – expansion of media distribution and migration – to illustrate various processes involved in that relationship.

Nicol suggested several theoretical models for analyzing the influence of improved telecommunications on the 'spatial economy', outlining how improvements in telecommunications might redefine intraurban and interurban patterns through technological and organizational effects. For his purposes, communication was defined as 'all means instrumental to two-way information exchanges', thereby excluding what he called 'purely informational' media or 'mass media' (Nicol, 1984a). Because here the primary concern involves the mass media, particularly radio, television and general-interest periodicals, I should say that these media have in fact been shown to be less than purely informational in both intent and effect. The entertainment, agenda-setting, 'watchdog', profit-making and political functions of the press certainly indicates a much more complex role in society than mere information transmission as Nicol suggested. It is true, however, that mass media are quite distinct from, say, postal or telephone service in that the former are generally unidirectional: messages flow only one way, usually downward through vertical social structures and, spatially, outward from urban centers.

Communications theories – individual effects from stimulus–response to agenda setting

Communications theory has undergone radical changes in the past several decades on both micro- and macrolevels. First, on the level of individual effects, the media were initially credited with powerful influence. For example, the stimulus–response theory was taken seriously, especially by those fearing the potential effects of television advertising. According to the early 'hypodermic needle' theory, with well-designed messages, perhaps including subliminal suggestions, the media could literally change an individual's behavior. Powerful effects theories fell from favor when 'active audience' models gained empirical support. These models credited audiences with selectivity in exposure, perception and retention, and suggested media were more likely to reinforce existing attitudes and behaviors than to change them (Klapper, 1960; Lazarsfeld, 1948). Development communications went through similar permutations. The most pertinent 'powerful effects' theory in development was offered by Lerner in the late 1950s. He suggested that the communications media foster 'empathy' – the ability to imagine oneself in another's situation. This ability, lacking in traditional societies, he claimed, would lead to individual self-improvement, a dynamic economy and a democratic system of government. Literacy and mass media played a pivotal role in the stages of political development:

1 Urbanization and economic concentration
2 Increasing literacy as urban immigrants are forced to survive in a new society
3 Exposure to flourishing mass media, leading to rising expectations
4 Development of 'empathy', or the ability to imagine self-improvement through emulation, and
5 Development of opinions and therefore the rise of political participation.

By 1976 Lerner himself reassessed the role of empathy, concluding ironically that more realistic stages of media effects might be (1) rising expectation, (2) rising frustration and (3) military takeover (Lerner, 1976). Development demands far more than mere emulation, and structures of communications comprise but one important component among many. The media, at best, many now claim, can reinforce existing beliefs and 'trigger' individuals to do what they were likely to do in any event.

Agenda setting represents a tentative return to a theory of powerful effects. According to proponents of agenda setting, media do not tell people what to *think*, but rather what to think *about* (McCombs and Shaw, 1972). Depending on the strength of this effect, media may be able to persuade rural people to become involved in urban issues and adopt a cosmopolitan perspective on the world. The pervasiveness of agenda setting is subject to ongoing debate in communications literature.

Communications theories – aggregate effects modernization or imperialism?

Besides an effect on the individual level, communications structures also have been said to affect development on a national or regional scale. By development I refer to intentional structural and scalar changes on a regional or national scale, thereby encompassing economic growth and improvements in social well-being. More specifically, to adopt Nicol's definition of regional development, it is (a) the expansion of output and income within the region, (b) the integration of the regional economy with diversified employment and entrepreneurship and (c) improvement of the efficiency of the system of cities within the region. Development communications refers in this chapter to the causal role of mass-oriented one-way media in fostering those types of regional change outlined above.

Nordenstreng and Schiller (1979) suggest that communications development literature has evolved through several distinct stages during the past several decades. The first stage of development communications began in the 1950s and was based on what is now considered a simplistic model of development: diffusion of technology outward from an industrial core, leading to 'modernization' of the traditional or backward rural areas. Lerner's *The Passing of Traditional Society* (1958) and Schramm's *Mass Media and National Development* (1964) are representative of the literature of the time. Because 'backwardness' was considered a product of isolation, structures of communications were considered the key to development. During this 'first generation' of literature a single Western model of development was considered applicable to all less developed societies (Nordenstreng and Schiller, 1979). The idea that growth in one sector leads inevitably to growth in other sectors gave way in the early to mid 1970s to a second, self-critical and less Western ethnocentric generation of literature. Many of the classical communications theorists, such as Schramm, Lerner and Rogers, refuted much of their own earlier theory which they recognized had not been sufficient in explaining Latin American, African or Asian experiences. Classical approaches to development which relied heavily on communications linkages had not led to widespread improvement in living conditions nor, generally, to accelerated economic growth.

One recurring argument was that the traditional Western media were inappropriate in developing countries because their one-way flows of information were not sensitive to local needs. Furthermore, goals of development other than Western modernization were gaining legitimacy among scholars and planners, making the design and implementation of policies for development more difficult. Schramm, a leading communications scholar, said in 1976:

> To put it simply, things are not as simple as had been assumed, and the generality sought by the old paradigm may not now be possible. Back to the old drawing board! (Schramm, 1976, p. 48).

By the late 1970s this retrenchment evolved into a third generation of communications development literature, which shifted focus to the international and global context within which national development occurs. Its main departure from earlier works was in its emphasis of external constraints in national development. The new body of literature benefited from increased contribution by less developed countries, whose representatives also participated in collective communications research efforts such as the critical MacBride Commission report of 1980 (MacBride, 1980). North–South communication was no longer assumed to spread benefits to less developed countries; instead, there were charges of cultural imperialism, cultural dependency, international domination through information flows and even conspiracy (Nordenstreng and Schiller, 1979; Schiller, 1973, 1976, 1979).

During the early 1970s considerable criticism was leveled at the existing international structure of information and communications flows. UNESCO was the main forum for the debate, in which the United States supported the 'Free Flow Doctrine' and many Third World countries called for a vaguely defined 'New World Information Order' (NWIO, also called a New International Information or New World Information and Communications Order). Of the more than twenty specific complaints outlined in a 1978 UNESCO report (ICSCP, 1978), three primary ones deal with the imbalances in volume and direction of news flows, the content and deprecatory quality of news about Third World countries and inequitable access to communication technologies.[1] Suggested solutions for restructuring the information order include development of Third World news agencies, increasing restrictions of international (North-to-South) information flows and a general shifting away from a 'free marketplace of ideas' to a 'developmental' theory of the press. These suggestions will be discussed at greater length.

Third World regional news efforts

In response to the call for a New World Information Order, many regional news agencies in Latin America, Asia and Africa have attempted to compete with or contribute to news gathered and disseminated by the big four – Reuters, AFP, AP and UPI (Jakubowicz, 1985). The Caribbean News Agency (CANA), Agencia Latino-Americana de Servicios Especiales de Informacion (ALASEI), Latin American Regional News Agency (LATIN) and ACAN, a Central American news agency, are Latin American examples of such regional cooperation. Mostly after 1970 similar joint ventures have been formed throughout the Third World and two (Inter-Press Service (IPS), and Non-Aligned News Agencies' Pool (NANAP)), are global, linking developing countries worldwide. Their intent is to provide both the North and the South with coverage of Third World issues from Third World perspectives. For all their efforts, however, both the regional and global agencies have been unable to break into Western markets or even win widespread acceptance in developing countries' media. Jakubowicz suggests their failures are primarily attributable to political manipulation of indigenous news reports and to Western-style education of journalists, which perpetuates the news values of the developed world (Jakubowicz, 1985).

Another large-scale effort to promote Third World views began in 1980 with first publication of *South*, a monthly journal promoting social, political and economic sovereignty of Third World countries and presenting facts and issues not well covered in the First World press. Among the stated goals of the editors is to help eliminate the instruments and agents of exploitation through reduction of the power of multinational corporations and other foreign and domestic elite influences.

Ted Turner's Cable News Network (CNN) is another attempt to provide world news coverage, this time from a multinational media institution based in the United States. Launched in 1980, by 1987 it served 10,122 cable systems in fifty-four countries (both developed and developing) with twenty-four-hour daily news coverage. Similarly, in October of 1986 USIA's Worldnet Service expanded its coverage to Latin America, enabling televised two-way dialogues one hour every day (Broadcasting, 1986). These efforts to increase or improve global mass communications indicate a growing need to understand cross-regional issues that arise.

North and South perspectives – free marketplace of ideas?

The second issue of the NWIO concerns restrictions on news and its distributions. Although the national postures regarding communication flows in the industrial North or the less developed South are by no means homogeneous, some generalities can be drawn. From the First World perspective, uninhibited flow of communications is considered not only necessary in a free market economy but also essential in preserving a basic human freedom, written in United Nations Universal Declaration of Human Rights, Article 19:

> Everyone has the right to freedom of opinion and expression; this right includes freedom to hold opinions without interference and to *seek,*

receive and impart information and ideas through any media and regardless of
frontiers. (emphasis added).

The 'free marketplace of ideas' is an ideological cornerstone of democracy.
The news and information media are called 'partners in man's quest for the
truth', and unrestricted flows of information are said to allow truth and
goodness to win over evil and oppression.

These ideas are subject to increased criticism and suspicion, often sup-
ported by empirical evidence. For example, some claim the imbalance in
North–South information flows represents neoimperialism. Beltran and
Cardona (1979) showed that foreign media, non-national advertising and
foreign programming dominate the mass communications institutions in
Latin American countries. They suggested as well that these imbalances
have had significant impact on indigenous cultures, and concur with
Naesselund:

(a) The distribution of communication resources in the world is
strongly disproportional to the distribution of population and the
information needs of the people. Thus an imbalance of the potential.
(b) It is estimated that the total flow of communication from the
industrialized part of the world (with one-third of the total world
population) to the developing countries is 100 times the flow in the
opposite direction. Thus an imbalance in the flow of information. (c)
The fact that the media in many developing countries fail to diversify
their content sufficiently to give it some significance to all audiences
(in particular in rural areas) leads to an irrelevance of content to the
social and cultural problems encountered in those countries. (Naesselund, 1975, p. 3).

Nordenstreng and Varis (1974) showed empirically that what is called a
'free flow' of information is in fact a one-way flow, dominated by the
United States and the industrial world. Furthermore, it is not only the
volume but also the content of foreign messages that is considered inappropriate. Although much of the continent relied heavily on US news media
during 1972–81, nearly all of South America was virtually ignored in
network television. Those few references to Latin America were generally
superficial and usually concerned major crises (Larson, 1985).

Less developed countries claim to suffer from Western-dominated
programming on cultural, economic and political levels: Western customs
are said to erode traditional values; demand for foreign consumer goods
increases faster than does purchasing power, causing an unfavorable balance
of trade; and the 'revolution of rising frustration' brought about by unmet
demand for consumer goods and public facilities theatens political stability
(Lerner, 1976). It is important to note that the academic community is split
regarding the shift toward protectionism and international press restriction.

Radical scholars, like Schiller, Nordenstreng, and a growing group of Third World critics, reject the 'free flow' doctrine as a tool of imperialism. Others argue that the flow of communication should be less restricted, not more so, to encourage development. De Sola Pool makes this argument:

> Dependence occurs when advanced countries possess know-how and techniques that developing countries are not able to acquire for themselves at will. Independence is therefore promoted by unrestricted free flow of information between countries, so that the developing country can acquire for itself whatever intellectual and cultural products it desires at the lowest possible price. The freer the flow of information, the wider the developing country's range of choice and the sooner it can acquire for itself the ability to produce the same sort of information or programming at home. (Pool, 1979, p. 152).

Developmental theory of the media

Proponents of the NWIO commonly argue that Third World countries should use the press as a tool for development, rather than adopt the principles of the Western 'free press'. Although international flow of media in the free world has been largely unimpeded until recent times, press restrictions are not new in many developing countries. A 'developmental' approach to media is often adopted, under which competing and conflicting messages of a liberal press system are seen as counterproductive. Negative messages, even from indigenous media institutions, are often not published in the interest of national pride, strength and stability (Hatchen, 1981; Martin and Chaudary, 1983). As opposed to the traditional Western approach, under which it is widely believed that from among competing messages 'the truth will out', the developmental theory encourages the use of media as a direct tool of national policy.

Communication development approaches

Against this backdrop of a changing climate regarding North–South media flows and indigenous approaches to media control, we can turn to Latin American experiences in communications development. There are many philosophies on how media can be used for development. Three broad approaches toward communication for rural development are outlined by Rahim (1976): (a) the extension and community development approach, (b) the ideological and mass mobilization approach and (c) the mass media and education approach. Of these, all have been implemented in Latin America, though the latter is by far the most common. According to the extension and community development approach, people are eager to learn new things and are capable of putting innovations to immediate use. This is the oldest, most common and classically 'Western' approach, supported by most UN and US development assistance. Communications media broadcast new ideas and others propose new solutions to problems with the expectation that diffusion and adoption will take place quickly. In contrast, the mass mobilization approach, which is traditionally implemented by centrally planned societies, considers development a function of heightened political consciousness. In Cuba, for example, the media are used for ideological exhortation. The mass media and education approach, most often adopted in Latin America, emphasizes individual learning. Adherents of this approach claim that the main barriers to rural development are illiteracy, ignorance, passivity, traditional attitudes of conformity and lack of civic and community values (Rahim, 1976). By increasing literacy and making lessons in health, economics, vocations and math available to the rural poor, development there is expected to follow.

Having outlined several ideological and theoretical differences in the field of communications development, I will present an overview of communications structures in Central and South America and consider the experiences of several countries whose governments have implemented representative communications policies over the past several decades.

Mass media in Latin America

Before discussion of the mass media's role in the development of Latin America, some descriptive comments will place a discussion of mass media issues in context. By considering the many and diverse countries in Central and South America as a single entity, 'Latin America', as I do often in this chapter, I make a sweeping generalization that, admittedly, cannot withstand specific criticism. The physical environment alone is obviously heterogeneous, ranging from desert to tropical rainforest to mountain highlands. The many natural resources (e.g. oil, coal, bauxite, fertile soils and temperate climate) are not equally distributed. Likewise, social and economic statistics show great differences between countries. In 1980 Gross Domestic Product per capita ranged from $1,089 to more than five times that amount. Life expectancy in Bolivia was fifty years and infant mortality rate was 131; in most of the continent life expectancy was sixty to seventy years and the IMR less than half Bolivia's rate (*UN Statistical Yearbook*, 1982). There are several accepted regional powers as well, including Mexico, Venezuela, Cuba, Colombia, Brazil and Argentina.

But despite these variations, Latin American countries in many ways are quite alike. Three centuries of exclusive and strict colonial control by Spain and Portugal and the unifying influence of the Roman Catholic Church have given most of the continent a similar culture, governing system, national language and economic system (A. Morris, 1979). Many also share similar problems: high rates of land rental, illiteracy and poverty; unfavorable balance of trade, high inflation and unemployment, and large portions of the population not actively involved in the market economy (Blakemore and Smith, 1983).

Some claim the remaining infrastructures of colonial mercantilism have not lent themselves to indigenous development. For example, the dendritic transportation systems, converging on port cities, allow relatively little movement in the interior because of high transportation costs (Gore, 1984). The pattern of rail and roads was largely designed, under colonial control, to facilitate the extraction and exploitation of interior natural resources.

Another phenomenon shared by many Latin America countries is a primate urban distribution in which the largest city is many times larger than the second in size. In 1970 the index of primacy (population of the largest city divided by the combined population of the next two in cities) was 11 in Argentina, 9.2 in Chile, 10.7 in Peru, 5.8 in Mexico (A. Morris, 1981). United Nations demographic study predicted that by 1990 half of the urban residents in Central and South America will live in cities of 1 million or more (Rondinelli, 1983), while already more than 60 percent of the entire Latin American population is urban. Most policy makers consider the primate pattern to be unbalanced and would encourage decentralization, smaller cities and rural settlement. Besides the human costs of overcrowding, huge urban areas are said to be economically inefficient on a national level. Central city slums and barrios surrounding the giant urban areas represent great welfare costs, and concentration of capital in a single location drives up land rents, reducing the viability of land-intensive industries (Gwynne, 1985; A. Morris, 1979). It is noted that while *most* policy makers and scholars promote policies of decentralisation, some claim that primate cities represent the greatest marginal return for investment and therefore, in terms of aggregate efficiency, are not a problem (Fuchs, 1967; Mera, 1978).

Regardless of its desirability, urban primacy is a common theme in Latin America and an important issue that communications policies should take into account. Most media messages originate in the cities or industrial world; very few are indigenous to rural areas (UNESCO, 1975). This raises issues regarding the content of media messages and their effects on remote audiences. The associations between urban-originating mass media and internal migration will be investigated below, after a look at media structures and policies in selected Latin American countries.

In the mid 1970s UNESCO surveyed media availability in member nations, counting newspaper circulations, magazines, radios, televisions and cinema seats. There was significant variation between countries. Uruguay, for example, had more than 500 radios, 125 televisions and a daily newspaper circulation of 250 for every 1,000 people. In Bolivia the corresponding figures were 87 radios, 9 television sets and 41 daily newspaper copies. (See

Maps 13.1, 13.2, and 13.3 for average densities of television sets, radio receivers and daily newspaper circulation. Table 13.1 shows major daily newspapers throughout Latin America.) Urban and rural distribution of mass media also differed widely. Because of the cost of receivers and unavailability of electricity in many rural areas, television broadcasting was largely an urban service. In geographically smaller countries *potential* television coverage was nearly complete in 1975, while in Mexico, Brazil, Argentina and elsewhere, the most remote 30 to 50 percent were not even reached by broadcast airwaves (UNESCO, 1975).

Like television, newspapers were also concentrated in urban areas, where advertisement and circulation revenues could cover production and distri-bution costs. Aggravating this urban–rural disparity are adult illiteracy rates approaching or surpassing 50 percent in rural regions, especially

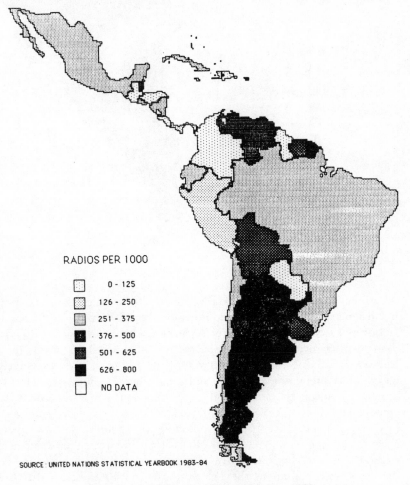

RADIOS PER 1000

	0 - 125
	126 - 250
	251 - 375
	376 - 500
	501 - 625
	626 - 800
	NO DATA

SOURCE : UNITED NATIONS STATISTICAL YEARBOOK 1983-84

Map 13.1   Radios per 1000 population, 1982

Map 13.2   Televisions per 1000 population, 1982

where the Spanish or Portuguese language are not spoken (UNESCO, 1984). In Guatemala, Paraguay, Peru and Ecuador, about 50 percent of the populations spoke Quechua, Guarani or Aymara as a native tongue; in Bolivia, more than 60 percent. No newspapers were published in any of these languages in all of Latin America (UNESCO, 1975). Radio, being a relatively inexpensive, literate-independent, and long-range medium, was already the main source of information in most Latin American countries, especially in rural areas. Yet only Bolivia and Paraguay reported occasional broadcasts in an Indian language, leaving the poorest, again, largely excluded. Even governmental educational and cultural programs, designed to provide basic education, were usually in Spanish and available in or near the cities. Information in the *Europa Yearbook* 1987 indicates that this situation has improved only slightly if at all during the succeeding decade.

For the most part, national media policies promoted basic education. Commercial television stations, radio stations and newspapers throughout Latin America were required to participate in educational campaigns, and many government-controlled media were primarily educational and cultural in scope. Probably the most aggressive educational broadcast drive began in El Salvador with the Ministry of Education in 1969 (Mayo and Hornik, 1976). Designed to increase rural education, to provide a minimum of nine years of schooling for everyone and to attract industrial growth in coming years, by 1974 the program had helped more than 10,000 students through grades 7 to 9 in 1,179 television-equipped classrooms. The multimedia project incorporated television, textbooks, a weekly newspaper and other teaching aids. Annual cost of the program was more than $2 million. The El Salvadorian government committed 36

DAILY NEWSPAPER CIRCULATION
PER 1000

- 0 - 25
- 26 - 50
- 51 - 75
- 76 - 100
- 101 - 150
- 151 - 275
- NO DATA

Map 13.3   Daily newspaper circulation per 1000 population, 1979 and 1983

Table 13.1 Major newspapers in Central and South America (daily circulation more than 100,000).

Country	City	Newspaper	Circulation
Argentina	Buenos Aires	Clarin	480,000
	Buenos Aires	Cronica	520,000
	Buenos Aires	El Cranista	100,000
	Buenos Aires	Diario Popular	145,000
	Buenos Aires	La Nacion	210,000
	Buenos Aires	La Razon	180,000
Brazil	Rio de Janeiro	O dia	207,000
	Rio de Janeiro	O Globo	220,000
	Rio de Janeiro	Journal de Brasil	146,943
	Sao Paulo	Folhade Sao Paulo	230,000
	Sao Paulo	Jornal da Tarde	250,000
	Sao Paulo	O Estado de Sao Paulo	220,000
	Sao Paulo	Noticias Populares	150,000
	Porto Alegre	Porto Alegre	110,000
Chile	Santiago	El Mercurio	120,000
	Santiago	La Tercera de la Hora	300,000
	Santiago	Las Ultimas Noticias	150,000
Colombia	Bogota	El Espectador	215,000
	Bogota	El Tiempo	200,000
	Medellin Antioquia	El Columbiano	123,707
Cuba	Havana	Granma	700,000
	Havana	Juventud Rebelde	300,000
	Havana	Trabajadores	150,000
Ecuador	Quito	El Comercio	130,000
Mexico	Mexico City	Diario de Mexico	110,000
	Mexico City	Esto	400,200
	Mexico City	Excelsior	175,000
	Mexico City	El Heraldo de Mexico	209,600
	Mexico City	Novedades	190,000
	Mexico City	Ovaciones	425,000
	Mexico City	La Prensa	297,803
	Mexico City	El Sol de Mexico	295,000
	Mexico City	El Universal	181,375
	Toluca	Estadio	200,000
	Monterrey	El Norte	100,000
Peru	Lima	El Comercio	100,000
	Lima	Expresso	123,000
	Lima	Ojo	180,000
	Lima	La Republica	114,000
Venezuela	Caracas	Meridano	300,000
	Caracas	El Mundo	195,000
	Caracas	El Nacional	140,000
	Caracas	Ultimas Noticias	234,431
	Caracas	El Universa	140,000
	Caracas	2001	160,000

percent of the national budget in the early 1970s to the Ministry of Education.

The Dominican Republic launched a six-year radio literacy program in 1964, enabling 26,500 previously illiterate adults to read and write in Spanish. After 1970 the emphasis was shifted to school-equivalency programs for grades 1 through 8, with annual enrollments of 12,500. This program, like El Salvador's, incorporated print and broadcast media. Other Latin American countries implemented similar, albeit less comprehensive, media educational campaigns. In Nicaragua a radio mathematics course was offered in primary schools from 1973 to 1977. Bolivia devoted 13 percent of television air time to education and also used radio schools for both children and adult groups.

A secondary thrust of Latin American communication policies, particularly in cinema, has been protection of indigenous media industries. Mexico, Argentina and Brazil have shown how state support can bolster the nation's media. Schnitman (1984) described how Mexican efforts since the 1930s turned that country into the major film exporter in Latin America. Implementation of loan banking, tax exemptions and a requirement of all Mexican theaters to support indigenous films was followed by a series of nationalizations beginning in 1959. By 1970 the state owned 60 percent of all theaters and distributed 95 percent of films (Schnitman, 1984). That year the industry produced 124 features and 557 short films. Brazil, with the second-ranked film industry in Latin America, produced 70 feature-length films annually in the 1970s. It required showing of at least one national film for every eight foreign films, and at least one short indigenous newsreel at every showing (UNESCO, 1975). Argentina, with twenty-eight feature releases and forty short films in 1970, also implemented measures to bolster its film industry including screen quotas for local films, import quotas for foreign films, production loans, subsidies geared to a film's attractiveness, prizes for artistic and socially relevant films, and state distribution of films to other countries (Schnitman, 1981). Probably the most radical communication reform in Latin America was launched in Peru after the 1968 military takeover by the Velasco Alvarado government. Before that year Peruvian media were mainly commercial, owned by a small number of elite families and carrying mostly imported programs and news. The reforms were designed to regain independence from the industrial nations in media facilities and programming, and to more equitably distribute media access and control. The government expropriated 25 percent equity of radio stations and 51 percent of television stations. A 1969 Press Law required the publishers to show 'respect for ... the demands for national security and defense', and the Telecommunications Law of 1971 required broadcast industries to operate in the service of Peru's socioeconomic development. Sixty percent of radio programming and 100 percent of advertising was required to be indigenous. In 1974 all major newspapers were expropriated and their ownership and management were transferred to labor and educational groups, rural organizations and professional and service associations.

These reforms were short-lived, however. Most of the media were quickly returned to previous owners, and by 1980 the new communications order was dismantled but for national control of a few key broadcast stations.

Atwood and Mattos (1984) observed that Peru's attempt at radical media reformation did bring about some important changes: for the first time a structured set of regulations was established for the management of media operations; exposure to cultural, educational and national programming during the reforms increased; and dependence on imported programming declined. The reform failed, however, in decentralizing broadcast ownership and control, transferring print media control to representatives of the lower and middle classes, and stimulating popular support for the reform itself.

Finally, the Cuban experience illustrates the Soviet use of the press and broadcast media as tools for ideological motivation. In 1961, 'The Year of Education', Cuba's new Prime Minister Fidel Castro launched a literacy campaign. Tremendously successful by official figures, the campaign led to a drop in illiteracy from 24 to 4 percent in a single year (Hedebro, 1982). This was accomplished by mobilizing the entire society with ideological exhortation under the slogan 'the people teach the people'. The media were used to encourage national commitment, but were not themselves seen as independent factors for social change.

> [The media] are used, where appropriate, to contribute to the fulfill-
> ment of development objectives that are formulated for the society as a
> whole . . . There is no communication policy for development, but
> there is a development philosophy in which communication repre-
> sents one of several parts. It is not the most important factor, but it is
> absolutely necessary. (Hedebro, 1982, p. 86).

Hedebro has suggested that centrally planned communications systems like those of Cuba, China and Tanzania, often provide for two-way vertical flows across social strata and also horizontal communication within each social class. This, combined with ideological message content, it is argued, facilitates more highly coordinated development efforts.

A bipartisan model of mass media and development

Recalling the model presented earlier (Figure 13.1) which shows various broad relationships between mass media and development, I will expand on these concepts by drawing on a variety of literatures. Scholars defending modernization theories of national and international development and those supporting radical and dependency theories often ignore the validity of the other's arguments, which results in models showing communications media as either a necessary link for the diffusion of innovations, or as a tool of repression. Although it may appear that the fundamental disagreement of modernization theorists and radical scholars is ideological, many of their ideas are in fact complementary. In Figure 13.2 I have modeled processes of urban–rural (or core–periphery) repression together with those of modernization, in a bipartisan interpretation of the underlying processes. I have expanded the highlighted portion of Figure 13.1 – the relationship between an expanding urban-based communications media and internal migration. These relationships are especially important in Latin America for two reasons. First, in much of the developing world the mass media are not available where poverty, illiteracy, geographical isolation and language barriers have blocked their diffusion. For these reasons potential growth for existing urban-based media structures in Latin America is extensive as well as intensive; by improving and expanding channels of distribution, translating programs into native languages or improving literacy, present media structures can expand into rural areas. Both market and governmental forces are behind this expansion. Second, rural-to-urban migration in Latin America is a pressing issue. The growth of cities in Latin America during the past two decades has been unprecedented, and fueled to a large extent by immigrants from rural areas.

What is the effect on the spatial economy when urban newspapers and magazines, and radio and television broadcasts reach the isolated rural communities? I argue that, particularly in Latin America where conditions in the rural and urban areas are very dissimilar, the expansion of urban-originating mass media messages into the hinterland has a variety of effects on the rural economy, the urban economy and on internal migration.

First, I should emphasize that the model I expand in Figure 13.2 is specific to the effects of *mass* media, *originating in the urban core*. Remote radio stations, rural community newspapers, or media not designed for mass consumption are not the focus of this model. I also exclude foreign-produced programming which Varis points out is sometimes appreciable especially in regards to entertainment broadcasting (Varis, 1984). The relationships between media and migration illustrated in this model are not deterministic – societies in different circumstances will not respond uniformly to similar patterns of media distribution. The net result on migration flows will depend to a large extent on the relative strengths of the various relationships depicted (and those of which I am unaware or have omitted for clarity), some of which encourage and some of which discourage rural–urban migration. Finally, this model applies to media expansion in the absence of other significant rural investments.

As Figure 13.2 shows, as distribution areas of urban media expand (e.g. through installation of another radio transfer tower, rural satellite dish for television signals, or outward expansion of newspaper circulation) they

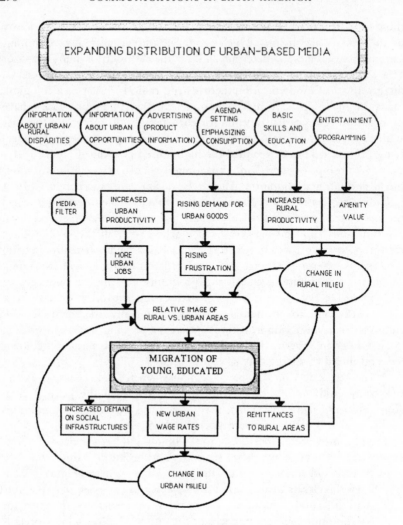

Figure 13.2   Mass media and internal migration

bring to heretofore virgin areas a mixed package. Through urban-based agenda setting (urban-value bias in media content), product advertising and basic education, commercial media can stimulate demand for consumer goods and expand market ranges. Unfortunately, rising demand for consumer goods among rural peoples outpaces their increasing productive capabilities and purchasing power. Urban-based manufacturers are better equipped to meet the new demand because of agglomeration economies, capital investments and economies of scale. This can stimulate urban production, create jobs, raise wages and increase the large and growing rural–urban disparity. Life expectancy, infant survival rates, per capita disposable income, rates of consumption and other measures of well-being are usually much higher in urban than in rural areas (Rondinelli, 1983; World Bank, 1983). Education levels are higher in Latin American cities, with urban adult illiteracy from 6 to 15 percent while the rural rate in most Latin American countries is between 30 and 55 percent (UNESCO, 1984). Wages and per capita income are equally disparate, with the urban worker receiving substantially higher wages even after the cost of living is considered (Mera, 1978). With a flow of urban messages, the media inform rural people of the disparate conditions, thereby encouraging migration to the cities. Furthermore, the media's depiction of the quality of urban life is often overstated. The press in the United States is accused of 'boosterism', suppression of unflattering messages in the interest of furthering development (R. Morris, 1981; Hochberg, 1980). Third World countries are even better known for their restricted publication of negative information in the interest of progress (Hatchen, 1981; Martin and Chaudary, 1983). The real disparity between rural and urban conditions is likely to appear even greater in the urban media. The images presented to the rural peoples indicate that city life is somehow better – better perhaps than it actually is. The net result is what Lerner calls a 'revolution of rising frustration'. When sought-after commodities are unavailable in remote areas or not affordable with rural wages, migration to the city may occur.

As mentioned earlier, the intent of Latin American media policies is most often to provide literacy, math and other basic skills to the rural peoples. The most educated are the most likely to emigrate, in search of higher wages and a better life in the cities. The more educated have also been shown to make more use of the mass media for information pertinent to their migration (although personal contacts are by far the most important source of such information for all groups (Levy and Wadycki, 1973)). When the young, skilled population moves to the urban areas, they often leave the rural areas more impoverished than ever.

> One negative effect of a successful education campaign is to draw off some of the best human resources and leave the area worse off than before. (McAnany and Mayo, 1980).

But not all effects of urban mass media are detrimental to rural areas. Radio, television, newspapers, cinema and magazines also provide diversion and entertainment to the media public. If we dismiss for the moment charges by some that new media supplant traditional and more beneficial channels of communications and erode cultural values (Hatchen, 1981; Kato, 1976; Lerner, 1976), we can consider both the education and amenity (entertainment) effects as improving the quality of rural life. An important, though indirect, countervailing force on rural out-migration is remittance from former rural–urban migrants, which in some cases can represent 10 percent of a village's income (World Bank, 1983). And finally, although the opportunities are often few, rural education can encourage efficiency and entrepreneurship there as well.

As the model shows, expansion of urban-based media into rural areas effects change in both rural and urban areas. Specific conditions will determine net effects. For instance, if investment capital or natural resources in rural areas are available, increased rural productivity may result from enhanced education, improving rural opportunities and reducing the attraction of city life. If, on the other hand, rural people respond dramatically to the lure of consumption goods and services and seek urban wages, rapid migration may result. In such a case the cliché, 'How are you going to keep them down on the farm once they've seen the city?' may be more appropriate than ever. This is not to say that media expansion or even rural-to-urban migration itself should be discouraged. Whether or not the primate cities in Latin America are too large is subject to some debate and beyond the scope of this paper. Through the schematic diagram I hope to point out various ways in which expansion or urban-based mass media might contribute to regional development and the migration phenomenon. I suggest that any combination of Figure 13.1's components can be similarly expanded for use by scholars and policy makers alike.

Conclusion

The study of communications in development has been frustrated by the complexity of both processes. Much of the work, such as that on diffusion, agenda setting, empathy, political change, etc., has concentrated on a specific function of the media without an overall framework for studying all of these influences as they affect policy goals. Furthermore, apparently conflicting theories of the media's role during the development process, and of development itself, have resulted in models emphasizing only negative or positive impacts of mass media. In addition, in any country both governmental and market forces are present; both must be recognized in a model of media effects. Figure 13.1, while admittedly simplistic, shows a range of relationships which should be incorporated in such a model. Each structural media adjustment – changes in program content, geographically extended distribution, changes in ownership or control, or introduction of new technologies – in turn, may affect social, political, economic, psychological and spatial conditions. Figure 13.1 offers a framework for study. Figure 13.2 shows how that framework can be used not only to juxtapose a variety of dynamic processes but also to combine key elements of apparently incompatible literatures. What is now needed is further investigation into the intricacies of each 'path' of change, as I have attempted to do in Figure 13.2 for expansion of distribution and rural–urban migration. Existing areas of study fit well into this framework. For example, agenda setting literature focuses on the ownership-control/political economy and individual psyche links; diffusion literature and radical scholars have covered a variety of paths, albeit with different perspectives and conclusions. Effects of new technology have also received considerable attention in recent literature, and, at the highest level, competition between market forces and governmental regulation as means of promoting development has long been debated. Case studies can be placed squarely in the framework of this model as well. Depending on the specific country or region, one portion of the model may be more important. In this example, designed for Latin America generally, expansion of media and rural–urban migration was particularly pertinent because of the rapid urban growth rates commonly found there. In other regions of the world other portions of the overall model of effects (Figure 13.1) may be more important. Finally, before a national policy could be based on this model a thorough analysis of the specific conditions there would be necessary – individual nations may differ dramatically on these issues.

Notes

1 Bullion points out that despite heightened publicity after 1970, many of the concerns surrounding the NWIO had their roots well before World War II (Bullion, 1982).

References

Atwood, Rita, and Mattos, Sergio (1984), 'Mass media reform and social change: the Peruvian experience, in Gerbner, G. (ed.), *World Communications: A Handbook* (New York and London: Longman), pp. 309–21.

Beltran, Luis Ramiro, and de Cardona, Elizabeth Fox (1979), 'Latin America and the United States: flows in the free flow of information, in Nordenstreng and Schiller (eds), op. cit., pp. 33–64.

Blakemore, Harold, and C. T. Smith (1983), *Latin America: Geographical Perspectives* (New York: Methuen).

Bullion, Stuart James (1982), 'The New World Information Order Debate: How New?', *Gazette*, 30, pp. 155–65.

Europa Publications (1987), *The Europa Yearbook 1987* (London: Europa Publications Limited).

Fuchs, Victor R. (1967), *Differentials in Hourly Earnings By Region and City Size, 1959.* (New York: Columbia University Press). National Bureau of Economic Research Occasional Paper, no. 101.

Gilling, E. (1975), 'Telecommunications and economic development: inter-county comparisons of the catalytic effect of the telephone services on development', unpublished MBA thesis, McGill University, 1975. From Nicol, op. cit., p. ii.

Gore, Charles (1984), *Regions in Question* (London: Methuen).

Gwynne, Robert N. (1985), *Industrialization and Urbanization in Latin America* (London and Sydney: Croom Helm).

Hatchen, William A. (1981), *The World News Prism* (Ames, Iowa: Iowa State University Press).

Hedebro, Goran (1982), *Communication and Social Change in Developing Nations* (Ames, Iowa: Iowa State University Press).

Hochberg, Lee (1981), 'Environmental reporting in boomtown Houston', *Columbian Journalism Review*, May/June, pp. 71–4.

Jakubowicz, Karol (1985), Third World news cooperation schemes in building a New International Communication Order: do they stand a chance?, *Gazette*, 36, pp. 81–93.

Kato, Hidetoshi (1976), 'Global instantaneousness and instant globalism – the significance of popular culture in developing countries', in Schramm and Lerner (eds), op. cit., pp. 253–8.

Klapper, J. T. (1960), *The Effects of Mass Communication* (Glencoe, Ill.: Free Press).

Larson, James F. (1984), *Television's Window of the World: International Affairs Coverage on the U.S. Networks* (Norwood, NJ: Ablex).

Lazarsfeld, P. F., Berelson, B. B., and Gaudet, H. (1948), *The People's Choice* (Chicago: University of Chicago).

Lerner, Daniel (1958), *The Passing of Traditional Society* (Glencoe, Ill.: Free Press).

——(1976), 'Technology, communication, and change', in Schramm and Lerner (eds), op. cit., pp. 287–301.

Levy, M., and Wadycki, W. (1973), 'The influence of family and friend upon geographic labour mobility: an international comparison', *Review of Economic Statistics*, 55, pp. 198–203.

MacBride, Sean, *et al.* (1980), *Many Voices, One World: Communication Today and Tomorrow* (Paris: Unesco).

McAnany, Emile G., and Mayo, John K. (1980), *Communication Media in Education For Low-Income Countries: Implications for Planning.* (Paris: Unesco International Institute for Educational Planning), Fundamentals of Educational Planning No. 29.

McCombs, Maxwell, and Shaw, Donald (1972), 'The agenda-setting function of the mass media', *Public Opinion Quarterly*, 36, pp. 176–87.

Martin, L., and Chaudhary, Anju Grover (1983), *Comparative Mass Media Systems* (New York: Longman).

Mayo, John I., Hornik, Robert, *et al.* (1976), *Education Reform With Television: The El Salvador Experience* (Stanford, Calif.: Stanford University Press).

Mera, Koichi (1978), On the urban agglomeration and economic efficiency, *Economic Development and Cultural Change*, 21, pp. 309–24.

Morris, Arthur S. (1979), *South America* (New York: Harper & Row).

——(1981), *Latin America* (London: Hutchinson).

Morris, Roger (1981), 'Buffaloed by the energy boom', *Columbia Journalism Review*, Nov./Dec., pp. 46–52.

Naesselund, G. R. (1975), 'Declaration Inaugural. In Seminario sobre Medios de Comunicacion Social para Periodistas y Productores de Radio y Television, Mexico, D.F., Mexico, 3–4 de Julio de 1975 Anexo III' (Paris: Unesco).

Nicol, Lionel Y. (1984a), 'Communications, economic development, and spatial structures: a review of research', *Working Paper No. 404* (Berkeley, Calif.: Institute of Urban & Regional Development, University of California).

——(1984b), 'Communications, economic development, and spatial structures: a theoretical framework', *Working Paper No. 405* (Berkeley, Calif.: Institute of Urban & Regional Development, University of California).

Nordenstreng, Kaarle, and Schiller, Herbert I. (1979), *National Sovereignty and International Communication* (Norwood, NJ: Ablex).

Nordenstreng, Kaarle, and Varis, T. (1974), 'Television traffic – a one-way street?', Reports and Papers on Mass Communication, No. 70 (Paris: Unesco).

Pool, Ithiel de Sola (1979), 'Direct broadcast satellites and the integrity of national cultures', in Nordenstreng and Schiller (eds), op. cit., pp. 120–53.

Rahim, Syed A. (1976), 'Communication approaches in rural development', in Schramm and Lerner (eds), op. cit., pp. 151–62.

Rondinelli, D. A. (1983), *Secondary Cities in Developing Countries: Policies for Diffusing Urbanization* (London: Sage).

Rogers, Everett M., and Antola, Livia (1985), 'Telenovas: A Latin American success story', *Journal of Communications*, 35, 4, pp. 24–35.

Saunders, Robert J., Warford, Jeremy J., and Wellenius, Bjorn (1983), *Telecommunications and Economic Development* (Baltimore and London: Johns Hopkins University Press).

Schiller, I. (1973), *The Mind Managers* (Boston: Beacon Press).

——(1973), *Communication and Cultural Domination* (White Plains: M. E. Sharpe).

——(1979), 'Transnational media and national development, in Nordenstreng and Schiller (eds), op. cit., pp. 21–32.

Schnitman, Jorge A. (1981), 'Economic protectionism and mass media development: film industry in Argentina, in McAnany, E., Schnitman, J., and Janus, N. (eds), *Communication and Social Structure* (New York: Praeger), pp. 263–86.

——(1984), International competition and indigenous film industries in Argentina and Mexico, in Gerbner, G., and Siefert, M. (eds), *World Communications: A Handbook* (New York and London: Longman), pp. 164–74.

Schramm, Wilbur (1964), *Mass Media and National Development: The Role of Information in the Developing Countries* (Stanford, Calif.: Stanford University Press).

Schramm, Wilbur, and Lerner, Daniel (eds) (1976), *Communication and Change: The Last Ten Years – And The Next* (Honolulu: The University Press of Hawaii).

UNESCO (1975), *World Communications* (Epping, Essex: Gower Press, Unipub, the Unesco Press).

——(1984), *Statistical Yearbook*.

Varis, T. (1989), 'The international flow of television programs', *Journal of Communications*, 34 (1),. pp. 143–52.

World Bank (1983), *World Development Report* (Oxford, New York: World Bank University Press).

'Worldnet opens doors to south', *Broadcasting*, 111, 16, pp. 52–4.

14 Information and urban growth in the periphery of the global village: New Zealand and the international information economy

PIP FORER & NIGEL PARROTT

The information economy

A well-documented development in Western economies during the last decade has been the changing balance between economic sectors. In traditional terms this has been cast as a growing dominance of the service over the primary and secondary sectors. However a more radical, and in some ways more functional, recasting of statistics by Porat (1977) has concentrated attention on job content, in particular whether employment involves handling information as opposed to physical products. On this basis employment is defined as either information or non-information related. In a period when employment in the information sector is growing this approach has attracted considerable attention.

Porat provided a robust methodological foundation in his work for the United States Bureau of Commerce which has subsequently been recast by others for application elsewhere. Most of these studies have been nationally based. Regional analyses using Porat's framework have been less common but some do exist (Hepworth, 1986; Hepworth *et al.*, 1987; Parrot and Forer, 1986). In addition writers such as Openshaw and Goddard (1987) have called attention to the research challenges posed by developments in the information sector, particularly in terms of regional patterns of change and locational factors that encourage its growth or spatial concentration.

The central issue of this chapter is the behavior over time of a national information sector both in a global context and at the subnational level. The particular focus is on the New Zealand urban system. In part the justification for this is a healthy geographic curiosity to understand how *national* changes articulate into *regional* adjustments. There are however broader questions of how the New Zealand system, or any national system, fits into the global pattern of information usage. With developments in long-distance communications the New Zealand information sector is increasingly linked to other systems – across the Tasman Sea, across the Pacific and in a global context. Employment in New Zealand in part reflects demand for information-processing services from Australian cities, while many high-order information services in New Zealand are supplied from Australia, Japan or the West Coast of the United States. This is a global process that has gathered pace significantly over the last two years.

Sequential or spatial restructuring?

An important question that has to be addressed is the true nature of the shift into information economies. In raw terms many observers have commented on the move towards information-based employment. Table 14.1 is a typical empirical description used to illustrate this shift. It presents the traditional view of sectoral employment, plus an estimate of employment in the information sector. Such tables are commonplace and are interpreted by many authors as showing a shift in economic activity away from the physical production of goods typical of the Primary and Secondary sectors. The consequent growth in predominantly information-based activities, even with these two sectors, has led to various labels being given to economies with high information-related employment. These include terms such as the postindustrial society and the information economy.

To explain this shift two contrasting interpretations are available. The

Table 14.1 International changes in workforce composition (in descending order of employment in the information sector).

Country		Information	Service	Industry	Primary
United States	1950	30.5	19.1	38.4	12.0
	1970	41.1	24.1	31.5	3.3
UK	1951	26.7	27.5	40.4	5.4
	1971	35.6	27.0	34.2	3.2
Sweden	1960	26.0	26.8	36.5	10.7
	1975	34.9	29.8	30.6	4.7
New Zealand	1956	25.6	21.7	36.2	16.4
	1971	33.7	20.6	34.0	11.6
W. Germany	1950	18.3	20.9	38.3	22.5
	1978	33.2	25.9	35.1	5.8
France	1954	20.3	24.1	30.9	24.7
	1975	32.1	28.1	29.9	9.9
Japan	1960	17.9	18.4	31.3	32.4
	1975	29.6	22.7	33.8	13.9
Australia	1947	22.7	28.1	31.9	17.2
	1971	29.7	30.5	30.7	9.1

Source: OECD 1981.

first seeks to explain recent and future developments in terms of a sequential shift in economic activity that is in many ways an extension of the stages of economic growth perceived by Rostow (1960). In these scenarios the information economy can be seen as a logical extension from, or adjustment to, the age of mass consumption. Toffler's 1980 work, *The Third Wave*, perhaps captures the inherently sequential view of development implied by much of this work. The basic tenet is that, given time, all societies will emerge into a postindustrial age that features high material welfare levels supported through an information-based economy: i.e there is an internal logic to economic activity which will turn many of us into information operatives. The evidence for this is taken to be the current growth of information-based employment.

A feature of this literature is a lack of concern for comparative spatial analysis. The interaction between different nations within the interntional system, or between regions and cities within countries, is largely ignored. The way in which these linkages affect relative levels of development is poorly considered.

Yet in the development literature sequential theories of economic development have so far proved to offer only partial truths even when a somewhat wider appreciation of the significance of interdependence is present. In response structural analyses have been called for. In these the interdependence of economies looms large, particularly through the international division of labor. Most writing on the international division of labor has concentrated on the development of a global manufacturing economy, often counterpointing the growth and prosperity of the Newly Industrializing Countries (NICs) with the pain of deindustrialization in many developed economies. Frobel, Heinrichs and Kreye pioneered this perspective (Frobel, Heinrichs and Kreye, 1980) but the literature is now well developed. There is however a developing international division of *information* labor, a phenomenon which has emerged through continued growth in communications technologies. Massey (1984) has suggested the separation of functions in capitalist economies which has fostered this and some work has emerged on the facilitative factors responsible (Langdale, 1985; Schiller, 1984). In general this aspect of the international division of labor has been less well researched, yet potentially has considerable implications for global patterns of development.

In the case of manufacturing, internationalization of production has had profound effects on some countries and led to a more critical analysis of economic development. In many cases the mechanism for national manufacturing change has not been an internally induced sequential development through time so much as an internationally induced spatial reallocation of activity. This reallocation has stripped manufacturing jobs from many developed economies (thereby inevitably swelling the *proportion* involved in services) and caused an enlarged industrial sector in other countries, especially those with cheap labour. However many of these latter countries retain a structural dependence on metropolitan markets. They require continued support from metropolitan capital and continued cost advantage relative to metropolitan production techniques to retain their manufacturing base. In such a global manufacturing system a key question is to what extent is development a locally sequential process and to what extent does it reflect the spatial implications of a global process?

The same question can be posed of information activities: do they improve communications and does the increasing homogeneity of the commercial culture permit their internationalization. Table 14.1 reflects the possible effects of a global concentration of information-based employment due to international specialization of labor as well as providing evidence of a universal, sequential process of economic development. Significantly, all the countries in Table 14.1 are developed countries and almost all of the prophets of the Information Society write from the perspective of large metropolitan-based economies.

Clearly, it can be argued that both the process of sequential development and the international division of labor might have the same effect on metropolitan economies. On the one hand, if the information economy relates to level of development then the wealthy economies might be expected to enter the Third Wave first. On the other hand, a burgeoning information sector could reflect processes of another kind. A concentration of control functions into core economies, which would follow from the growth of multinational enterprises (MNEs) would be part of this. Another component would be the repatriation of funds from the profits of overseas operations of such enterprises which would buoy the core economy and support additional employment which, if Engels' Law holds true, would most likely be in the information and service sectors rather than in manufacturing. Together these aspects of the functional specialization of the global economy can create local economies that have enriched information and service sectors.

These two interpretations suggest very different processes and invite very different policy responses. They suggest a need to look more closely at the information sector in national ecoomies outside of the core OECD nations and at the regional impacts of the growing information sector. To keep comparison meaningful, however, the need is for an economy with underlying similarities and strong links to the metropolitan nations. New Zealand provides such an example.

To consider developments in the information sector in New Zealand we will firstly need to discuss the methodology used by Porat to define the sector. We will then need to consider the relevant background to the New Zealand economy, which is an unusual one, involving many recent adjustments. Thirdly we will need to consider what happened to the information sector in New Zealand between 1971 and 1986. Finally we will need to return to the principal question of just what processes lie behind these adjustments.

Defining employment in the information sector

Foundation work on the contribution that the information sector makes to employment and value added derives from Porat's 1960s study of the American economy (Porat, 1977). The essence of his approach is to categorize work not by normal sectoral affiliations but by whether or not it involves the handling of information – the manipulation of incorporeal data. Porat's definitional procedures permitted reaggregations of existing statistics to yield data on the information sector and in the case of the United States showed that '46 percent of the [US] Gross National Product in 1967 was bound up with information activity'. Also '41 percent of the work force holds some sort of information job earning 53 percent of labor income'. Using Porat's approach it is also possible to suggest what proportion of employment in traditional economic subsectors is engaged in information-handling activity.

Many workers have updated and sought to improve Porat's analysis. Porat's approach has often been marginally modified, most notably in publications of the OECD which has adopted its own standard definition (OECD, 1980). In most studies figures have been only marginally affected by alternative sectoral definitions and have clearly indicated a strong growth in employment and product value in the information sector. In New Zealand Michael Conway employed a local modification in his Commission for the Future working paper that examined changes in the intercensal time period 1971–6 and reported strong growth, thus supporting the evolving stereotype of the information economy. This was passed into popular mythology in New Zealand by Scott (1984) who claimed that New Zealand was already half-way to completing the transition to being an information economy.

The general basis of all the methodologies for measuring the information sector stem from Porat's definition of information workers 'as those whose income originates primarily in the manipulation of symbols and information'. While this broad definition has been generally accepted, it has been interpreted in a variety of ways by different analysts. The problem lies in identifying which occupations are primarily engaged in the production, processing or distribution of information. Within certain occupations this poses no problem; for example Porat compares a computer programmer with a carpenter:

> Both are skilled workers earning roughly the same salary. Both require a certain amount of education before they can function productively, and both use attention, concentration and applied knowledge in their respective tasks. However, the programmer's livelihood originates with the provision of an information service (a set of instructions to a computer), while the carpenter's livelihood originates with the construction of a building or piece of furniture – non

information goods. The former sells information as a commodity; the latter sells a tangible physical product. (Porat, 1977, Vol. 1, p. 27)

The distinction between these occupations is clear-cut, but with certain occupations the distinction is not so readily apparent. In the case of a factory foreman, his or her time could be shared equally between manipulating information and involvement in the manufacturing process. Ideally, a time budget should be calculated over the course of a year to assess the percentage of time spent manipulating information. However, such data do not exist, except in a few cases, so commentators on occupational trends have made less precise judgments on how these workers should be assigned to the information sector. These judgments have depended on variations in personal interpretation as well as the employment categories used in different countries and at different time.. For this reason no overall agreement exists between writers on which group of occupations constitutes the definitive information sector.

Despite this, if we look at the work of four writers (Toffler, Stonier, Porat and Conway) a number of occupations are common to each of their definitions of the information sector. These core occupations are found in Stonier's definition which identifies five types of information operative (Stonier, 1983, p. 43) and to some degree are mirrored by the four classes employed by the OECD definition of the sector:

1   Organization operatives – key managers, whose role it is to coordinate the various facets of modern production systems (land, labor, capital, machinery, energy and material inputs coupled to equally complex transport, communication and distribution systems).
2   Transmission of information operatives – secretaries, telephone operators, postal workers, journalists and others working in the mass media, educators of all sorts, technical salesmen.
3   Operatives in information storage and retrieval – filing clerks, librarians, computer programmers.
4   Information producers. A relatively small number of people are involved in producing new information, or patterns of information – scientists, artists, statisticians, architects and designers.
5   Information processors and analysts. These operatives apply information in order to solve specific problems – lawyers, doctors, counsellors and accountants.

All definitions of the sectors also include a number of occupations that are hard to classify and it is here divergence between authors occurs. While a good case exists for establishing an international standard of classification for information occupations, none yet exists and for our purposes we shall use three definitions of the information sector. We shall use the OECD definition (OECD, 1980) to allow international comparison. When looking at New Zealand trends we shall use Conway's definition (Conway, 1981) so as to enable comparison with his findings. With respect to an exploratory look at trends in five urban centers between 1971 and 1981 we shall use a variant on Conway's definition. This reflects the fact that published data on occupational trends in urban areas are aggregated in the census such that Conway's definition cannot be reproduced exactly.

The New Zealand context

As a member of the OECD New Zealand is generally recognized as one of the more developed of the world's trading economies. However, it also displays unique traits which need to be understood if we are to appreciate fully the patterns of change in its information sector.

The New Zealand economy is frequently typified as small, fragile and distant. It is perched apparently on the most distant rim of the Pacific (Fig. 14.1) and frequently overshadowed by neighboring Australia (from a distance viewed as culturally similar to which it is increasingly linked by functional ties). New Zealand's 3.2 million total population is equivalent only to a modest city in most OECD countries, but is highly urbanized. Until the 1970s New Zealand was tied closely to the economy of the United Kingdom with an economy based on guaranteed market access for agricultural products. It exhibited a highly regulated economy and a local manufacturing sector that enjoyed extremely high levels of protection (Franklin, 1978; Gould, 1982). While Latin America drew ideas and innovations from other countries, particularly the United States (Grey, 1984), its outlook was polarized on its past colonial ties with Britain.

In the context of such an economy New Zealand's tertiary sector was sizable and among other things performed a variety of government administrative roles including the support of an elaborate regulatory environment for trade and finance. The small size of the New Zealand economy meant limited sources of large scale domestic capital emerged and this encouraged considerable government involvement in all sectors of the economy, especially those requiring large capital works. This influence extended into financial management. Extremely rigid exchange controls persisted throughout most of the 1970s resulting in the maintenance of a quite small and very introspective financial employment sector. Other outcomes were a relatively low level of foreign ownership and a low participation of New Zealand capital in overseas enterprises.

New Zealand's urban and regional structure reflected this polarized trade and protected manufacturing. The service of a hinterland was a dominant function of all the metropolitan centers, in Wellington's case augmented by its role as capital and in Auckland's as the (largely domestic) commercial center, a hierarchy which still reflected the legacy of provincial government. While differences in size of the main metropolitan centers existed,

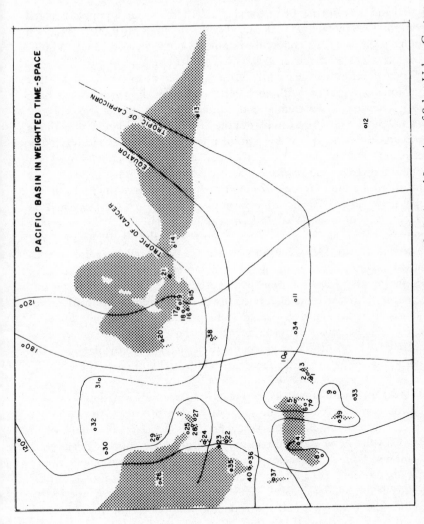

Figure 14.1 New Zealand: Location within the Pacific Basin and Location of Select Urban Centers

there remained a relatively even level of development and appreciable commercial influence outside of Auckland and Wellington as a reflection of the early investment dominance of the South Island (Johnston, 1968).

A combination of the oil shock of 1974 and the loss of markets consequent upon Britain's accession to the European Community began a process which has transformed that economy and the role that international capital and the information sector play in it. This change is popularly dated from the 1976 budget of the National government of Sir Robert Muldoon but has been accelerated under the nominally more socialist Labour administration which took power in 1984.

These developments have involved a far closer, more direct and broader functional integration with the world economy. In part this has been precipitated by recent trading and political realities, but it also reflects a gradual process in which this part of the Pacific rim has integrated with the growing Pacific economy. Even in the mid 1970s New Zealand was not as peripheral as it seemed. An analysis of air transport at that time, for instance, revealed the degree to which Australasia's isolation was functionally misleading. Figure 14.2 indicates the relative accessibility by air for forty nodes in the Pacific using a simple index of accessibility based on travel times (Shimbel, 1953). In terms of the needs of personal contacts for international businessmen the South West Pacific already enjoyed favored treatment. Figure 14.3 shows this even more graphically and uses a well-known technique of mapping time distances to illustrate the relative positioning of the nodes in time space (Forer, 1978). These figures represent the result of three decades of airline growth that altered spatial relationships and made wider functional integration possible. Significantly for later arguments, the effect of the influence of the Australasian centers, particularly Sydney, in distorting the 'natural' access pattern highlights the emergence and nature of the international hierarchy of cities in the Pacific region.

This relative ease of contact, and later the ease of information flows, has reciprocal links to trade and the adoption of new practices. An illustration of this is the seed of innovation of new products from overseas. In the 1960s and 1970s the New Zealand consumer culture was, quite literally, years behind developments in major markets. Models of new equipment would typically become available with a lag time of twelve to twenty-four months from overseas release. This delay was reflected not just in the products themselves but in the surrounding culture of trade magazines and consumer expectations. By the mid 1970s new innovations were arriving much more quickly. By the 1980s some new products arrived within days of global release. An example which can be widely appreciated is that of Apple microcomputers: the Apple II took almost two years to be marketed in New Zealand after release, the Apple Macintosh eighteen days. In general the shrinking adoption lag times have had particular significance in terms of information technology with delays moving from years to weeks.

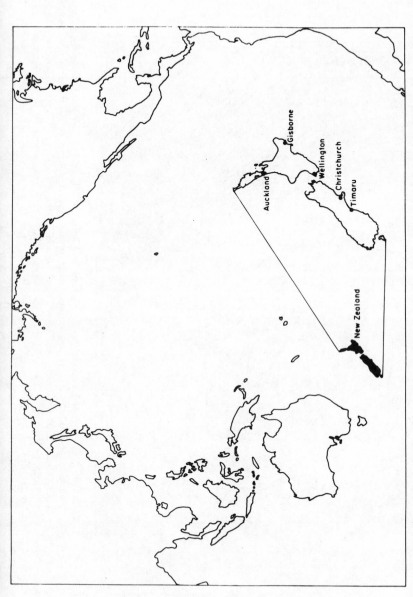

Figure 14.2 Airline access within the Pacific Basin 1975. The isolines are drawn for accessibility amongst forty nodes in terms of flight times between nodes inversely weighted by frequency of flights. The value of 100 represents average accessibility amongst the nodes. Individual node values are displayed by the node's location.

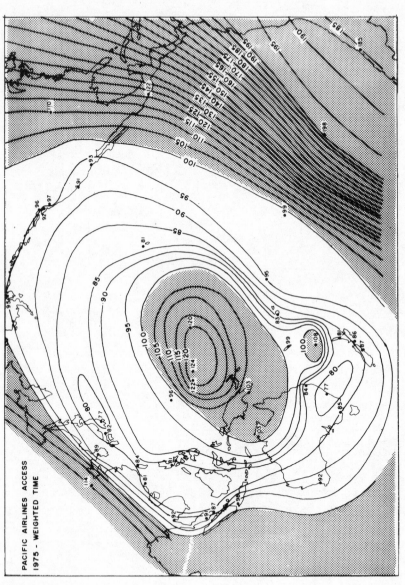

Figure 14.3 The Pacific Basin mapped in weighted time-space. The points in Figure 14.2 have been mapped using their separation in time-space, and sample lines of latitude and longitude have been superimposed to show distortions resulting from differential access to the system. Numbers by the nodes indicate the city code number. Nodes are: 1 Christchurch, 2 Wellington, 3 Auckland, 4 Perth, 5 Sydney, 6 Melbourne, 7 Brisbane, 8 Darwin, 9 Nandi, 10 Noumea, 11 Papeete, 12 Easter Island, 13 Santiago, 14 Acapulco, 15 Los Angeles, 16 San Francisco, 17 Portland, 18 Seattle, 19 Vancouver, 20 Anchorage, 21 Chicago, 22 Manila, 23 Hong Kong, 24 Taipei, 25 Seoul, 26 Beijing, 27 Tokyo, 28 Osaka, 29 Guam, 30 Truk, 31 Majuro, 32 Ponape, 33 Norfolk Island, 34 Pago Pago, 35 Bangkok, 36 Singapore, 37 Djakarta, 38 Honolulu, 39 Port Moresby, 40

New Zealand has embraced the computer culture very heavily and even though its communications infrastructure still lags behind overseas developments the faster diffusion of products has allowed the widespread acceptance of microcomputing into its business sector, which has one of the highest penetration rates of information technology in the world. Such integration has also been a precondition for the emergence of a New Zealand software industry serving world markets. The LINC development center in Christchurch, which created one of the major commercial computing environments adopted by the multinational Unisys corporation, is a case in point.

This growing functional proximity reflects the fact that from 1976 onward successive New Zealand governments have begun more positively to address the need to dismantle the protectionist barriers surrounding the economy and to integrate with the world trading system. They have also begun the process of disposing of state assets and privatizing and deregulating many traditional government activities. By 1987 this process began to extend to postal and communications services. All of this has been combined with a dramatic deregulation of trade and finance. In particular restrictions on the international flow of capital, which were quite draconian, were abolished in a matter of months in 1984–5. This presented significant room for growth in the financial sector. New Zealand's position as the first financial trading market to open on any business day, along with growing movements of some functions and capital from Hong Kong, offered appreciable opportunities in some small but strategic areas of information handling.

Many aspects of this pattern are not unfamiliar. The underlying move towards a lower state involvement and reliance on the free market bears comparison with many other OECD members' economic responses, especially those of Great Britain. However the context differs dramatically: a small, protected economy still largely driven by agricultural trade poses different problems than declining and traditionally manufacturing based economies.

All of these changes have thrown up major research questions for geographers, an agenda for which has been proposed by Britton and Le Heron (1987). The key implications for the information sector since 1971 have been clear. Government restrictions on new jobs in the state bureaucracy in the late 1970s – especially a 'sinking lid' policy in force from 1976 onwards – caused significant contraction in one area of the information economy, and one with a distinct regional bias. At the same time the private sector has moved to provide employment in areas of new demand caused by the same process.

The information economy in New Zealand 1971–81

In terms of sequential versus structural explanations of development the New Zealand economy developed slowly in wealth terms between 1971 and 1986, but experienced significant changes in its integration with the world economy. It is in the light of these changes that the information sector deserves close attention. Census data for 1971 and 1976 have already been used to show that New Zealand had rapid growth in its information sector between those years, from which it was argued that more change was to come as an inevitable transformation of the economy occurred (Scott, 1984). However the record of 1976–81 can be placed alongside these earlier data to confirm whether this 'irreversible' trend did continue. The following paragraphs outline an analysis of the complete period from 1971 to 1981. The quite startling conclusion is that from 1976 to 1981 the information sector *contracted* as a proportion of total employment. The regional distribution and the nature of employment within the sector also underwent some modification.

Two definitions of the information economy are used to examine these trends in New Zealand. Conway's definition is one which is effectively based on occupational definitions using seven classes. The slightly more conservative OECD definition is also used, which divides occupations into four main groups according to their role in the information economy. These are: information producers, information processors, information distributors and information infrastructure occupations. Their use permits comparison with Hepworths' *et al.* study of the United Kingdom in 1981. Tables 14.2 and 14.3 summarize trends using these definitions.

From the table it can be seen that the *total* number of people employed in the information sector in New Zealand grew between 1971 and 1976 and

Table 14.2 Changes in New Zealand information sector employment using Conway's definition of the information sector (only subsector totals plus major categories or changes in each group are shown).

Occupation	Numbers employed			% increase	
	1971	1976	1981	71–76	76–81
Professional Technical	94,158	121,506	125,673	29.0	3.4
Administrative, Managerial	28,403	40,908	45,993	44.0	12.4
Clerical	171,112	196,709	203,091	15.0	3.2
Sales	40,111	40,429	39,711	0.8	−1.7
Service	4,146	4,530	4,770	9.3	5.2
Agriculture	932	1,212	1,395	31.3	15.1
Production	33,995	41,910	42,105	23.3	0.4
Total Information Sector	372,848	447,214	462,738	19.9	3.4
Total Classified Workforce	—	—	—	12.7	6.8
% in Information Sector	33.7	35.9	34.7	—	—

Source: New Zealand Census of Population and Dwellings 1971, 1976 and 1981.

Table 14.3   Changes in New Zealand information sector employment using the
OECD definition of the NZ information sector.

Occupation	Numbers employed			% increase	
	1971	1976	1981	71–76	76–81
Information Producer	50,569	62,324	64,497	23.2	3.4
Information Processor	227,431	270,083	277,523	18.8	2.2
Information Distributor	51,758	65,286	70,632	26.1	8.2
Information Infrastructure	40,404	48,826	43,536	6.0	1.7
Information Sector Total	370,162	440,519	456,188	19.0	3.6
% of total workforce	33.4	35.3	34.2	—	—

Source: as per Table 14.2.

between 1976 and 1981. If we take Conway's definition first, then between
1971 and 1976 the information sector grew at a rate of 19.9 percent, while
the total workforce expanded by only 12.7 percent. This fits with popular
myths and supports ideas of a burgeoning information economy.
However, between 1976 and 1981 growth was more modest at 3.47
percent, which was *below* the rate of growth for the total workforce (6.89
percent). That is to say that the proportion employed in the information
sector actually contracted. In 1971 the information sector accounted for 34
percent of the total workforce, this figure rising to 36.5 percent in 1976 and
dropping back to 35.5 percent in 1981. Using the OECD definition for the
sector a similar picture emerges with a rapid increase in numbers between
1971 and 1976 and more modest growth and proportional shrinkage to
34.2 percent of the workforce during the second period. This figure should
be compared with the figure of 45.2 percent for the United Kingdom in
1981, an economy that was only two percentage points higher in this sector
in 1971 (see Table 14.1). In general the sector has not developed in New
Zealand to the same extent as in some other trading partners.

The reasons for this slowing in the rate of growth of employment in the
information sector are not clear, though they could well be tied to the slow
rate of economic growth between 1976 and 1981 and to increased productivity in the information sector. Further clues on what is happening in the
sector at large may be given by considering the subsectors. Taking
Conway's classification of occupations first, Table 14.4 summarizes the
percentage share of the information sector and growth between 1971 and
1976, and 1976 and 1981 for each of these groups. This shows that two
groups (professional plus technical and clerical dominate while service and
agriculture groups make only a small contribution to growth. Between
1971 and 1981, two groups, professional plus technical and administrative
plus managerial have increased their share of the information sector by
several percent, while clerical and sales groups have declined by a similar
amount. The remainder have been fairly constant.

If we look at the percentage increases of these groups over the two

Table 14.4   Percentage share of the information sector (Conway's definition).

Occupational group	% share of information sector			% increase	
	1971	1976	1981	71–76	76–81
Professional, technical	25.2	27.1	27.1	29.0	3.4
Administration, managerial	7.6	9.1	9.9	44.0	12.4
Clerical	45.9	43.9	43.8	15.0	3.2
Sales	10.7	9.0	8.5	0.8	−1.7
Service	1.1	1.0	1.0	9.3	5.2
Agriculture	0.2	0.2	0.3	31.3	15.1
Production	9.1	9.3	9.0	23.3	0.4
Overall	—	—	—	19.0	3.4

Source: as per Table 14.2.

five-year periods, several points emerge. Firstly the national trend of substantial increases between 1971 and 1976 and much smaller increases during the second period is followed, with the exception of sales and service groups. Secondly two groups, administrative plus managerial and agriculture, experienced sustained growth over the two periods. However, it should be pointed out that the number of people employed in these occupations is relatively small – about 10 percent of the information sector. Also of note is the number of occupational groups that shed workers between 1976–1981. Significant among these are (1) physical scientists, (2) architects, planners and engineers, (3) surveyors, draftpersons etc., (4) telecommunication technicians, (5) stenographers and typists and (6) production supervisors. In each case, with the exception of telecommunication technicians, their numbers are still greater than 1971 figures. These changes may reflect the shifts in the economy which saw a move into more intensive agriculture and a downturn in construction and linked developments.

Clearly the information sector is not expanding as a uniform whole. There is significant growth in some areas, more modest growth elsewhere and decline in a number of areas, some surprising. A similar picture emerges if we look at the OECD definition. Table 14.5 summarizes for each of these OECD functional groups, their percentage share of the information sector and their percentage growth rates between 1971–6 and 1976–1981. From this we can see that information processors dominate the sector with approximately a 60 percent share. Between 1971 and 1981 the situation was relatively stable: information distributors and producers only slightly increased their share of the information sector at the expense of information processors and information infrastructure occupations. As with Conway's definition, the percentage increases for the groups over the two five-year periods conform to the national trend. However, during the first period information distributor and producer categories grew most rapidly with 26.1 percent and 23.2 percent increases exceeding the information sector growth rate or 19.9 percent. In the second period, growth was more modest with only information distributors expanding significantly at 8.2 percent.

The rate of growth for information processes and information infrastructure occupations was particularly low at 1 percent. One suggestion that emerges is that the links between information producers and distributors and information processors may warrant further investigation to establish whether a lead–lag relationship exists between the subsectors or whether they function largely autonomously. Once again comparison with Hepworth et al.'s findings is valuable, even though they do not provide data on change over time. The final column of Table 14.5 shows that both countries' information sectors are dominated in employment terms by information processors, fundamentally clerks and sales. Both have roughly 60 percent of their information employment in this category. However, the UK has a much larger representation within the (generally higher-order and more strategic) Information Producers (23 to 14 percent) and substantially less in the Distributor and Infrastructure categories. This may be some evidence of comparative specialization.

Urban trends

If there are structural changes within the information sector are there also spatial shifts in where the information sector is expanding? To answer this we have examined information sector changes in five selected New Zealand urban areas: Auckland, Wellington, Christchurch, Gisborne and Timaru. Auckland is the main commercial center with the largest share of corporate headquarters and a workforce over one-third of a million. Wellington, the political capital, is a city with the next-largest share of corporate headquarters and a major share of government employees. It is functionally tied to Lower Hutt, a separate accounting unit with a more

Table 14.5 Percentage share of the information sector (OECD definition).

Occupational group*	% share of information sector			% increase	
	1971	1976	1981	71–76	76–81
Information Producer (23.5)	13.6	14.1	14.1	23.2	3.4
Information Processor (61.2)	61.4	61.3	60.8	18.7	2.7
Information Distributor (9.3)	13.9	14.8	15.4	26.1	8.2
Information Infrastructure (6.0)	10.9	9.7	9.5	6.0	1.5
Overall	—	—	—	19.0	3.5

*Percentage share figures for UK 1981 in parentheses.
Source: as per Table 14.2 and Hepworth, Green and Gillespie (1987).

provincial and manufacturing economic base, to form a settlement of around 300,000 inhabitants. Christchurch is of a similar size to Greater Wellington, but reliant on a regional role as the major center of South Island. The other two centers are small provincial cities of around 20,000 to 30,000 people, both dependent on agriculture and provincial activity and both having suffered a relatively high degree of economic depression.

As noted earlier, working with subnational figures requires using a somewhat wider and less precise variation of Conway's definition of the information sector. In 1981 this definition adds an extra 10,434 workers to the national figures for the information sector, an acceptable revision of about 3 percent. Looking first at the percentage share of the national information sector found in these five urban areas, we note that Auckland accounts for approximately 29 percent of the national information sector and between 1971–81, increased its share of the sector by several percent. Wellington and Christchurch have a 9 percent share each and their share has declined slightly. The two provincial centers account for fractionally under 1 percent of the national total and their national share remained fairly constant (Table 14.6).

If we look at Table 14.7, which shows the information sector as a percentage of all employment in the area, we see that the information sector dominates employment in Wellington taking nearly a 60 percent share in 1981. This is on a par with the concentration in London, which reaches 58 percent (Hepworth et al., 1987) but proportionately to the national average is a much greater concentration. In other urban centers the share mirrors more closely the national average. In each center except Gisborne the information sector's percentage share of employment in the area increased over time, with the largest rises, 3 and 2 percent respectively, going to Auckland and Wellington. Only Auckland has a growth rate for the proportion of information employees that is higher than the national average for the two periods 1971–6 and 1976–81 (Table 14.6). During the first period growth in the other four urban centers was lower than the national average. However, during the second Wellington actually declined by 2.4 percent although the national rate of growth was exceeded

Table 14.6 Percentage share of the national information economy by selected urban area.

Urban area	1971	1976	1981	% increase 71–76	% increase 76–81
Auckland	26.8	28.1	28.9	26.5	6.7
Wellington	10.2	9.1	8.6	8.4	−2.4
Christchurch	10.7	9.8	9.9	10.0	4.5
Gisborne	0.9	0.8	0.8	10.7	5.2
Timaru	0.9	0.8	0.8	7.3	3.7

Source: as per Table 14.2.

Table 14.7 Information economy as a percentage of all employment by selected urban area.

Urban area	1971	1976	1981
New Zealand	34.0	36.5	35.5
Auckland	37.7	40.3	40.7
Wellington	56.0	58.6	58.6
Christchurch	37.1	36.7	38.8
Gisborne	33.2	34.4	32.4
Timaru	33.0	34.0	34.6

Source: as per Table 14.2.

in both Christchurch and Gisborne. Clearly growth in the information sector has been concentrated in Auckland. This is borne out by the fact that between 1971–6 the sector nationwide expanded by 20.37 percent with 35 percent of this expansion, 28,850 new jobs, located in Auckland. Similarly between 1976–81 the sector as a whole grew by 4.02 percent with 47 percent of this growth, 8,659 jobs, in Auckland. Growth in the other four urban centers has been of a much smaller order.

Having established the overall trend for the information sector in these urban areas over the ten-year period, it is interesting to ponder if they have shown different sectoral growth patterns. Initial analysis reveals considerable variation between centers. For example, in the case of the managerial group during the first period, this grew at a rate of 74.29 percent in Auckland, 50.53 percent in Gisborne, 20.35 percent in Wellington, while it declined by 10.63 percent in Christchurch and 17.78 percent in Timaru. During the second period, a number of occupational groups increased their numbers exclusively in the three larger centers of Auckland, Christchurch and Wellington. These include: (1) economists, (2) statisticians, (3) mail clerks, (4) authors plus journalists, (5) life scientists, (6) composers plus performers. In two occupational groups, sales managers and clerical supervisors, the reverse occurred with these centers seeing disproportionately lower growth. The general tenor of the changes is a numerical and functional concentration of the information sector into the major, non-government urban node.

Effects, causes and emerging processes

The facts to date show a reordering and relative contraction of the information sector in New Zealand. The reasons behind this change are far from clear however. Certainly some doubts are cast upon the suggestion that growing information sector employment is a guaranteed outcome of advanced economies. Several possible reasons for the stagnation can be advanced. One is that the information sector may have been economically buoyant in the whole period but became relatively less labor-intensive as time progressed. Another is that gross numbers in the information sector reflect relative levels of affluence, with affluent economies being able to sustain higher levels of information sector services. Certainly in the period in question the New Zealand economy lost ground compared with many of its traditional partners and its terms of trade were noticeably eroded. In a more localized perspective some links can be suggested between developments in New Zealand and Australia. During most of the period 1971–86 the Australian economy demonstrated more growth and offered an enhanced material lifestyle to some New Zealanders. The result has been a strong flow of long-term migrants across the Tasman Sea. At some points these flows have been noticeably linked to local conditions, for instance in terms of architects and draftspersons migrating in the face of reduced building activity. In the 1980s computer industry specialists were widely reported to be beating a similar path.

Related to these specific issues must be the relationship between economies, both in the system of the South West Pacific and in a global sense. With a wider participation in international capital investment the New Zealand information sector has become part of an international division of labor. At present this is most marked in the financial and international currency exchange areas, but there is a growing emphasis elsewhere. Trans-Tasman transaction processing of financial records is practiced by several companies in one or the other direction, while one scheme exists to provide a global archive repository for corporate computer records in what is seen as a highly stable and secure environment.

Related to this is the degree to which the urban centers have become part of a global hierarchy of cities (Ettlinger and Archer, 1987). Within Australia Sydney has strengthened its position as a dominant center, followed by Melbourne and Brisbane, the latter rampantly growing within the free market financial structure of Queensland. Sydney has become the provider of many higher-order service functions to New Zealand, while some observers see Auckland as in turn challenging the second level Australian centers as an Australasian service provider. For various reasons Auckland has shown some attractions to, and growth from, the international financial sector. In particular it has grown in attractiveness as a politically stable financial base for operations in the Western Pacific.

At present it is difficult to unravel these different factors. The most recent census (1986) has not yet been published at sufficient disaggregation to allow time for further analysis. What we would expect to find is that the shift from public to private employment in the information sector has continued. The period to 1986 was one where the economy continued to be deregulated and where New Zealand continued to have difficult trading conditions. The financial liberalization of the early 1980s spurred dramatic growth in financial activity, most marked in the Auckland and Wellington landscapes. The New Zealand currency was floated in this period and foreign exchange dealings became liberalized. Consequently sectors associated with financial management and foreign exchange enjoyed a boom, one aided and abetted by the bull stock market in full flood until October 1987. At the same time the manufacturing sector experienced a brief-lived growth in employment based on a low-valued currency, which also encouraged a boom in tourism, an industry in which information-related jobs score highly. However, the removal of agricultural subsidies as part of the same liberalization meant a decline in the agricultural sector and related employment. These factors between them may well have proportionately swollen the information sector. The 1986 data should provide a clearer understanding of the degree of these changes.

They have themselves however been overtaken by events since March 1986 (census night), and especially since October 1987. This period has seen even larger impacts on state bureaucracies due to privatization measures. These have been quite significant, involving the loss of jobs of thousands of state employees in the cities, regional centers and small towns. Some of these jobs and functions have been picked up by the private sector, but others have been simply shed. These changes have not been solely centered on information jobs, but they have been significant casualties and some changes have had strong impacts on certain communities' information employment. Of these losses the most publicized ones have been those caused by the rationalization of the network of post offices, which had disproportionately severe effects on small townships.

The global stock market adjustments of October 1987 have had far more profound effects on the urban financial sector in New Zealand. The New Zealand stock exchange suffered among the highest value losses of any exchange, and this was most marked in corporations with a high proportion of management and financial functions, in which the exchange was rich. Staff shedding and corporate regrouping can be imagined to have had most significant effects in Auckland, although there have been difficult adjustments in Christchurch and Wellington. They have been compounded by difficult conditions in manufacturing, which has shed jobs as its levels of protection have waned.

Given the complex interactions between job-shedding by an automating and contracting manufacturing sector, an inarguably restructuring international economy and the availability of new technologies we cannot hope to find a simple answer to the behavior of the information sector. Part of the answer may lie in the measure of the sector used here (jobs): other measures (such as information handled or value added) might yield different results. Nonetheless the apparent trends revealed profoundly question the concept of the information society as a universal stage of development, query its employment potential and raise the question whether it may not be a state which is functionally and spatially far more limited than some writers would acknowledge. It is true that when the figures for the latter 1980s become available they may show that the contraction in the sector from 1976 to 1981 was a brief aberration. They may show a growth in the sector which supports the idea of the information economy as a stage of development. More likely they will show a sector which is proportionately smaller, or just as significantly, restructured towards the low-level service end of the information activity spectrum. Some evidence exists for this in the comparisons between the situation in New Zealand and the United Kingdom in 1981. To get a clearer picture there is a growing need for international comparison. Sadly, there is also a growing disparity between the pace of change and the speed at which data become available to sustain comparison.

References

Britton, S., and LeHeron, R.' (1987), 'Regions and restructuring in New Zealand: issues and questions in the 1980s', *New Zealand Geographer*, 43, 3, pp. 129–39.

Conway, M. (1981), 'Information occupations: the new dominant in the New Zealand workforce', in *Commission for the Future* Network New Zealand Working Papers (Wellington: Government Printers).

Ettlinger, N., and Archer, J. C. (1987), 'City-size distributions and the world urban system in the twentieth century', *Environment and Planning A*, 19, pp. 1161–74.

Forer, P. C. (1978), 'Time–space and area in the City of the Plains', in Carlstein, T., Parkes, D., and Thrift, N., *Timing Space and Spacing Time*, vol. 2 (London: Edward Arnold, 1978), pp. 99–118.

Franklin, S. H. (1978), *Trade, Growth and Anxiety: New Zealand Beyond the Welfare State* (Wellington: Methuen).

Frobel, F., Heinrichs, J., and Kreye, O. (1980), *The New International Division of Labour* (Cambridge: Cambridge University Press).

Grey, A. H. (1984), 'North American influences on New Zealand's landscapes', *New Zealand Geographer*, pp. 66–77.

Gould, J. (1982), *The Rake's Progress: The New Zealand Economy Since 1945* (London: Hodder & Stoughton).

Hepworth, M. (1986), 'The geography of technological change in the information economy', *Regional Studies*, 20, pp. 407–24.

Hepworth, M., Green, A., and Gillespie, A. (1987), 'The spatial division of information labour in Great Britain', *Environment and Planning A*, 19, pp. 793–807.

Johnston, R. J. (1968), 'Commercial leadership as an urban function: some international comparisons', *Proceedings of the Fifth New Zealand Geographical Society Conference* (Christchurch: New Zealand Geographical Society), pp. 153–7.

Langdale, J. (1985), 'Electronic funds transfer and the internationalisation of the banking and finance industry', *Geoforum*, 16, pp. 1–13.

Markusen, A. R. (guest editor) (1983), 'Silicon landscapes: high technology and job creation', *Built Environment*, 9, p. 1.

Massey, D. (1984), *Spatial Divisions of Labour* (London: Macmillan).

OECD (1980), *OECD Publication DSTI/ICCP/80.10 (first revision)* (Brussels: OECD).

Openshaw, S., and Goddard, J. (1985), 'Some implications of the commodification of information and the emerging world information economy for applied geographical analysis in the United Kingdom', *Environment and Planning A*, 19, pp. 1161–74.

Parrott, N., and Forer, P. (1986), 'The information sector in New Zealand 1971–81', *New Zealand Geographer*, 42, 1, pp. 25–30.

Porat, M. V. (1977), *The Information Economy: Definition and Measurements* (Washington, DC; US Department of Commerce), 9 vols.

Rostow, W. W. (1960), *The Stages of Economic Growth* (Cambridge: Cambridge University Press).

Schiller, H. (1984), *Information and the Crisis Economy* (Norwood, NJ: Ablex).

Scott, N. (1984), 'The information society is here', *New Zealand Listener*, 11 August 1984, pp. 60–2.

Shimbell, A. (1953), Structural parameters of communication networks', *Bulletin of Mathematical Biophysics*, 15, pp. 501–7.

Stonier, T. (1983), *The wealth of information: a profile of the post-industrial economy* (London: Thames–Methuen).

Toffler, A. (1980), *The Third Wave* (London: Collins).

PART IV

Political, social and cultural aspects
of communications expansion

15 *Fishing in muddy waters: communications media, homeworking and the electronic cottage*

JOHN R. GOLD

> One may not doubt that, somehow, good
> Shall come of water and of mud;
> And, sure the reverent eye must see
> A purpose in liquidity.
>
> Rupert Brooke, 'Heaven'

Anyone familiar with recent history of futures research will readily appreciate the traumas that the business of forecasting has suffered since the heady days of the 1960s. The OPEC oil embargo and ensuing recession swept away the confident expectations born of the brief postwar era of, seemingly, never-ending economic growth. Many assumptions about economic, social and political trends were scrutinized and found to be wanting; many cherished icons of social progress needed to be discarded.

The subsequent postmortems revealed many deficiencies in forecasting strategy (Freeman and Jahoda, 1978; Linstone and Simmonds, 1977), including, *inter alia*, premature preoccupation with methodology at the expense of conceptual understanding and inadequate regard for the true complexity of the relationship between technology and society. With regard to the latter, forecasters had basically emphasized the direct impact of technology on society. The subtleties of actual social response to technology and the extent to which that technology would bring real social change somehow never fully entered the analysis. On the surface, this critique produced a lasting reappraisal of previous beliefs about redemption through technology (e.g. see Goudzwaard, 1984; Lyon, 1986; Norman, 1981), yet it is plain that the warning has been only partly heeded.

It is a fundamental premise of this paper that the belief in social transformation through technological advance is still very much alive. Indeed its most prevalent contemporary expression comes from the work of those who maintain that developments in electronic communications media, combined with advances in computing, will play a central role in remolding the economic, political, cultural and spatial patterning of society. Writers such as James Lee Smith (1972), James Martin (1978), and Peter Laurie (1980) put forward a view of a postindustrial society transformed by an 'information revolution'. They portrayed an optimistic picture of electronically interconnected or 'wired' societies, in which technology opened up possibilities for economic regeneration, for greater leisure time due to decentralized working patterns, for cultural enrichment from ubiquitous, multichanneled cable networks, for increased choice through electronic banking and shopping, and even for enhancement of democracy by instant electronic referenda.

Availability of space does not permit a full explanation of why this blissful state of affairs has so far failed to arrive; why the so-called 'information revolution' may have transformed the workings of the military, political and economic elites of society but has left the lives of the masses largely untouched, save for word-processing in the office and video-recorders in the home and classroom (Traber, 1986; see also Webster and Robins, 1986). Rather, the aim here is to take one specific version of the idea of the 'wired nation', to analyze its current empirical accuracy and to examine in greater detail the imagery that underpinned it.

The particular version that will be considered here comes from the writings of the American social commentator and futurist, Alvin Toffler. In his book, *The Third Wave*, Toffler (1980, p. 204) noted with characteristic assertiveness that: 'Hidden inside our advance to a new production system is a potential for social change so breathtaking in scope that few among us have been willing to face its meaning'. The scenario that could result was termed 'the electronic cottage'; in which there could be 'a return to cottage industry on a new, higher, electronic basis, and with it a new emphasis on the home as the center of society'.

The essay that follows has four main sections. The first introduces Toffler's exposition of 'electronic cottage' society. The second briefly considers empirical evidence from Great Britain on the extent, in practice very limited, to which homeworking of the sort described by Toffler and others has come about. The third section turns its attention to the conceptual failings of the electronic cottage scenario. It argues that this scenario, as a representative example of what will be termed 'technological utopias', effectively perpetuates the weaknesses of that genre with respect to its treatment of the society-technology relationship. The final section surveys the foregoing material and argues, despite its empirical and conceptual shortcomings, that the underlying principles behind the electronic cottage scenario are likely to persist unless steps are taken to address the relationship of society and technology more effectively.

The 'electronic cottage'

There is, of course, nothing particularly new in saying that cheap and efficient information transfer through electronic communications media has important implications for social and spatial organization. The age of electronic communications began in the nineteenth century, ushered in by Morse's invention of the electro-telegraph (1844), Bell's invention of the electromagnetic telephone (1876) and Marconi's experiments in wireless telegraphy (1895–9). By the end of that century, transatlantic telegraph cables were already in place, a telephone service existed between London, Paris and other European capitals, and wireless telegraphy linked Britain and continental Europe. Indeed the idea that such media would dramatically reduce the frictions of distance was already being articulated by the turn of the twentieth century by writers as diverse as H. G. Wells, Marcel Proust, Sir Halford Mackinder, and Filippo Marinetti (Kern, 1983, pp. 211–40).

That argument was to be further embellished as the century wore on, but perhaps reached its acme in the 1960s with the work of Marshall McLuhan (1962, 1964). McLuhan's concept of the 'global village' provocatively suggested new social forms that were effectively created, by media innovations. Older patterns of communication required clustering of people and activities at specific points; the new electronic media made this condition unnecessary. Extending this contention, developments in communications technology could be said to obviate the need for cities:

> With instant electric technology, the globe itself can never again be more than a village, and the very nature of (the) city as a form of major dimensions must inevitably dissolve like a fading shot in a movie. (McLuhan, 1964, p. 366)

McLuhan's ideas may have done more to provide inspiring slogans than to supply the foundation for research, but there is no doubt that his work contributed to a growing interest in the possibilities of urban and industrial decentralization through the availability of electronic communications media. Such research grew substantially in the 1970s (Gold, 1974; Gold and Barke, 1979). Peter Goldmark (1972), for example, suggested the development of a 'new rural society' with the diffusion of the urban population into low-density settlements in rural areas interconnected by telecommunication networks. Harkness (1976) and Nilles et al. (1976) produced scenarios of the dispersed city forms possible through 'telecommuting' – the substitution of telecommunications for purposes that otherwise required use of transport. Martin (1978, p. 193) foresaw the emergence of 'virtual cities', in which the normal functions of the city would no longer be geographically concentrated, but 'scattered across the earth . . . connected electronically'.

Alvin Toffler took these ideas further and gave a more explicit view of the type of society that might result. Originally a political correspondent in Washington and Associate Editor of *Fortune* magazine, Toffler first attracted public attention with his book *Future Shock* (1970), which speculated upon the possible hidden social costs of rapid and sustained technological change. In *The Third Wave* (1980), Toffler sought to place the social implications of technological change in a wider historic perspective. The book contained the same emphasis on change and dislocation as in his earlier work, but at its heart lay an enthusiastic endorsement of new 'high' technology and an alluring vision of the social Nirvana that could follow the adoption of that technology.

Borrowing an analogy from Norbert Elias (1978), Toffler conceived of cultural history as a succession of three waves; each imposing its patterns on the one before. The First Wave of change came with the spread of agriculture and the Second with the Industrial Revolution. According to Toffler, a Third Wave had become apparent from around 1955 onwards. It would be based on high technology, with a new production system centered on information-handling and processing. Places of employment would be transformed, with smaller work units, the adoption of energy-efficient technology and the decentralization of many sectors of industry. The Third Wave, however, would have implications beyond simply the sphere of economic activity. Rather, it would set in train forces that would rapidly supersede the social, economic and political patterns associated with factory-based industrialism. In Toffler's (1980, p. 24) words:

> The Third Wave brings with it a genuinely new way of life based on diversified, renewable energy sources; on methods of production that make most factory assembly lines obsolete; on new, non-nuclear families; on a novel institution that might be called the 'electronic cottage'; and on radically changed schools and corporations of the future. The emergent civilization writes a new code of behaviour for us and carries us beyond standardization, synchronization, and centralization, beyond the concentration of energy, money and power.

The notion of the 'electronic cottage' created a scenario by which the home was reestablished as the workplace for millions of people – a position which it had last occupied during the preindustrial First Wave. Toffler pointed to various celebrated examples in which the employees in high-techology industries, from managers down to secretarial and other support staff, were operating for at least part of the working week from home. He recorded a growing trend towards 'telecommuting' and held that both commercial logic and wider political and environmental pressures were also militating in the same direction. He argued too that it was even possible that labor movements might arise 'demanding that all work that *can* be done at home be done at home' (Toffler, 1980, p. 213).

Given the place that work routines occupy as an organizing feature of everyday life, it was considered inevitable that the home and family life itself would be revolutionized by this trend towards homeworking. Toffler (1980, pp. 214–17) suggested four possible sets of consequences that could result. The first was termed the *community* impact. Homeworking would lead to greater community stability, especially because changing jobs would not necessitate physical relocation, merely plugging into a different computer network. With less residential turnover could come greater commitment to a locality and to community life. In this way, the electronic cottage could help restore a sense of belonging and 'touch off a renaissance among voluntary organizations'.

Secondly, there was the *environmental* impact, seen primarily in terms of implications for energy production and provision. Instead of having to supply vast amounts of power to centralized urban-industrial complexes, electronic cottages would spread out energy demand and create a pattern that might be better met by small-scale, home-based generators. This could make it economically viable to employ such environmentally desirable energy sources such as solar and wind power, which, in turn, would have beneficial consequences for levels of atmospheric pollution.

The third set was the *economic* impact. Quite apart from generating a new pattern of employment, the Third Wave could have major implications for relationships within industry since

if individuals came to own their own electronic terminals and equipment . . . they would become, in effect, independent entrepreneurs rather than classical employees – meaning, as it were, increased ownership of the 'means of production' by the worker. We might also see groups of home-workers organize themselves into small companies to contract for their services or, for that matter, unite into cooperatives that jointly own the machines. All sorts of new relationships and organizational forms become possible (Toffler, 1980, p. 215).

Finally, there was the *psychological* impact of the electronic cottage. In place of the contractual relations which characterize the industrial workplace would be found the strengthened interpersonal relationships and affective bonds of the homeworking environment. It was recognized that homeworking itself could lead to feelings of isolation, but these could be countered by dividing hours of employment between the home and local work-centers in order to preserve contact with a wider group of employees.

Such, then, were the many and varied implications linked to the electronic cottage. Although hazy as to the practical details of social organization, Toffler effectively sketched a radically different society to that which applied in the late 1970s. The electronic cottage would bring a return, on a new basis, to a golden preindustrial age without the attendant drawbacks of the First Wave. Society would be comprised of enterprising, self-reliant individuals, living in stable communities, in harmony with Nature. The key to this new world was technology, which supplied the preconditions and dynamo for social transformation.

Taken as a whole, the electronic cottage scenario managed to embrace an extraordinary *mélange* of the popular bandwagons of the time. The advent of telecommuting is seen as providing an opportunity to address simultaneously environmental conservation, the nature of relations in the workplace, community ownership of the means of production, emerging concerns with a 'human scale', anxieties about personal lifestyles, the strengthening of family, the resuscitation of neighborhood and the energy crisis. In doing so, the author stressed at periodic intervals that what was on offer here was simply 'a possibility – a plausibility, perhaps – to be pondered' (Toffler, 1980, p. 217), but populism is not born of equivocation. In an important statement, that has been eagerly quoted with approval by many other writers (e.g. Knaap *et al.*, 1987, pp. 4–5), Toffler (1980, p. 217) left the reader in no doubts about the likely consequences of his scenario:

> it is worth recognizing that if as few as 10 to 20 per cent of the work force as presently defined were to make this historic transfer over the next 20 to 30 years, our entire economy, our cities, our ecology, our family structure, our values, even our politics would be altered almost beyond recognition.

The 'electronic cottage' quickly became a standard term in the literature (e.g. Deken, 1981), adopted by a large number of popular science authors who have generally treated Toffler's speculations as established fact. To them must also be added various commentators who, while sharing the broadly evangelical tone of his writings, developed the theme of homeworking in accordance with their own social visions. James Robertson (1983), for instance, argued that the work patterns of factory-based industrialism might eventually be seen as only a brief interlude in the evolution of industrial organization. Robertson envisaged a resumption of home-based patterns, and the possibility of gaining a way of life to which he gave the acronym SHE (Sane, Humane, Ecological). Charles Handy (1984) argued for the death of place-based ('gathered') industrial corporation and the advent of new dispersed organizations. He foresaw the possible emergence of a contractual fringe of self-employed professionals and consultants, working from home, able to take on tasks according to their chosen working hours and to arrange their lives in a manner that asserted their independence and self-reliance.

Viewed collectively, these writings indicate the appearance of a broad pattern of expectation linked to telecommuting. It is considered inevitable that the decentralizing processes, already seen at the present day, will continue apace. The new society will consist of a population interlinked by communications technology but, with their new-found freedom of residential location, essentially able to locate wherever they found most conducive to living the good life. The scenario is certainly plausible, but the empirical evidence that this pattern is emerging remains far from conclusive, as a brief case study of the new homeworking in Great Britain makes clear.

New homeworking in Britain

Homeworking traditionally has had a poor image in Britain. It is usually associated with the systems of low-paid outwork practiced in textiles, packaging and other industries, whereby a widely dispersed, non-unionized labor force work long hours in poor conditions with no job security. Nevertheless, it was apparent that a new type of homeworker had appeared by the early 1980s and Ursula Huws (1984), in a report for the London-based Low Pay Unit, provided a profile of them. From a sample of seventy-eight homeworkers employed in high-technology industries, Huws found that three-quarters of the sample were highly skilled professionals – mainly consultants, computer programmers, analysts and project managers – with only 10 percent being clerical operators (i.e. operating word processors). Almost all those interviewed were found to be female with young families, who, despite possessing skills that were in short supply, might not have been able to participate in the labor market without homeworking.

Leaving aside independent consultants and freelance writers who are not part of wider groups, most of these new homeworkers would be employed by one of the four major networks shown in Table 15.1.

The oldest and largest of these is an independent holding company, F International. Established by Stephanie ('Steve') Shirley in 1962, F International now employs 1,000 homeworkers in Britain (with subsidiaries in Denmark and Holland). A second and much smaller network, established originally in 1982 by the British Government's Department of Trade and Industry (DTI) but now administered by the Manpower Services Commission's Disablement Advisory Service, was set up as a limited pilot project specially aimed at creating employment for the disabled. The remaining two networks are rather different in that they are wholly-owned subsidiaries of large information-technology conglomerates. Only around 280 of the 20,000 employees of International Computers Limited (ICL) work for its two homeworking networks, CPS (Contract Programming Services) and PMS (Product Maintenance Support). The proportion is approximately the same for Rank Xerox (UK). While these networks are viewed by both companies as worthwhile operating experiments in their own right (see, for example, Judkins et al., 1985), with possible commercial spin-offs, there is no doubt that both companies see homework-

Table 15.1 An overview of formal British homeworking programs.

Organisation	F Int	ICL	Rank Xerox	DTI/MSC
Founded	1962	1969	1981	1982
Terms of employment	Self-employed, contracted to FI	Employees	Self-employed, to RX	Employees contracted
Participants	1000	260	60	100
Reasons cited for forming network:				
Employment	√	—	—	√
Skills	√	—	—	—
Retaining workers	—	√	—	—
Cutting o/hs	—	√	√	—
			√	
Work undertaken:				
Computer consultancy	√	—	√	—
Project management	√	√	—	√
Applications development	—	√	√	—
Software management and tech. writing	—	√	—	√
Clerical (w/p etc.)	—	—	—	√

Source: Adapted from Kinsman (1987, p. 68).

ing primarily as a potential answer to specific present-day intraorganizational problems. In the case of ICL, homeworking offered a means to recruit and retain skilled staff who might not otherwise have been available for work; in the case of Rank Xerox, there was an element of searching for means to cut overheads and staff levels without losing skilled staff.

All of these networks owe their rationale to telecommunication technology, although they vary markedly in the extent that everyday operations require the interconnection of operatives with one another or with the parent company. The ICL networks embody a high level of connectivity with other parts of the organization. The upper echelons of the homeworking staff are currently equipped with ICL's OPD (One-per-Desk) hardware, a work station with a built-in telephone and microcomputing facilities originally based on the Sinclair QL. Others are equipped with answerphones and, increasingly, freestanding microcomputers with modems. F International typically makes more modest use of data communications (Kinsman, 1987, p. 68), but there are indications that this situation is changing. The company, for example, is currently extending provision of electronic mail facilities to its homeworkers. As with ICL, extension of facilities is proceeding from the top down. All managers and members of the company's sales force have now access to electronic mail, with that facility now being afforded to around 15 percent of the ordinary panel members. The availability of electronic mail offers the company new ways of dividing up and managing projects and gives panel members a greater opportunity to participate in and contribute to strategy. The DTI/ Manpower Services Commission scheme, which principally distributes discrete tasks for members to work on in their own homes, makes more limited use of data communications.

These networks have been widely discussed as the precursors of the new order, but the evidence again needs careful examination. Although these schemes have aroused great interest, it is noticeable that the available studies (e.g. Kinsman, 1987; Forester, 1987; Francis, 1986), keep working and reworking the same examples. The truth is that progress has been slow. F International has had a highly chequered history, being several times close to bankruptcy in the 1970s and adversely affected by forays into the international market (particularly in the USA between 1978 and 1985), which overextended the company. The ICL and Rank Xerox schemes have only grown at a modest rate, remaining isolated offshoots of larger concerns. The DTI/MSC scheme has succeeded in bringing employment to a disadvantaged group but is still very small. At the present time, perhaps significantly, few proposals for new networks are in prospect. The prospects of homeworking as presently conceived in itself acting as a significant spur to further decentralization or significantly reducing commuter flows seem extremely remote.

These doubts are reinforced by the adverse comments that are ranged against certain aspects of homeworking. Reservations have been expressed about the true quality of the working environment that the new homeworking supplies. The evidence here is contradictory. Reports stemming from the computer industry itself or from authors excited by the prospects afforded by the new homeworking suggest that staff are happy with their working regime and that employees 'rarely feel isolated in the ways predicted' (Forester, 1987, p. 164). Others would take issue. The report by Huws (1984), mentioned earlier, observed that there were two sides to the story. While it was true that homeworking opened up opportunities of employment for some workers who might not otherwise have been able to participate in the labor market, Huws noted that their average pay levels, opportunities for in-service training, conditions of job security and fringe benefits were significantly worse than those provided for on-site staff. In addition, the new homeworkers often complained of isolation, stating that they missed the everyday contacts of working in an office. Even Forester (1987, p. 165), who predicted that homeworking will increase considerably, was forced to conclude that this 'grim picture of high-tech homeworking hardly accords with some of the more utopian visions of life in the Electronic Cottage'.

In both quantitative and qualitative terms, then, the British experience shows only scant support for the idea that the 'electronic cottage' scenario is likely to have a significant impact on social and spatial organization. To date, only a small number of homeworker networks have been founded and these have primarily offered the chance for highly skilled married women with young families to reenter the labor market. The numbers of those actually opting out of conventional jobs and adopting telecommuting in pursuit of an enhanced lifestyle at present seems miniscule. While many homeworkers are genuinely content with the resulting working patterns and conditions and champion the cause of homeworking, others find it harder to function on this independent basis. It is hard to avoid the conclusion that proponents of telecommuting may well have overemphasized the positive elements of working at home (flexibility, lifestyle decisions, avoidance of the physical stresses of commuting) relative to the negative aspects of homeworking (isolation, poorer terms of employment) and the *positive* aspects of actually working in an office (everyday face-to-face contacts, social relationships and a feeling of being in touch with what is 'going on').

Two broad arguments are frequently made against the points raised above. First, it is maintained that rates of adoption in other countries have been higher (Kinsman, 1987) and that development in Britain has been hampered by the high costs of telecommunications, particularly for specialist teleconferencing and information services. Secondly, it can be argued that it is facile to expect such a dramatic shift in working patterns, population and lifestyles to occur overnight: that progress has been made and that the 'takeoff' point *will* arrive in the near future.

Yet, having presented these arguments, a nagging doubt remains. The type of predictions made by Toffler and his supporters are not without precedent. There have been various other occasions in the last hundred years in which it has been predicted that a particular technology holds prospects for a major reconstruction of society. The precise technology to which this expectation has been attached may have changed over time, but the essential logic of associating radical social change to given advances in technology has remained remarkably constant. This is a point which requires further consideration, since examining the construction of past, and failed, visions of the future does give us some basis for assessing the true nature and plausibility of the 'electronic cottage' scenario.

The construction of futures

In a reflective essay on the nature of the 'information revolution', Hamelink (1986) pointed to the importance of 'myth': the creation of a powerful story, told and retold, through which phenomena in society are explained. To Hamelink (1986, pp. 7–8), the 'myth' in question offers an explanation of the world of the late twentieth century and presents the normative implications of its historical interpretation. It suggests that the 'information revolution' is the most significant historical development of our time: a revolutionary transition to a fundamentally different age.

This transition is the shift from an industrial society to an information society. Related descriptions of this new society include 'post-industrial', 'post-capitalist', 'post-ideological' and 'post-protestant' . . .

The most important aspect of the myth is the notion that we are entering a radically different stage of human history. The information society is a 'post-society': it means a break with previous values, social arrangements and modes of production.

This is a useful analysis. It clearly identifies the arguments of those who have equated technological progress in the field of information technology with a qualitative improvement in human life and feel that current technology has brought human society to a critical juncture from which it is possible to move forward to a radically different, and better, future. Yet the analysis also needs to be placed in its wider historical frame. In many ways, the advent of the 'information revolution' merely gave an opportunity for old and discredited arguments to be dusted off and reused.

An example helps to clarify matters. In the 1920s and 1930s, modernists, particularly in architecture, were claiming that advances in technology offered the chance not just to reconstruct cities but also society. The plans and drawings produced by the Modern Movement during this period, therefore, not only depicted bold schemes for linear cities and for gleaming cities of towers, they also showed a new society living a blissful new leisured, convenient and mobile lifestyle amidst their new surroundings and dwellings (usually rented apartments).

The starting points for the Modern Movement's formulations were the innovations in building technology: new building materials (steel, glass and reinforced concrete), sophisticated constructional techniques and secondary energy sources (especially electricity). These advances gave the Modern Movement the possibility of producing structures and cities unlike anything that had been previously attempted, but modernists clearly recognized that their ideas would be adopted only if it could also be shown that society would also want the built environment so created.

The resulting strategy, as has been shown elsewhere (Gold, 1984, 1987) was essentially twofold. First, the Modern Movement produced highly selective interpretations of social, cultural and economic trends in order to argue that radical change in living styles was actually taking place. Secondly, they drew upon their own values in order to argue that such radical change *should* take place. The result was the production of an extremely evocative image of 'modern society'; a society that would be rational, progressive, and filled with 'creative' individuals more than able to grasp the potential of technology to transform their lives. It would only be the passage of time and the experience of the vision realized that would cast doubt on these assertions.

This example could be complemented by others: it is, for example, equally possible to formulate the same case with regard to writings in the 1960s about the supposed impact of universal mobility in creating new placeless communities (Gold, 1985). Nevertheless, the example of the urban vision of interwar modernism can serve to provide a parallel with the scenario of the 'electronic cottage'. When compared, it is apparent that while these scenarios portrayed quite different future environments, both suffered from the same conceptual weaknesses. They began with a known pattern of technology and proceeded, by dint of selective assumptions about social adoption of technology, to devise idyllic pictures of future societies and their living environments. Both, in fact, are representative examples of a genre of thought that may be termed 'technological utopianism': a style of thought that pursues, through the agency of technology, a comprehensive vision of an ideal society radically different from, and, it is believed, better than that which obtains at the time.

The history of 'technological utopianism' itself extends back to the late eighteenth century and the ensuing years has yielded an enormous variety of different forms (Kumar, 1987; Manuel and Manuel, 1979; Segal, 1985). Some, like those produced by the Modern Movement, portrayed visions of a future world unlike anything that had so far existed; a world that embraced the new metropolitan scale made possible by the machine. Others, like Toffler, have effectively retreated back into the Garden; evoking an image of a world in which population decentralization through technology permits reversion to living patterns of a bygone golden age. The strategy of these, and other variants of technological utopias, may be different, but the net result is essentially the same: envisaging technological progress as the mechanism by which radical change in living patterns is produced, even if technological utopians tend to be hazy as to how those transformations would come about (Segal, 1985, pp. 23–4).

Conclusion

It is said that nothing ages faster than yesterday's vision of tomorrow. The public indignation over the failings of modernism led to a profound reappraisal of the Modern Movement's concepts of the future city and a growing appreciation of the divergence between vision and subsequent reality. The current mood of nostalgia for the optimism of the 1960s has not masked the fact that the icons of that period – such as the pursuit of mobility, unfettered consumerism, the alternative culture – have similarly proved unrealized.

Naturally, it is too early to say whether the scenario of the 'electronic cottage' will go the same way. With the enormous worldwide interest in computer and communications technology, there is great interest in the type of scenario put forward by Toffler. By any estimate, too, Toffler's notion of the 'electronic cottage' has done us a considerable service by focusing attention on the potential significance of telecommuting and by linking together many apparently diverse themes into a single scenario. It contains many ideas that deserve to be pursued in greater detail. Nevertheless, the previous section has indicated the one-sidedness and weaknesses of the analysis, with the section before that revealing that the empirical evidence as yet gives scant support to any major structural change in social and spatial patterns due to homeworking. Certainly no significant pressure groups have developed to campaign for this change. One can also see that the circumstances to which Toffler was responding have themselves already changed: for example, with the falling oil prices of the late 1980s, it is harder to retain that sense of urgency for finding alternative, energy-conserving settlement patterns that was a recurrent theme in Toffler's vision.

Yet it is important not to conclude merely by assessing the likely accuracy or otherwise of the electronic cottage scenario in isolation. As shown, the electronic cottage scenario is representative of a genre of thought that, in varying guises, has run through twentieth-century writings about technology and society; another assembly of a limited range of technological variables from which premature generalizations are made. If we are to make better sense of the patterns that are arising, however, it is essential to move beyond simply looking for the impacts of technology and start showing the role of human agency in the process: why people choose technology and how they choose to use it. Only by doing so will it be possible to move beyond seductive utopian images of social redemption through technology towards an understanding that reflects the true complexity of the relationship between technology and society.

Note

This paper was written and researched in 1987 while on sabbatical as an Official Academic Visitor in the Department of Geography, London School of Economics. I would like to convey my thanks to them for their generous hospitality and assistance and to Joanne Stead and Frank Webster for bringing valuable sources to my attention at that time.

In the time since this essay was written, much new literature has appeared on the technological and social aspects of electronic homeworking (e.g. Ahrentzen, 1989; Antonelli, 1990; Ernste and Jaeger, 1989; Lyon, 1988; Miles, 1988a), although the adoption of such working practices has clearly continued to disappoint their advocates and enthusiasts (e.g. Shirley, 1989). The reader interested in further comment on Toffler's 'electronic cottage' scenario may be recommended to an exchange of views that appeared in the journal *Futures* in the summer of 1988

(Forester, 1988; Miles, 1988b). In a subsequent paper (Gold, 1990), I have developed further the idea of 'wired society' scenario as an expression of utopian literature. The basic conclusions of the present paper, however, remain unaltered.

References

Ahrentzen, S. (1989), 'A place of peace, prospect, and . . . a PC: the home as an office', *Journal of Architectural and Planning Research*, 6, pp. 271–88.

Antonelli, C. (1989), 'Induced adoption and externalities in the regional diffusion of information technology', *Regional Studies*, 42, pp. 31–40.

Deken, J. (1981), *The Electronic Cottage* (New York: Morrow).

Elias, N. (1978), *The Civilizing Process: The Development of Manners*, trans. E. Jephcott (Oxford: Basil Blackwell).

Ernste, H. and Jaeger, C. (eds) (1989), *Information Society and Spatial Structure* (London: Belhaven).

Forester, T. (1987), *High-Tech Society: The Story of the Information Technology Revolution* (Oxford: Basil Blackwell).

——(1988), 'The myth of the electronic cottage', *Futures*, 20, pp. 227–40.

Francis, A. (1986), *New Technology at Work* (Oxford: Oxford University Press).

Freeman, C. and Jahoda, M. (eds) (1978), *World Futures: The Great Debate* (Oxford: Martin Robertson).

Gold, J. R. (1974), *Communicating Images of the Environment*, Occasional Paper 29, Centre for Urban and Regional Studies, University of Birmingham.

——(1984), 'The death of the urban vision?', *Futures*, 16, pp. 372–81.

——(1985), 'The city of the future and the future of the city', in King, R. (ed.), *Geographical Futures* (Sheffield: Geographical Association), pp. 92–101.

——(1987), *Modernism and the Urban Imagination*, PhD dissertation, Centre for Urban and Regional Studies, University of Birmingham.

——(1990), 'A wired society?: utopian literature, electronic communication and the geography of the future city', *National Geographical Journal of India*, 35 (in press).

Gold, J. R. and Barke, M. (1979), *Telecommunications and Urban Space*, Public Administration Series 362, Monticello, Ill., Vance Bibliographies.

Goldmark, P. G. (1972), 'Tomorrow we will telecommunicate to our jobs', *The Futurist*, 6, pp. 55–8.

Goudzwaard, B. (1984), *Idols of our Time* (Downers Grove, Ill.: Inter-Varsity Press).

Hamelink, C. J. (1986), 'Is there life after the information revolution?', in Traber, M. (ed.), *The Myth of the Information Revolution: social and ethical implications of communication technology* (London: Sage), pp. 7–20.

Handy, C. (1984), *The Future of Work: A Guide to a Changing Society* (Oxford: Basil Blackwell).

Harkness, R. C. (1976), 'Innovations in telecommunications and their impact on urban life', in Golany, G. (ed.), *Innovations for Future Cities* (New York: Praeger), pp. 21–53.

Huws, U. (1984), *The New Homeworkers* (London: Low Pay Unit).

Judkins, P., West, D., and Drew, J. (1985), *Networking in Organizations* (Farnborough: Gower).

Kern, S. (1983), *The Culture of Time and Space, 1880–1918* (London: Weidenfeld & Nicolson).

Kinsman, F. (1987), *The Telecommuters* (Chichester: John Wiley).

Knaap, G. A. van der, Linge, G. T. R., and Wever, E. (1987), 'Technology and industrial change: an overview', in Knaap, B. van der, and Wever, E. (eds), *New Technology and Regional Development* (London: Croom Helm), pp. 1–20.

Kumar, K. (1987), *Utopia and Anti-Utopia in Modern Times* (Oxford: Basil Blackwell).

Laurie, P. (1980), *The Micro Revolution* (London: Fontana).

Linstone, H. A., and Simmonds, W. H. C. (eds) (1977), *Futures Research: New Directions* (Reading, Mass.: Addison-Wesley).

Lyon, D. (1986), *The Silicon Society* (Tring: Lion).

——(1988), *The Information Society: Issues and Illusions* (Cambridge: Polity Press).

McLuhan, H. M. (1962), *The Gutenberg Galaxy: The Making of Typographic Man* (London: Routledge & Kegan Paul).

——(1964), *Understanding Media: The Extensions of Man* (London: Sphere Books).

Manuel, F. E., and Manuel, F. P. (1979), *Utopian Thought in the Western World* (Oxford: Basil Blackwell).

Martin, J. (1978), *The Wired Society: A Challenge for Tomorrow* (Englewood Cliffs, NJ: Prentice-Hall).

Miles, I. (1988a), *Home Informatics: Information Technology and the Transformation of Everyday Life* (London: Frances Pinter).

——(1988b), 'The electronic cottage: myth or near-myth', *Futures*, 20, pp. 355–66.

Nilles, J. M., Carlson, F. R., Gray, P., and Hanneman, G. J. (1976), *The Tele-communications-Transportation Tradeoff* (New York: Wiley).

Norman, C. (1981), *The God that Limps: Science and Technology in the Eighties* (New York: Norton).

Robertson, J. (1983), *The Sane Alternatives: a choice of futures*, revised edition (Ironbridge: the author).

Segal, H. P. (1985), *Technological Utopianism in American Culture* (Chicago: University of Chicago Press).

Shirley, S. (1989), 'Information technology', *Royal Society of Arts Journal*, 137, p. 758.

Smith, R. L. (1972), *The Wired Nation* (New York: Longman).

Toffler, A. (1970), *Future Shock* (London: Bodley Head).

——(1980), *The Third Wave* (London: Pan).

Traber, M. (ed.) (1986), *The Myth of the Information Revolution: Social and Ethical Implications* (London: Sage Publications).

Webster, F., and Robins, K. (1986), *Information Technology: A Luddite Analysis* (Norwood, NJ: Ablex).

16 Socio-spatial implications of electronic cottages

BRIAVEL HOLCOMB

The spatial reconvergence of home and paid employment facilitated by telecommuting technologies has different implications for men than for women. This chapter explores the preliminary and potential impacts on women's lives as 'homework', once largely confined to manufactured piecework and regulated by the federal government since the 1940s, becomes increasingly prevalent. The microcomputer, modem and other telecommuting technological innovations have created the potential for electronic cottages to revolutionize the spatial patterns and conditions of paid employment (Toffler, 1980). Technological advances have nurtured flexitime and flexispace to the point at which, theoretically, much clerical work could be done at any time of the day or night and in any place with electrical and telephone connections. Because of gender differences in occupations, workforce attachment and commuting patterns, the new technology is likely to have greater impact, both positive and negative, on women (Christensen, 1987).

The spatial separation on a grand scale of reproductive and productive work (of domestic labor and childrearing versus paid employment outside the home) is a relatively recent phenomenon accompanying the Industrial Revolution and the rise of the factory system. While men, women and children all worked long hours in the mills and factories of the mid nineteenth century, it was not until the rapid expansion of cities in the twentieth century that longer commutations to work accompanied the evolution of spatially separate 'domestic spheres' (Mackenzie and Rose, 1983). The growth of suburbs and land use zoning in the mid twentieth century consolidated a pattern in which paid employment was concentrated in central and industrial districts while domestic labor was segregated into residential neighborhoods.

The 1934 congressional ban on home-based industrial work, part of an effort to revive the national economy, covered a hundred different industries. While the ban, part of the National Recovery Act, was found to be unconstitutional the following year, in 1942 a second ban was imposed on seven garment industries and other home industries were strictly regulated (Gerson, forthcoming). These regulations, supported by union leaders, the US Labor Department and many state governments reduced (though by no means eliminated) home work. The immediate post World War II years were also the halcyon housewife era. Women should be wives and mothers, should not be employed outside the home except perhaps in voluntary work. The spatial separation of 'home' and 'work', while always far from complete, may have reached a zenith in this period.

In the last twenty-five years a new trend has modified this pattern. As women have entered the paid labor force in escalating numbers, although the spatial distinction of home and work has been largely maintained, now workers of both genders commute and spend part of their days in each sphere. Nearly two-thirds of all adult women are now gainfully employed and mothers are almost as likely to be in the labor force as women who do not have children under 18 years old. In 1960 only 30.4 percent of American mothers were employed. By 1986 the figure was 62 percent. While women have entered the paid labor force they continue to assume the major share of reproductive or domestic work. Numerous surveys document continued gender inequality of time spent in childrearing and home maintenence. Many women in the labor market are single heads of households and are responsible for virtually all domestic work.

This is the context into which technological advances, especially the microcomputer and telecommuting, became available about a decade ago. Two-thirds of women work outside the home, 40 percent of them in clerical occupations. Ninety-seven percent of clerical workers are female. The new possibilities for collapsing commutation distances from miles to feet and for scheduling working hours with maximum flexitime are just beginning to be realized. Meanwhile, federal bans on some home work have been lifted and others are under discussion (Christensen, 1986). This chapter explores the advantages and drawbacks of home waged work using microcomputing technology. If the phenomenon of the electronic cottage proliferates, there are significant implications for the spatial distribution of employment, commuting patterns and urban structure.

The recency of telecommuting complicates estimates of the numbers of workers involved, but it is clear that the full potential of this mode of employment is far from realization. While over a fifth of US homes had microcomputers by the mid 1980s, it has been estimated that nearly 90 percent will be so equipped by 1990 (National Academy of Sciences, 1983). Most home computers are used for playing video games or for tasks previously done by typewriters and calculators. However, one recent estimate was that between 13 and 14 million people regularly perform paid work at home using personal computers. An AT&T study estimated that as many as 25 million people work out of their home using computers (Zeldes, 1986). In 1986, between two and three hundred companies had some type of home-based work arrangement. These included large corporations such as J. C. Penney, Citibank and Xerox as well as small office service firms. Christenson notes the likelihood that 'the strongest force accounting for home-based work is the increasing number of small businesses. From 1977 to 1982 the total number of sole proprietorships increased 21 percent, whereas the number for women increased 46 percent. Women entrepreneurs are the fastest growing segment of American small business' (1986, pp. 4–5).

Microcomputer and telecommunication technology remove locational inhibitors from a wide variety of tasks, but in examining the implications of this it is essential to distinguish between two general categories of work. One is the creative 'brain' work of the programmer, scientist, author or stockbroker who may find the home environment freer of distraction and more conducive to mental activity than the traditional workplace of office or laboratory. The other is the routine and sometimes 'mindless' work of the clerk who punches in the extremely repetitive data of insurance claims or inventories, or types other peoples' manuscripts. In the absence of empirical data, it seems reasonable to assume in contemporary America that the majority of the first group of micro users are men (since males predominate in most professional and managerial occupations) while women are the preponderance of the latter (since most clerical workers are female). While there are fascinating issues related to the impact of this technology on professional occupations (see, for example, Cooley's, 1981, discussion of the Taylorization of intellectual work), the concern of this paper is largely with the latter group. Through examination of the advantages and drawbacks of this kind of employment in the home, implications both for policy and the changing geography of work and home are explored.

Among the tasks which can be performed in 'remote offices' or electronic cottages are word processing, programming, data entry, catalog sales, insurance claims processing, keypunching programs, transcribing dictation from automated tape recorders, handling warranty claims, ticket sales and numerous other routine operations. A growing number of US companies, including Blue Cross/Blue Shield, Control Data, Aetna Life and Casualty, AT&T, American Express and General Motors, and some governmental agencies such as the US Army and the General Services Administration, employ people who work at home with micros (*Business Week*, 1981; Pratt, 1984). It is important to note that these are tasks which could equally well be performed at conventional offices. While a few rely on interactive telecommunication technology, in most cases the work done by homeworkers still entails regular trips to a central office to collect raw data and return completed work. A very small portion of this potential fertility of the marriage between microcomputers and telecommunications has yet been realized. Few electronic cottages are linked to central offices only by telephone and modem. However, the reasons for retaining the umbilical cord of occasional commuting or of supervisory home visits are more social than technical. There are few data as yet on the spatial distribution of electronic cottages but the technology already exists to completely free much clerical work from locational constraints and to enable offices to be as 'remote' as the worker wishes.

Benefits of working with microcomputers in flexiplace

There are numerous advantages to both employee and employer of working at home. Our quintessential model is a mother of young children who has a modicum of education, is appropriately imbued with the work ethic and has a spouse employed outside the home. She desires both the material and psychological rewards of paid employment, but constraints on her time and mobility from her maternal role severely curtail her employment opportunities. In times past she may well have opted for producing goods at home. The grandmothers of today's word processors knitted stockings by the score. There is a long history, laced with exploitation, of home working for women. Alternatively, both in the past and present she may prefer to market products and become an Avon Lady or sell Tupperware at parties. However, microelectronics offers employment which is slightly better paid and has more social status than knitting, without the psychological and social effort entailed in selling cosmetics or plastics. In her comparison of office versus home-based clerical workers, Gerson found that the strongest difference between the two groups was not whether they were mothers of young children, or in the numbers of hours worked or average income, but rather in the values they expressed. Homeworkers were more likely to express traditional values, placing emphasis on religions and little on career advancement, while office workers agreed more with statements advocating shared household labor, equal employment opportunities for women and general gender egalitarianism (Gerson, forthcoming).

Computerized clerical work offers employment in flexiplace with flexitime. The freedom to schedule one's paid working hours facilitates both the combination of more roles and enables one to maximize other opportunities. Some women begin work early in the morning, completing a few hours before the children awaken; others prefer the evening when the children are in bed. Most people, especially women, do many things at once. Hence working at home can be combined with machine laundering, waiting for and supervising repairpersons, being available for emergencies and even, though less comfortably, supervising children, none of which is easily possible with a central office job. The time flexibility of home computerwork is a major advantage to many users, including professionals who may find they are most productive at certain nonconventional hours of the day or night.

Work at home reduces the cost in time and money of dressing for and traveling to work and of lunches. Not having to be 'presentable' and eliminating the time of commuting and the socially compulsory lunch hour can add several hours to the day. Pratt (1984) reports that her sample of home employed computer users estimated their monthly money savings at $200 on food, $100 on transportation, and $100 on clothes. While these figures seem high (how many women spend, let alone save, $10 on lunch a day and $100 a month on clothes?) estimates must be difficult. When electronic work is combined with childcare, these costs are also saved.

Work at home is usually 'piece work', that is, paid by the amount produced rather than by the time spent working. This can have both pros and cons for the employee, an advantage being that a worker has more control over the amount she produces and she may opt to work faster and more hours in order to generate more income. Many such jobs are part time, so offer more variation in the amount of time devoted to work than does a standard office job. Indeed, the autonomy offered by working at home can be a major motivation for seeking a change in the location of one's job. It is being increasingly recognized that the opportunities for autonomy, for working at the pace and in conditions which allow for individual variation (Bach versus Heavy Metal accompaniment, coffee or wine refreshment, breaking for a telephone chat or a catnap), and without the oppression of direct supervision by hierarchical superiors, is a major determinant of job satisfaction and is related to productivity.

For some women, computer work at home means the difference between being employed and not. Employment carries more than financial rewards. Working at home can (though by no means always does) enhance a woman's social status. The locational flexibility offered by this employment assists people who are half of two-job couples in which the spouse is either required to make frequent residential moves or to live in places remote from centers demanding clerical work. Telecommunications innovation has theoretically homogenized US space to the point where most clerical jobs performed at home could be done in a cabin located in a mining boomtown in Wyoming or a retreat for transcendental meditation in Vermont. That in practice most teleworkers are probably ensconced in suburbia negates neither the theory nor its romance.

While for reasons previously noted this discussion has focused on women, another group of home workers requires mention. Those whose physical handicaps impair mobility to the point that journeys to work are implausible find new opportunity in microcomputer related employment. So also do semi-retirees who enjoy a few hours of work a day. Working at home can be especially attractive to people who prefer to minimize their social contacts and work alone. Not having to participate in office politics, and perhaps even avoiding such unwanted advances as sexual harassment, can be an incentive for some. Paid employment within the home can entitle the employees to a home-office tax deduction and, if the electronic equipment is purchased by the user, to a depreciation deduction also. Many firms, however, either lend or lease equipment used by homeworkers, thus (again more in theory than practice) home teleworkers have part-time use of the capital investment of their employers for personal gain (e.g. the freelance typist of manuscripts on a micro used mainly for processing insurance forms for a major corporation). Again in theory, and perhaps in practice, home work enables the worker to sell her labor to the highest bidder without locational constraints. If a firm cross-town, or cross-country is willing to pay higher rates than one closer, one is free (perhaps) to accept the offer.

The employer also benefits. A major advantage is that of saving office space. Given high rents in many central and even suburban office locations, added to the costs of heating, cooling and other services, the savings in work space can be considerable. Most home employment, paid by piecework rates, does not offer standard benefits such as sick leave, vacations and pension contributions, costs which are saved by the employer. When a company owns (or leases) expensive mainframe computers, extending their use over twenty-four hours increases utilization of that investment. Homeworkers often prefer to work at hours when a central office is closed, both because of reasons discussed earlier and because access to the mainframe is easier. When the mainframe computer is 'down' during regular office hours, the employer continues to bear the cost of idle on-site workers, while those at home must absorb the inconvenience themselves. One estimate of the savings to employers of work done at home versus at a central office is 20 percent (Keller, 1984).

There are labor supply advantages to companies using homeworkers. There is the potential for an expanded labor supply in employees released from time and space constraints. Considerable fluctuations in demand for labor can be accommodated by employing part-time, off-site workers. Since most wages are based on production, the inefficient worker or those who waste time on the job are remunerated accordingly. Retention rates for trained personnel can be higher when work location is flexible and can accommodate some residential mobility. Absenteeism and tardiness become irrelevant. 'Workoholism', a trait which may be detrimental to the addict but beneficial to her employer, has been suggested as an occupational hazard of home work in which one rarely escapes the potential for labor!

The drawbacks of home micro-electronic employment

With so many advantages as discussed above, why then are companies, labor unions and employees often reluctant to support work at home? Is this new technology an avenue to liberation for women who seek to combine reproductive with productive labor, or is it merely the most recent form of female labor exploitation? Like advances in household technology from the Hoover and Maytag to the food processor and microwave, do microcomputers merely 'facilitate' the assumption of more roles by women and raise the standards of acceptable performance? Will home electronic employment simply reinforce the pink-collar clerical occupational ghetto and weaken efforts to organize labor?

For both individuals and their employers there are disadvantages to home employment. Chief among them for some employees are the lack of benefits (mentioned earlier) and the absence of compensation for the 'downtime' which humans, as well as computers, require. Others miss the social interaction which employment at a central office offers. Such interaction has not only emotional benefits but can enhance one's career prospects and productivity. To be remote from the informal contacts which foster the communication of useful information for job performance, about openings in the hierarchy, or simply psyche soothing, can be difficult. While 'networking' among home computer users is part of the conventional wisdom, the screen and earpiece have their limitations in the roles of consoler and confidante. The social isolation of home work could be particularly hazardous to the handicapped since their invisibility may reduce public awareness of their needs.

For all home computer users, but perhaps especially women, there can be heavy psychological costs in the lack of spatial separation of home and work. Combining waged work and parenting in the same space and time may be financially functional but emotionally and physically difficult. Anyone who has attempted to combine even marginally mental work with the care of a truculent toddler will appreciate the potential trauma. Indeed, Gerson's 1985 study comparing 106 home-based workers with 260 office-based workers found that homeworkers were almost as likely to use some type of childcare as were office-based workers and that the homeworkers had higher childcare expenses than did office workers (Gerson, forthcoming). Nearly half the women in Olsen's sample of 'cottage keyers' reported deterioration of health habits such as eating and smoking more and drinking more coffee (Olsen, 1982). There is also the inconvenience of sharing one's living space with cumbersome equipment which may require expensive maintenance. For non-working household residents, the presence of a homeworker striving to concentrate can be as irritating as interruptions are to the computer operator. The contradiction between the guilt of ever present work (one could always pop downstairs and process a few more claims) versus the perception by acquaintances that since one is not getting dressed and leaving the house by 9.00 a.m. one is not, therefore, 'working', may require adjustment. Nemy also wrote half humorously about the homeworkers' vulnerability to chatty phone callers and visitors. While the relative social isolation of home work is attractive to some people, it is anathema to others. Particularly vulnerable in this respect are the disabled. If handicapped workers are isolated in the home, their needs will be less obvious to the public. For women who regard escape from domesticity as a *sine qua non* of independence, home centered work lacks appeal. Workers who are at home alone must add heating and cooling costs to those of employment, and some telecommunicators require the expense of a second phone. The labor supply flexibility which advantages employers is bought at a cost to employees who are not paid if there is no work for them to do.

The potential for exploitation of labor which led over forty years ago to a federal legislative ban on commercial home knitting might equally be applied to microcomputer home employment. Labor union organizing is extremely difficult when employees are spatially dispersed and willing to accept very unconventional work conditions (Keller, 1984). When fringe benefits and job security are absent, employees have few bargaining chips. As noted earlier, the great majority of clerical workers either in conventional offices or at home, are female. Clerical work is one of the most gender-segregated occupations and has become increasingly 'deskilled' as tasks have been compartmentalized into smaller units requiring fewer personal decisions and with increasing opportunities for managerial supervision of productivity (Duncan, 1981; Glenn and Feldberg, 1977). Work done on microcomputers can be routinely checked for productivity as measured by time logged on, number of keystrokes, proportion of errors, and so forth. The operator becomes, in some ways, an extension of the machine, and work done at home is immune to neither this depersonalization nor the potential for remote monitoring. A Congressional Office of Technology Report (1987) estimates that 7 million computer workers are now being monitored, most of them clerical workers. The study found that such monitoring adds stress and can be regarded as invading the privacy and dignity of workers. Much clerical computer work is numbingly boring and uses none of the operator's creative and imaginative potential. Indeed, deviations from routines can cause costly errors and even computer downtime.

In some ways, women who work at home are simultaneously spatially and temporally liberated and held captive in a work hierarchy which denies opportunity to solitary workers for organizational advancement. There are definite limits to upward mobility unless one participates in the corporate atmosphere. Homeworkers are more dispensable and replaceable than their on-site peers. They expect, and get, less. Technological innovation is transforming not only the techniques of production but the role of labor (Mattera, 1983). There are even fears of child labor exploitation . . . perhaps justified by the relative technical sophistication youths raised on computer games may have acquired! Opportunities for team work and for building cooperative relationships between workers are diminished by solitary job conditions.

The primary disadvantage to the employer may be the diminishment of control over the worker. If and when homeworkers have access to 'sensitive' material or contact with customers, the possibilities of security lapses or alienation of consumers is obvious. If workers are on salaries or hourly wages the ability to monitor productivity may be lessened. It is more difficult to socialize homeworkers into corporate expectations and to train them for new and complicated procedures. In brief, home computer employees could, but by no means necessarily do, represent a threat to corporate image and loyalty. There are also issues for employers regarding insurance for off-site workers and equipment, complications which could escalate if occupational hazards associated with exposure to visual display terminals are demonstrated.

Geographical implications of home microelectronic employment

Given the advantages and drawbacks of this kind of employment as discussed above, it seems inevitable and probably desirable that the trend towards work at home with computers will grow. The speed and extent to which it does so will depend on numerous factors such as the price of equipment, fluctuations in the economy affecting the demand for labor, improved computer education, energy and office space costs, as well as the unionization of clerical labor, the attitudes of women to home employment, and so forth. The executive board of the Service Employees International Union passed a resolution in 1982 calling for a federal ban on electronic home work, and other unions have also opposed the trend (Mattera, 1983). Recent Congressional hearings on the regulation of home employment and considerations by local planning boards about appropriate zoning regulations to accommodate this growing phenomenon are evidence of the complexity of adopting laws which will minimize the social costs of home employment (Herbers, 1986; Noble, 1986; Zeldes, 1986). Nevertheless, many people, especially women, welcome the opportunities for employment that this technology affords. The phenomenon has interesting implications both for employment and labor policies and for geography.

In some senses, the recent mushrooming of electronic cottages is one more step in a longstanding trend towards the decentralization of office work. Escalating costs of office production in downtown locations (including high rents, congestion and increasing commutation) in the last twenty years, together with the suburbanization of the workforce, has led many companies to either relocate headquarter offices in suburbia, or to maintain a smaller central office with many routine functions being moved to 'back offices' in peripheral locations. Nelson argues that this migration of office work is in response to the abundance in suburbia of placid female clerical labor – women who have relatively high educational levels but are willing to accept modest wages with few benefits because their domestic commitments offer little time/space flexibility (Nelson, 1984). These same women are candidates for home clerical employment since even the constraints of a short commute and absence from the home are eliminated. Thus a growth in home computer employment could mean a continued decentralization and dispersal of clerical employment, theoretically to the point where the clerical home labor force is completely free of employment-related locational constraints. In that case, a measure of residential locational freedom is also gained in that people who are not living in households geographically limited by employment of at least one member can choose to maximize other locational criteria such as amenity or rent.

Conceivably, a rapid growth in electronic home employment could ultimately lead to a declining demand for central and back office space which would be an unwelcome trend for cities whose revitalization has been spurred by commercial office space construction, and those (such as Denver) which already have a glut of such space due to overbuilding in boom periods. Greater Los Angeles has more self-employed people (not all microcomputer users) than any other city in the US. 'One result of this office-at-home phenomenon is already clear: the metropolitan area does not have as much office space as would be expected for a city of its size and wealth. Most major metropolitan areas have at least 20 square feet of office space per person. In greater New York, the bastion of the organization man, the figure is 28 square feet. In greater Los Angeles the figure is a mere 15 feet' (Lockwood and Leinberger, 1988, p. 36). On the other hand, a diminution of commutation would reduce energy consumption, air pollution and traffic congestion (Schiff, 1979). The average work commuting trip in the USA in 1983 was 8.3 miles one way (US Department of Transportation, 1985). While the average for women is likely to be slightly less (since various surveys have found that women work closer to home than do men, e.g. Hanson and Johnston, 1985; Madden, 1981) each commuting trip eliminated saves energy and conserves environmental quality.

Finally, an increase in home teleworking could lead to changes in space at the domestic level. Already more homes are being specifically designed to incorporate an office in which complicated and cumbersome equipment can be housed and where the worker has a degree of privacy and quiet. A recent competition to design a 'New American House' had guidelines which required, among other elements, space for professional work within each unit. The winning design provided for a modicum of auditory and visual supervision of children on the second floor while working on the third (Leavitt, 1985). Federal tax deductions for home offices encourage the dedication of a separate space within the home for this function.

There are numerous issues related to the expansion of employment in microelectronics which require further study. The more critical of these concern the ways in which the new technology changes the conditions of work and the relationships between employer and employee, factors of particular concern to women (Boneparth and Stoper, 1983; Zuboff, 1982). To date, the expansion of high technology has done little to advantage women. A recent report on this issue noted that '[h]igh tech may produce integrated circuits, but it does not necessarily produce an integrated work force or eliminate the female–male earnings differential' (Gathright, 1985). While women were 2 percent of all engineers in the computer industry, they form 92 percent of all data-entry operators. A major explanatory factor in gender wage differentials is gender occupational segregation. Johnston (1987) has demonstrated that women with convenient spatial access to employment are disproportionately in female-typed occupations and that geographical factors contribute to gender occupational segregation. Ironically, home telework may reinforce this pattern and lead to even more intense gender differentials in occupation and income.

Olson and Primps (1984) predict that the expansion of home computer work will be slow. If, however, there is progress in solving problems of employment conditions, home design and technological barriers to complete remote workplace integration, then this form of work could expand rapidly and significantly alter geographical space. The changes will not be confined to the USA. Already, Britain has more computers per capita than any other country and advanced industrialized countries such as Japan, Canada and West European states are likely to experience similar trends (Zimmerman, 1986). There is also the possibility that just as the garment industry has migrated to places with cheap skilled labor such as Mexico or Taiwan, so too will routine clerical work. Currently, the relative cheapness of home-based labor helps US companies compete internationally, but if homeworkers become more organized and are able to improve wages and benefits, the spectre arises of the same exports of jobs as has occurred in other manufacturing industries (McKay, 1988). Electronic cottages, while collapsing the space between home and work, have geographical implications at the intraurban, regional and international scales.

References

Boneparth, Ellen, and Stoper, Emily (1983), 'Work, gender and technological innovation', in Diamond, Irene (ed.), *Families, Politics and Public Policy* (New York: Longman), pp. 265–78.

Business Week (1981), Issue on microelectronics, 6 January, pp. 94–103.

Christensen, Kathleen (1986), 'Impacts of computer-mediated home-based work on women and their families', *Office Technology and People*.

——(1987), *Women and Home-based Work: The Unspoken Contract* (New York: Henry Holt).

Congressional Office of Technology Assessment (1987), *The Electronic Supervisor: New Technology, New Tensions* (Washington, DC: US Government Printing Office).

Cooley, Mike (1981), 'The Taylorization of Intellectual Work', in Levidow, Les, and Young, Bob (eds), *Science, Technology and the Labor Process* (London: Blackrose), pp. 46–65.

Duncan, Mike (1981), 'Microelectronics: five areas of subordination', in Levidow, Les, and Young, Bob (eds), *Science, Technology and the Labor Process* (London: Blackrose), pp. 172–207.

Gathright, Alan (1985), 'High-tech fields not providing big gain for women, study says', *Home News*, 9 June, E16.

Gerson, Judith (forthcoming), *At Home and in the Office: A Comparison of Home- and Office-Based Clerical Workers* (Chapel Hill, NC: University of North Carolina Press).

Glenn, Evelyn Nakano, and Feldberg, Roslyn L. (1977), 'Degraded and deskilled: the proletarianization of clerical work', *Social Problems*, 25, 1, pp. 52–64.

Hanson, Susan, and Johnston, Ibipo (1985), 'Gender differences in work-trip lengths: explanations and implications', *Urban Geography*, 6(3), pp. 193–219.

Herbers, John (1986), 'Rising cottage industry stirring concern in US', *Times Argus* (Montpelier, Vt), 15 May.

Johnston, Ibipo (1987), *A Geographic Perspective on Occupational Segregation: A Case*

Study of the Workforce in Worcester, Massachusetts, 1980, PhD dissertation, Massachusets Clark University.

Keller, Bill (1984), 'Unions battle against jobs in the home', *New York Times*, 20 May, pp. 1, 32.

Leavitt, Jacqueline (1985), 'A new American house', *Women and Environments*, 7, 1, pp. 14–16.

Lockwood, Charles, and Leinberger, Christopher B. (1988), 'Los Angeles comes of age', *The Atlantic Monthly*, 261, 1, pp. 31–56.

McKay, Roberta (1988), 'International competition: its impact on employment', in Christensen, Kathleen (ed.), *The New Era of Home-Based Work: Directions and Policies* (Boulder, Colo.: Westview Press).

Mackenzie, Suzanne, and Rose, Damaris (1983), 'Industrial change, the domestic economy and home life', in *Redundant Spaces* (London: Academic Press) pp. 155–200.

Madden, Janice Fanning (1981), 'Why women work closer to home', *Urban Studies*, 18, pp. 181–94.

Mattera, Philip (1983), 'Home computer sweatshops', *The Nation*, 2 April, pp. 390–2.

National Academy of Science (1983), *Office Workstations in the Home* (Washington, DC: National Executive Forum).

Nelson, Kristin (1984), 'Back offices and female labor markets: office suburbanization in the San Francisco–Oakland SMSA', unpublished PhD dissertation, University of California, Berkeley.

Noble, Kenneth B. (1986), 'House Committee reports people employed in home are exploited', *New York Times*, 23 July, A16.

Olson, Margrethe (1982), 'Remote office work: implications for individuals and organizations', *Women and Environments*, 5, 3, pp. 10–13.

Olson, Margrethe and Primps, Sophia B. (1984), 'Working at home with computers: work and nonwork issues', *Journal of Social Issues*, 40, 3, pp. 97–112.

Pratt, Joanne (1984), 'Home teleworking: a study of its pioneers', *Technological Forecasting and Social Change*, 25, pp. 1–14.

Schiff, F. W. (1979), 'Working at home can save gasoline', *Washington Post*, 2 September, c1, c2.

Toffler, Alvin (1980), *The Third Wave* (New York: Morrow).

US Department of Transportation (1985), *1983–4 Nationwide Personal Transportation Study* (Washington, DC: US Government Printing Office).

Zeldes, L. (1986), 'Home-based businesses touch off zoning debate', *Sunday Star*, Chicago, 1 June.

Zimmerman, Jan (1986), *Once Upon the Future: A Woman's Guide to Tomorrow's Technology* (New York: Pandora Press).

Zuboff, S. (1982), 'New worlds of computer-mediated work', *Harvard Business Review*, 60, 5, pp. 142–52.

17 Spatial implications of religious broadcasting: stability and change in patterns of belief

ROGER W. STUMP

The origin of religious broadcasting in the United States during the early 1920s marked the start of a revolution that has transformed the process of religious communication. In its impact on religious patterns, this revolution has been compared to Guttenberg's printed Bible and the Protestant Reformation (Armstrong, 1979; Cotham, 1985). The significance of religious broadcasting derives from the enthusiasm with which it has been adopted. Two months after becoming the first commercial radio station on the air, KDKA of Pittsburgh transmitted the world's first religious broadcast on 2 January 1921. The event was considered a great success and led to an immediate and continuing growth in religious broadcasting. This trend has been most pronounced in the United States, but has diffused to other areas of the world as well. The establishment of powerful international stations around the globe has been particularly important in transmitting religious programming to Latin America, Africa and Asia. The arrival of television in the late 1940s, and its enhancement by satellite and cable technology in the 1970s, added a new dimension to the realm of religious broadcasting, especially in the United States (Hadden and Swann, 1981).

Through its rapid expansion since the 1920s, religious broadcasting has developed a significant presence throughout the world. Christian organizations have been especially active in this development. By 1980, Christian radio and television shows were broadcast by over 1,400 Christian stations and a much higher number of commercial and government stations; and together these programs reached nearly a billion listeners during an average month (Barrett, 1982, pp. 18–19). Besides Christians, only Muslims have made much use of electronic broadcasting in religious communication. Islamic broadcasting has been limited, however, to radio programs dealing with the Koran, transmitted across the Islamic world from Egypt, Libya and Saudi Arabia (Browne, 1982).

The growing presence of religious broadcasting has important implications for the spatial distribution of religious belief. The electronic media have enabled Christian evangelists to reach vast audiences, including many listeners who, because of their isolation or government policies, might never have an opportunity for direct contact with a Christian missionary. Religious broadcasting also serves to confirm the faith of existing believers, a significant function given the growing influence of secularism during this century. Religious broadcasting may thus have varied effects on spatial patterns of belief, in some places contributing to change, in others reinforcing stability. Its influence in a specific area will depend on many factors, including the existing beliefs of listeners, their access to religious broadcasting, and the nature of their social and religious environments.

The purpose of this chapter is to identify major trends in the influence of religious broadcasting on world religious patterns, and to develop a model characterizing the nature of these trends and their interrelationships. In examining the effects of religious broadcasting, the discussion focuses on three basic processes in religious communication: evangelization, or disseminating knowledge of a particular faith; conversion, or convincing listeners to adopt that faith; and confirmation, or maintaining the faith of existing adherents. These processes correspond to stages in the diffusion of any innovation, although the third is especially important in the context of religion. In the diffusion of many innovations, after a brief period of confirmation the innovation's utility guarantees its continued use. In the case of religion, however, an individual may require regular reinforcement to maintain a consistent level of religious commitment (Davis, 1976). Confirmation of belief thus represents a particularly important function of religious communication.

The following discussion begins with a brief survey of world patterns of religious broadcasting. It then addresses broadcasting's role in each of the above processes of religious communication. This part of the discussion refers most often to Protestant evangelical broadcasting, the leading force worldwide in this phenomenon. The discussion finally presents an empirical model that consolidates the patterns examined in previous portions of the paper. This model provides both a systematic assessment of the spatial and spiritual effects of religious broadcasting and a framework for further research regarding this issue.

Patterns of religious broadcasting

The United States has been the dominant force in religious broadcasting throughout the latter's history. This dominance reflects the American origin of religious broadcasting, the rapid development of a sophisticated broadcasting infrastructure in this country, and the commitment of many American religious groups to evangelism. Religious broadcasting has grown most rapidly in the United States since World War II. From 1945 to 1960, an average of ten radio stations per year began weekly transmissions of at least fourteen hours of religious programming in the United States; and from 1972 to 1987 the number of radio stations broadcasting at least

Table 17.1   Patterns of Christian broadcasting, 1975.

Region	Percent Christian	Christian stations		Listeners and viewers	
		N	Per 100 million population	N (millions)	Percent of population
Anglo American	89.7	780	329.3	197.6	83.4
Latin America	94.1	436	134.5	165.4	51.0
Europe	86.7	56	11.8	250.2	53.0
USSR	35.8	0	0.0	20.7	7.7
Africa	42.5	6	1.5	55.7	13.9
East Asia	1.6	14	1.4	80.7	8.0
South Asia★	7.4	32	2.6	131.3	10.5
Oceania	89.5	9	42.2	11.9	55.9
World	33.2	1,333	33.6	913.4	23.0

★Includes all of southwest, south and southeast Asia.
Source: Barrett, 1982.

twenty hours of Christian programming a week grew by more than twenty per year. During the late 1970s, new Christian radio stations appeared at the rate of one per week. Over 1,000 radio stations in the United States currently offer some religious programming, and over 70 percent of these broadcast at least twenty hours of such programming every week (Armstrong, 1972, 1979; Barrett, 1982; *Broadcasting Cablecasting Yearbook*, 1987).

Television has also contributed to the expanding presence of religious broadcasting in the United States (Tweedie, 1978). The success of a handful of seminal figures in the 1950s laid the foundation for an explosion in religious television during the 1960s and 1970s, aided by the development of satellite and cable technology. By the 1980s, commercial television in every major broadcasting market in the United States carried at least some religious programming. Over thirty independent Christian stations broadcast religious material virtually full time, and cable systems carried hundreds of additional all-religious channels (Armstrong, 1979). The leading organization in religious cable television, the Christian Broadcasting Network, is now the fifth largest cable service in the country, transmitting to cable systems serving over 32 million subscribers (*Broadcasting Cablecasting Yearbook*, 1987).

Radio and television stations located in the United States continue to account for a majority of religious stations worldwide (Figure 17.1; Table 17.1). Nonetheless, religious broadcasting has developed a significant presence outside the United States over the past forty years, particularly through the medium of radio. This trend reflects in part the growth of local religious broadcasting in other countries, through the creation of independent religious stations and the transmission of religious programs by

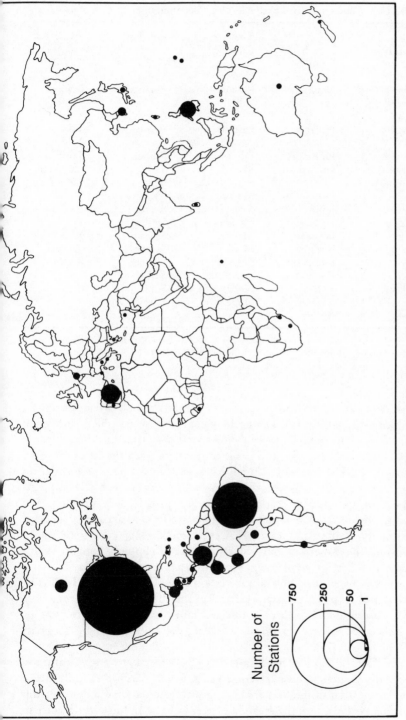

Figure 17.1 Christian radio and television stations, 1975
Source: Barrett, 1982

Number of Stations

750
250
50
1

Table 17.2   Languages used by the leading Protestant international broadcasters*

Afrikaans	Farsi	Kirghiz	Oromo	Swatow
Albanian	French	Korean	Pashto	Swedish
Amharic	Fulfulde	Kuki	Pedi	Tagalog
Amoy	German	Kurdi	Polish	Tamil
Arabic	Gujarati	Kuruk	Portuguese	Telugu
Armenian	Hakka	Lahu	Punjabi	Thai
Assamese	Hausa	Lao	Quechua	Tigrinya
Bengali	Hebrew	Latvian	Rumanian	Tshwa
Berber	Hindi	Lingala	Russian	Tswana
Burmese	Hmong	Lithuanian	Santali	Turkish
Byelorussian	Hungarian	Macedonian	Serbian	Ukranian
Cantonese	Ibo	Malagasay	Shan	Umbundu
Catalan	Indonesian	Malay	Shangaan	Urdu
Chewa	Italian	Malayalam	Shona	Vietnamese
Croatian	Japanese	Mandarin	Sindhi	Wa
Czech	Javanese	Marathi	Sinhalese	Xhosa
Danish	Kabyle	Meithei	Slovak	Yoruba
Dari	Kannada	Ndevele	Slovene	Zulu
Dutch	Kanuri	Nepali	Somali	
English	Karen	Norwegian	Sora	
Estonian	Kazakh	Nupe	Spanish	
Faroese	Khmer	Oriya	Swahili	

* Trans World Radio, Far East Broadcasting Co., Adventist World Radio, ELWA (Sudan Interior Mission), HCJB (World Radio Missionary Fellowship).
   Source: World Radio TV Handbook, 1987.

commercial and government stations. Local religious broadcasting has grown most rapidly in Latin America, which contains roughly a third of all Christian stations and over three-fourths of those outside the United States. Brazil alone accounts for nearly half of the stations in the latter category (Figure 17.1). The Catholic Church has founded over 80 percent of Latin America's religious stations, which it uses to transmit religious and educational programming. Various Protestant groups also operate local stations, however, often as part of missionary efforts to expand evangelical Christianity's Latin American presence (Barrett, 1982).

Beyond the Americas, only Spain and the Philippines contain a substantial number of local religious broadcasting stations. As in Latin America, most are run by Catholic organizations and provide both educational and religious programs. The Far East Broadcasting Company, a major international broadcaster, operates nine local stations in the Philippines that are devoted to Protestant evangelical programming (Barrett, 1982, pp. 566, 631).

While local religious broadcasts have become increasingly common outside the United States in recent decades, in most areas their influence is secondary to that of international broadcasts. International religious radio originated in 1931 with the creation of Radio Vatican, the leading voice in

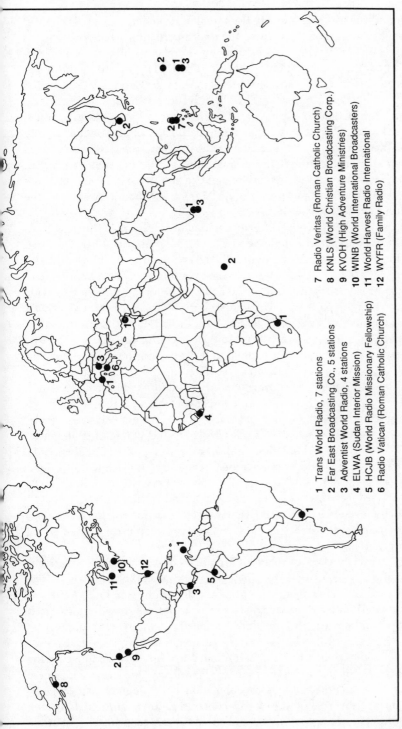

Figure 17.2 Major international Christian broadcasters, 1987
Source: World Radio TV Handbook, 1987

1 Trans World Radio, 7 stations
2 Far East Broadcasting Co., 5 stations
3 Adventist World Radio, 4 stations
4 ELWA (Sudan Interior Mission)
5 HCJB (World Radio Missionary Fellowship)
6 Radio Vatican (Roman Catholic Church)
7 Radio Veritas (Roman Catholic Church)
8 KNLS (World Christian Broadcasting Corp.)
9 KVOH (High Adventure Ministries)
10 WINB (World International Broadcasters)
11 World Harvest Radio International
12 WYFR (Family Radio)

Catholic broadcasting, and HCJB, a major Protestant station located in Quito, Ecuador. These remained the only international religious stations through World War II, but since then the number of international Christian broadcasters has steadily increased. The leading operations in international Christian radio now include Trans World Radio (TWR), the Far East Broadcasting Company (FEBC), Adventist World Radio (AWR), station ELWA, Radio Veritas and various stations located in the United States (Figure 17.2). Of the major international Christian stations, only Radio Vatican and Radio Veritas are affiliated with the Catholic Church; the rest are run by Protestant organizations, mostly of American origin. Several international Islamic stations also began operating during the 1960s and 1970s, transmitting from Egypt, Libya and Saudi Arabia to the Muslim world (Browne, 1982).

As the number of international religious broadcasters has increased, so has their potential for influence. The leading international Christian broadcasters transmit daily to virtually every nation in the world. TWR, FEBC, HCJB and Radio Vatican rank among the top twenty international broadcasters in terms of the hours of programming they provide each week (Browne, 1982). The leading stations broadcast, moreover, in a great diversity of languages. Radio Vatican and AWR transmit international broadcasts in over thirty languages, FEBC in nearly fifty languages (not counting those used in local broadcasts in the Philippines), and TWR around seventy. Together the five top Protestant international broadcasters offer programming in over a hundred different languages (Table 17.2). Radio Moscow, by comparison, broadcasts in about sixty languages, and Radio Beijing and the Voice of America in about thirty languages each (*World Radio TV Handbook*, 1987). Because of its linguistic diversity, international religious radio has a vast potential audience. In 1980 the languages used in international religious radio claimed 3.7 billion native speakers, about 85 percent of the world's population (Barrett, 1982). Based on this fact and the widespread availability of low-cost radios, the president of TWR has estimated that international religious broadcasting has the ability to reach eight of every ten people in the world (Mumper, 1986).

The actual audiences of religious broadcasts do not approach 80 percent of the world's population, but they are still sizable. The *World Christian Encyclopedia* estimates that, during an average month in 1975, local and international Christian broadcasts on secular and religious stations reach over 900 million listeners, or almost a fourth of the world's population (Barrett, 1982, p. 18). The same source estimates that by 1980 the average monthly audience had passed 990 million. Most listeners are found in the Americas, Europe and Oceania, where Christianity dominates and where the ratio of religious stations to population is the highest (Table 17.1). Nonetheless, the rest of the world contains about a third of Christian broadcasting's average monthly audience, comprised primarily of those listening to international broadcasts. The spatial extent of the actual and potential influence of religious broadcasting is thus nearly unlimited.

The functions of religious broadcasting

The goal of religious broadcasting is to disseminate a set of beliefs and strengthen their influence in the world. In meeting this goal, broadcasters have emphasized the three key processes in religious communication: evangelization, conversion, and confirmation of belief. Religious broadcasting's influence on these processes varies; it tends to be more effective in confirming belief and in evangelization than in conversion, which commonly depends on personal contacts. In places where contacts between current believers and potential converts are rare, however, religious broadcasts may emerge as a significant force in conversion. In any case, most religious broadcasters, and especially those with Protestant affiliations, consider all three of the above processes to be integral features of their mission.

Religious broadcasting's involvement in these processes has varied implications for spatial patterns of belief. Because it regularly reaches large and widely dispersed audiences, religious broadcasting has the potential to effect substantial changes in religious distributions. By spreading awareness of their beliefs and by converting new adherents, broadcasters may significantly expand the spatial extent of their religion's influence. By confirming the faith of their coreligionists, on the other hand, broadcasters may reinforce existing religious patterns. The effects of religious broadcasting on patterns of belief in a specific location will thus depend on the relative emphasis placed on evangelization, conversion and the confirmation of faith in the programming received by that location.

Religious broadcasting and evangelization

Use of the electronic media to spread awareness of their faith has been practiced almost exclusively by Christians. Other religious groups involved in broadcasting tend to direct their attention to existing adherents only. As a function of Christian broadcasting, evangelization has achieved its greatest importance in international radio transmissions to the non-Christian world, where access to other sources of information about Christianity is often limited (Hughes, 1984). Indeed, over 80 percent of the world's non-Christians in 1982 lived in areas at least partially closed to foreign Christian missionaries, including most of Asia, Eastern Europe, the Soviet Union and much of northern Africa (Barrett, 1982). For many non-Christians, radio thus represents the only means to acquire information about Christianity. Evangelization may also be a function of religious broadcasting in predominantly Christian areas, but its impact in such areas is generally small because few people are unfamiliar with Christianity.

Much evidence suggests that Christian broadcasters have had considerable success in increasing the awareness of their beliefs among non-Christians. The world's unevangelized population, those having little or no knowledge of Christianity, still included 1.4 billion people in 1980, almost all of them living in Africa, Asia or the Soviet Union. Nonetheless, the evangelized proportion of the world's population has risen sharply in recent years, from 61 percent to 68 percent during the 1970s alone (Barrett, 1982, pp. 18, 798). The contribution of Christian broadcasting to this trend cannot be determined precisely, but it appears to be substantial. Two-thirds of the growth in the world's evangelized population during the 1970s occurred in East and South Asia, areas that contain few Christian missionaries but represent key targets of international Christian broadcasting. In the mid 1970s, for example, India and China each received over twenty hours of Christian programming per day from the stations operated by FEBC, only one of several organizations broadcasting to these countries. At the same time, the evangelized proportion of both countries' populations increased by at least ten percentage points between 1970 and 1980. Religious radio's impact has been especially pronounced in China since 1975, when FEBC added AM transmissions to its existing schedule of shortwave broadcasts to the Chinese mainland. At that time China contained only 12 million shortwave sets, but 100 million receivers over which AM broadcasts could be heard. AM broadcasts thus increased FEBC's potential audience to about 90 percent of the Chinese population (Barrett, 1982, pp. 234, 377).

Estimates of the number of Christians and of the average monthly audience for Christian broadcasting offer additional evidence of the importance of the electronic media in evangelization (Barrett, 1982). In a country where Christian broadcasting's audience outnumbers the Christian population, the difference between the two equals the minimum number of non-Christians listening to Christian broadcasts. Division of this value by the total number of non-Christians yields the most conservative estimate of the proportion of non-Christians in the audience for Christian broadcasting. The larger this proportion, the more effective Christian broadcasts have presumably been in evangelizing a nation's population.

The widespread distribution of non-Christians listening to Christian broadcasts suggests that the latter have contributed significantly to the process of evangelization. By the mid 1970s, the audience for Christian broadcasting outnumbered the Christian population in most nations of South and East Asia and North Africa (Figure 17.3). Non-Christian audiences tended to be large in countries that contained or were close to Christian stations. For example, nearly a third of non-Christians listened to Christian broadcasts in Liberia, home of station ELWA. ELWA in fact has many listeners throughout northwestern Africa, the station's primary

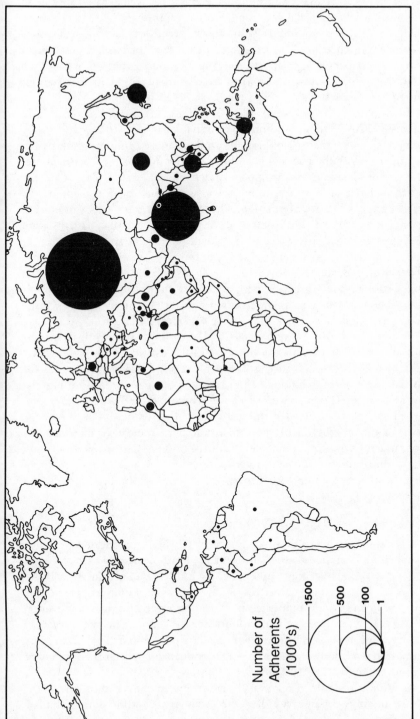

Figure 17.3 Isolated radio church adherents, 1970

Source: Barrett, 1982

Number of
Adherents
(1000's)

1500

500

100

1

target area. Christian broadcasting via television as well as radio appears to account for large non-Christian audiences in countries where access to both media is relatively common. Thus, about a tenth of South Korea's non-Christians listened to Christian broadcasts, a third of Japan's, and half of Taiwan's. Broadcasting's effect on evangelization has been more limited, on the other hand, in areas that are relatively distant from Christian stations or that contain few radio and television receivers per capita.

In absolute terms, non-Christians listening to Christian broadcasting in the mid 1970s were most common in India, Japan and China, where they numbered more than 50 million, 30 million and 20 million respectively. Comparisons at the national level between the number of Christians and the size of Christian broadcasting's audience indicate that, altogether, the latter includes at least 136 million non-Christians, or 5 percent of the world's non-Christian population. This percentage represents a minimum estimate, however, as it is based on the assumption that all Christians listen to Christian broadcasts. Since in fact not all Christians are likely to do so, the non-Christian portion of the audience for these broadcasts will be larger than estimated above. Moreover, the figure of 5 percent indicates the proportion of non-Christians listening at least monthly to Christian broadcasts. The number who listen less often but have become aware of Christianity through such broadcasts may be considerably larger.

In any case, broadcasting has clearly played an important role in spreading awareness of Christianity among non-Christians, in Asia especially. Christian broadcasting's success in this process has major implications for spatial patterns of religious belief and knowledge. By promoting the diffusion of Christian concepts to areas where they are largely unknown, broadcasters have expanded the spatial extent of awareness of one of the world's major religious traditions. In addition, they have increased the potential for significant change in the distribution of Christianity by laying the foundation for future conversions, either through the medium of broadcasting or by some other means.

Religious broadcasting and conversion

Conversion of listeners is a major goal of Christian broadcasters, and for evangelical Protestants especially, but in most situations broadcasting's influence on conversion appears to be rather weak (Hughes, 1984). Research into mass media's effects on the diffusion of innovations indicate that they can be very influential in spreading awareness of an innovation and in confirming the decision to adopt an innovation. The mass media are less effective, however, in actually persuading people to make that decision. Adoption tends instead to arise from the influence of personal contacts and communication (Davis, 1976).

Studies of the role of religious broadcasting in conversion replicate the above findings. In a survey of Pentecostals in Colombia during the late

1960s, for example, only 1 percent of the respondents indicated that they had been converted via radio, although by that time a Pentecostal station had been broadcasting in Colombia for about two years (Flora, 1976, pp. 65, 141). On a broader scale, using evidence from mass-evangelistic campaigns conducted via radio and television, the *World Christian Encyclopedia* estimates that such campaigns produce an average of one conversion per 10,000 listener- or viewer-hours (Barrett, 1982, p. 103). Because these campaigns are aimed specifically at gaining converts, rates of conversion among the audiences of other varieties of religious broadcasting should be even smaller.

Christian broadcasting thus appears to have relatively little impact on conversion since that process depends so heavily on personal communication. This pattern does not hold, however, when opportunities for personal contact between believers and nonbelievers are rare. In such cases, broadcasting may contribute conspicuously to the process of conversion. This contribution is most evident in isolated or inaccessible regions, like interior South America; in countries dominated by another religion, including the Muslim, Buddhist and Hindu nations of Africa and Asia; and in Communist states where the government is, or was, antagonistic to religion. For example, the Slavic Gospel Association, the largest producer of Russian-language Christian radio programs, estimates that international radio broadcasts are responsible for 80 percent of Russian conversions to Protestant churches (Mumper, 1986).

Broadcasting's contribution to the process of conversion is perhaps best exemplified by the phenomenon of the isolated radio church, a small community of believers created and maintained through the influence of Christian radio. Such communities develop as individual listeners begin to spread their enthusiasm for Christian broadcasts to friends and family. Because they evolve in isolation from organized Christianity, these communities are often unaware of the existence of specific denominations, and so form their own independent congregations (Barrett, 1982).

The number of radio churches in existence has been estimated from the frequency with which their members write to Christian radio stations. According to these estimates, approximately 74,000 isolated radio churches were active in over sixty different countries by 1970, with a combined membership of nearly 3 million (Figure 17.4). By the early 1980s, their membership had surpassed 4 million (Barrett, 1982, pp. 830–1). The Soviet Union contained by far the largest number of isolated radio churches, nearly 40,000 in 1970, with more than 1.5 million members. South and East Asia represented the other major focus of this phenomenon, with over a million isolated radio believers in 1970, almost two-thirds of these in India alone. Most of the Muslim nations of Southwest Asia and North Africa contain radio churches, but their membership has been limited by Islam's religious and cultural dominance. Radio churches have also arisen throughout Eastern Europe, despite the presence of established Chris-

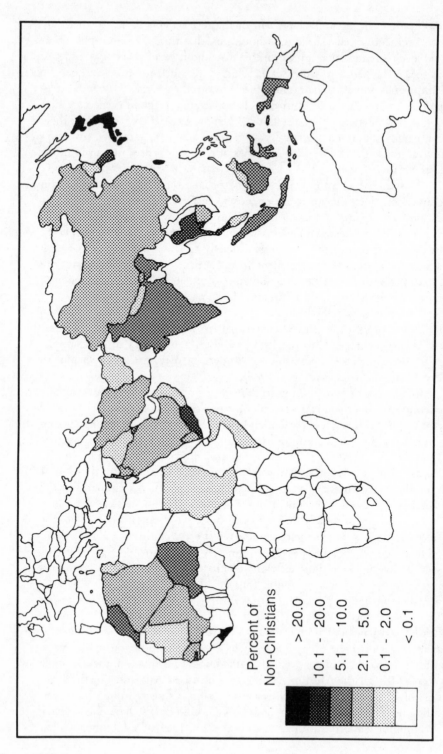

Figure 17.4 Proportion of non-Christians within the average monthly audience for Christian broadcasting, mid 1970s

Source: Barrett, 1982

Percent of
Non-Christians

> 20.0

10.1 - 20.0

5.1 - 10.0

2.1 - 5.0

0.1 - 2.0

< 0.1

tianity. Most adherents of radio churches there as elsewhere have adopted a Protestant orientation, however, and thus are not served by existing Catholic and Orthodox churches. Finally, nearly 9,000 radio church members are scattered throughout isolated regions of the Amazon basin.

In areas that have little direct contact with organized Christianity, then, radio may play an important role as a medium of conversion. As a result, broadcasting has helped spread Christianity to areas to which it would otherwise be unlikely to diffuse. Various factors may hinder the success of Christian broadcasting in this context. Social pressures inhibiting conversion to Christianity will be strong in many places where the latter is weak. The electronic media may in fact reinforce such pressures through broadcasts supporting another religion or through anti-religious programming. Since many areas lacking a strong Christian presence are remote and relatively undeveloped, their inhabitants may also have limited access to radio receivers and thus to Christian broadcasts. Despite the obstacles raised by such factors, however, Christian radio has succeeded in generating conversions in most non-Christian areas.

Thus, while broadcasting is less effective in the process of conversion than in other aspects of religious communication, it may still play a key role in the process. The changes in religious distributions brought about by broadcasting-based conversions are rarely large. Nonetheless, they may be highly conspicuous because they are concentrated in areas where Christianity's advance has been most limited.

Religious broadcasting and the confirmation of belief

In many ways confirmation of belief represents the most important function of religious broadcasting. Its importance derives in part from its pivotal role in religious communication generally. Missionaries and evangelists have commented widely on the need to reinforce the belief of recent converts. The decision to adopt a particular faith often occurs in an initial rush of enthusiasm, and the convert's commitment to a new set of beliefs may weaken as this enthusiasm fades. The convert's faith thus requires continuing support. Furthermore, the need for such support is not limited to recent converts; the persistence of religious commitment among long-time believers may also depend upon the regular reinforcement of their faith (Davis, 1976).

Confirmation of belief also represents the primary function of religious broadcasting because of the extent of its impact. Most members of religious broadcasting's audience share the broadcaster's faith. Two-thirds of the average monthly audience for Christian broadcasting, for example, inhabit the predominantly Christian regions of Europe, the Americas and Oceania (Table 17.1); and Christians living in non-Christian nations comprise perhaps another 20 percent. For these people, confirmation of belief represents Christian broadcasting's most significant attraction. Moreover, the audience for Christian broadcasting includes a majority of the world's Christians. Out of a total audience of 913 million in 1975, non-Christians accounted for at least 136 million, but probably no more than 200 million. Christian listeners should thus have numbered between 700 and 800 million, or between 50 percent and 60 percent of the total Christian population (Barrett, 1982).

Non-Christian religious broadcasting, moreover, is devoted almost exclusively to sustaining the faith of existing believers. The Koran programs broadcast within the Muslim world, for example, are intended 'to deepen and reinforce the faith of believers' (Boyd, 1982; Browne, 1982). The Voice of America, one of the international broadcasting services of the United States government, provides regular programming with a similar intent to adherents of Judaism and Russian Orthodoxy within the Soviet Union (Mainland et al., 1986). The Russian Orthodox broadcasts draw an estimated 20 to 30 million listeners each week from the Soviet Union's 50 to 80 million Orthodox believers (Spring, 1983).

Much evidence confirms the effectiveness of religious broadcasting in reinforcing belief. Most of this evidence is based on survey research conducted in the United States, but it presumably applies to other situations as well. A major survey of American religious television conducted in the early 1980s found, for example, that most regular viewers watch out of the desire to confirm or cultivate their own beliefs and because they are dissatisfied with the moral content of other forms of television programming (Gerbner et al., 1984). The similarity between an individual's beliefs and those presented in religious broadcasts was in fact found to be the strongest predictor of whether an individual would watch religious television. Other studies have replicated this finding (Gaddy, 1984; Stacey and Shupe, 1982).

Studies of the effects of religious broadcasting on local church involvement also suggest that the former plays an important role in reinforcing faith. In a 1980 survey of people who had watched religious television in the previous month, 27 percent said that their church involvement had increased in response to watching religious programs, only 7 percent that it had decreased. The remaining respondents observed no change in their involvement (Gallup Organization, 1981). A similar survey conducted in 1983 reported similar results. Among respondents viewing more than an hour of religious broadcasting in the previous week, 14 percent stated that this activity had increased their church involvement, only 3 percent that it had reduced their involvement (Gerbner et al., 1984, V–75).

That most respondents to these surveys observed no change in their church involvement could mean that religious broadcasting has little influence in this context. Interpretation of these surveys must take into account, however, the already high rates of church involvement found among religious broadcasting's audience. In the 1983 survey cited above, half of those who watched religious television attended church at least weekly, compared to a third of non-viewers (Gerbner *et al.*, 1984, V–47). The ability of religious television to increase church involvement among a significant number of its viewers, whose average rates of participation are already high, provides additional evidence of its power to strengthen its audience's faith.

Given its role in confirming belief, religious broadcasting may influence the spatial distribution of religion by reinforcing existing patterns. This effect of religious broadcasting may have especially important implications in situations where other factors work against the preservation of established religious patterns. Christian broadcasters direct a substantial amount of programming to the Muslim world, for example (Head, 1974). An important effect of Islamic broadcasting may therefore be to offset the influence of Christian broadcasting on its Muslim audience. Many areas of the world, and especially the Western nations traditionally dominated by Christianity, have become increasingly secularized during this century; the nonreligious, who made up less than 1 percent of the population of Europe and North America in 1900, now account for over 10 percent (Barrett, 1982). Religious broadcasting may serve to slow this process. Indeed, among Western nations the effects of secularization have been least pronounced in the United States, where the influence of religious broadcasting is greatest. Communist governments hostile to religion have also challenged the stability of religious patterns, but Christian broadcasting again appears to have reduced the impact of this trend in traditionally Christian regions.

A model of the effects of religious broadcasting

The patterns presented above reveal that religious broadcasting has varied effects on its audience. The nature of these effects in a given area depends on many factors, including the beliefs held by the area's inhabitants, the ease of access to religious broadcasting, and the influence of other social and religious factors on interest in religious broadcasts. Moreover, the varied influences of religious broadcasting on its audience have affected spatial patterns of religion in different ways, in some cases encouraging stability and in others leading to change.

A simple conceptual model is proposed here to synthesize the varied effects of religious broadcasting (Table 17.3). Although the model depicts trends in Christian broadcasting specifically, it can easily be generalized to fit other cases as well. Several important assumptions concerning Christian broadcasting and its effects are expressed in the model, drawn from the evidence presented above. Of these assumptions, the most basic and perhaps the most obvious is that the nature of Christian broadcasting's effects depends most directly on the religious character of its audience. More precisely, the effects of Christian broadcasts to Christian audiences

Table 17.3 A model of the effects of Christian broadcasting.

Audience:	Christians	Non-Christians	
Function:	Confirmation of belief	Evangelization	Conversion
Magnitude of effects:	Large	Moderate	Small
Direction of effects:	Stability	Change	Change
Religious effects:	Provides significant support for current adherents' beliefs	Creates moderate potential for change in religious beliefs	Converts small numbers in places lacking a strong Christian presence
Spatial effects:	Significantly reinforces existing religious distributions	Creates moderate potential for change in religious distributions	Produces small but conspicuous changes in religious distributions
Factors influencing effectiveness:	Hours of broadcasting, access to receiving sets, content of available programs	Hours of broadcasting, access to receiving sets	Hours of broadcasting, access to receiving sets, content of available programs, social effectiveness
Major areas of influence:	North and Latin America, Europe, Oceania	Asia, Africa	Eastern Europe, USSR, Asia, North Africa, isolated areas in Latin America

will differ significantly from those of broadcasts intended for non-Christians.

Broadcasts to Christians should have a greater impact than broadcasts to non-Christians, in part because the former have a larger total audience. This difference in audience size reflects the greater presence of Christian broadcasting in areas dominated by Christians, as well as the greater availability of radios and televisions in many Christian areas, which overlap extensively with the developed regions of the world. In addition, the impact of broadcasting directed at Christians will be enhanced by the greater receptivity of its audience to the broadcaster's message. In transmitting to non-Christian areas, broadcasters must overcome disinterest or even hostility toward their beliefs among their potential audience.

The *nature* of the effects of Christian broadcasting will also depend on its audience. Broadcasts intended for Christians serve primarily to enhance religious stability. This stability arises from the support that such broadcasts provide for their audiences' beliefs. Such support maintains religious commitment among members of the audience, and thus reinforces existing aggregate patterns in the spatial distribution of belief. Broadcasts directed at non-Christians, on the other hand, serve primarily to encourage religious change. Such change may derive from increasing awareness of Christianity through evangelization or from growth in the number of active Christians through conversion. The process of evangelization should affect more people than the process of conversion, because the former involves merely the spread of information rather than a more complex process of persuasion. Indeed, only in places where other forms of Christian communication are limited or nonexistent does broadcasting play an important role in conversion. Unlike the process of conversion, however, evangelization produces little actual change in religious beliefs. Broadcasting to non-Christians affects the spatial distribution of religion, then, by increasing the *potential* for religious change among moderately large evangelized populations, and by introducing small but conspicuous increases in Christian membership in areas dominated by other religions or by the absence of religion.

The magnitude of religious broadcasting's impact on its diverse audience will, of coruse, vary. In reaching both Christian and non-Christian listeners, the effectiveness of Christian broadcasts will first depend on their accessibility. The latter will in turn be a function of two variables: the hours of programming broadcast in the audience's language, and the number of radio and television receivers available to the intended audience. The ability of religious broadcasting to attract large audiences in the United States, for example, reflects both the great amount of programming and widespread ownership of radio and television sets in this country. Local and international stations also transmit a large amount of religious programming to and within Latin America, but lower rates of radio and television ownership have limited the size of regular audiences there. In Western Europe, on the other hand, the smaller amount of religious programming offsets the greater availability of receiving sets (Table 17.1).

The content of Christian broadcasts will also influence their effectiveness, especially in terms of the processes of confirmation and conversion. For a listener to receive religious support from a broadcast, the latter must espouse beliefs compatible with the listener's. The content of the religious programs available to an individual thus determines whether he or she will find a program useful in confirming his or her beliefs. Similarly, a religious broadcast will be successful in converting listeners only if it presents its message in an acceptable and convincing manner. In the case of evangelization, on the other hand, religious broadcasts intend to inform rather than to change or reinforce beliefs. The effectiveness of broadcasting in this context should thus depend less directly on the compatibility of the broadcaster's message and his or her listeners' beliefs.

Finally, because religious conversion may radically change the convert's social identity and position, the effectiveness of Christian broadcasts in this process may be diminished by various social forces. Such forces are likely to be most evident in nations containing little religious diversity, or where the government is hostile to religion. The religious homogeneity of the Islamic countries of North Africa and southwest Asia, for example, creates very strong social pressures against conversion to other faiths. This pressure may take on official status, as in Saudi Arabia where all other religions are officially prohibited. Sanctions against religion in former Communist states, including the restriction of religious activities, discrimination against religious minorities and the dissemination of anti-religious propaganda, also contributed to the difficulty of converting listeners through broadcasting (Barrett, 1982; Powell, 1975).

The combined influences of the above factors have produced important spatial variations in the nature and effectiveness of Christian broadcasting. In confirming the faith of existing adherents, religious broadcasting has been most effective in the world regions dominated by Christianity: the Americas, Europe and Oceania. Within these regions its influence has been most pronounced in the United States, where religious broadcasting began and is now most widespread. Christian broadcasting as a medium of evangelization has achieved its greatest efforts in Asia, and to a lesser extent Africa, where the growing availability of radio receivers has given international religious broadcasts access to very large unevangelized populations. In the context of conversion, Christian broadcasting has achieved its most conspicuous successes in former Communist nations, in portions of Africa and Asia dominated by Islam, Hinduism and Buddhism, and in isolated areas within Latin America. These areas tend to be inaccessible or inhospitable to Christianity, and thus lack other forms of Christian missionary activity. The success of broadcasting in these areas thus reflects its status as essentially their only contact with Christianity. In places where more traditional forms of missionary activity are common, broadcasting contributes less conspicuously to conversion.

Concluding remarks

The above model depicts the major effects of Christian broadcasting in evangelization, conversion and confirming beliefs, the three basic processes of religious communication. The model could be applied to other religions as well, but only Islam has developed a substantial presence in religious broadcasting, and its activities are limited to confirming the faith of existing believers. Nonetheless, the portion of the model relating to this function applies fairly well to Islamic broadcasting.

Although it identifies the major effects of religious broadcasting and many of the important factors related to them, the proposed model remains incomplete. Limitations in the available data concerning religious broadcasting have prevented consideration of at least four key issues. First, broadcasting's role in conversion is probably not limited to persuading non-Christians to adopt Christianity. Christian broadcasts may also encourage listeners to switch from one Christian denomination to another. Shifts in affiliation at this level may be insignificant at the world scale, since they are likely to involve small numbers of people moving between relatively similar denominations. Within some regions, however, such changes may have a marked influence on religious patterns. The expanding presence of Protestant broadcasting in Latin America in recent decades, for example, has been accompanied by rapid growth in Protestant proportions throughout the region. Examination of the connection between these patterns could provide useful insights into the effects of religious broadcasting on a regional scale.

A second issue concerns the effectiveness of broadcasting in preparing non-Christians for conversion. Broadcasters often cite this as an important function of their activities (Davis, 1976; Hughes, 1984); indeed, it is largely to this end that broadcasters have become widely involved in evangelization. A significant behavioral gap separates the act of becoming aware of a religion and the act of converting to it, however. To understand fully the effects of Christian broadcasting on non-Christians, and on the spatial distribution of the world's religions, it must be determined to what extent broadcasting helps close that gap.

The third basic issue not treated in the model concerns the relative effectiveness of radio and television in religious communication. The smaller cost of radio receivers and the greater range of radio transmissions has enabled radio to reach a larger audience worldwide, but in places where religious programming is available via both radio and television, one medium may prove to be more effective than the other. Opinions differ, however, concerning which the more effective will be. Some argue in favor of television on the basis that it creates among its audience an illusion of participating in the broadcast (Hadden and Swann, 1981, pp. 66–7). Others argue in favor of radio, asserting that it creates a sense of intimacy between the broadcaster and listener (Davis, 1976; Van Der Puy, 1984). It is perhaps most likely that radio will be more appropriate for some tasks, television for others. Assessment of the relative utility of each medium in different situations would help to clarify general trends in the effectiveness of religious broadcasting.

A final issue deserving further attention is the American role in religious broadcasting internationally. The leading Protestant organizations involved in international broadcasting are almost all American in origin; many local Protestant stations in Latin America are operated by denominations headquartered in the United States; and much of the religious programming broadcast over local and international stations outside the United States is produced either in the United States or by American organizations (Barrett, 1982, p. 720). The prominent contribution of Americans to religious broadcasting internationally raises important questions. Does American involvement influence the effectiveness of international religious broadcasting, either positively or negatively? How has this involvement affected the spread of American denominations to other parts of the world? Do American religious broadcasters contribute to the diffusion of American values not directly tied to religion? Has religious broadcasting aided in the diffusion of American cultural features generally? Answers to these questions would enhance our understanding not only of the implications of religious broadcasting as a cultural force, but of contemporary processes of transculturation as well.

In sum, religious broadcasting can have significant effects on religious belief and thus on the spatial distribution of religion. The model presented above identifies some of the major features of these effects, but many more remain unexamined. Further inquiry into the character and influence of religious broadcasting, locally, nationally and internationally, should provide important insights into patterns of cultural change and stability both within and beyond the realm of religious belief.

References

Armstrong, Ben (1972), *Religious Broadcast Stations in the United States* (Madison, NJ: National Religious Broadcasters).

——(1979), *The Electric Church* (Nashville, Tenn.: Thomas Nelson Publishers).

Barrett, David B. (ed.) (1982), *World Christian Encyclopedia* (Nairobi: Oxford University Press).

Boyd, Douglas A. (1982), *Broadcasting in the Arab World* (Philadelphia, Pa: Temple University Press).

Broadcasting Cablecasting Yearbook, 1987 (Washington, DC: Broadcasting Publications).

Browne, Donald R. (1982), *International Radio Broadcasting* (New York: Praeger).

Cotham, Perry C. (1985), 'The electronic church', in Allene Stuart Phy (ed.), *The Bible and Popular Culture in America* (Philadelphia, Pa: Fortress Press), pp. 103–36.

Davis, B. E. (1976), 'Mass media in missions', in George P. Gurganus (ed.), *Guidelines for World Evangelism* (Abilene, Tex.: Biblical Research Press), pp. 140–69.

Flora, Cornelia Butler (1976), *Pentecostalism in Colombia* (Rutherford, NJ: Fairleigh Dickenson University Press).

Gaddy, Gary D. (1989), 'Some potential causes and consequences of the use of religious broadcasts', in Bromley, David G., and Shupe, Anson (eds), *New Christian Politics* (Macon, Ga: Mercer University Press), pp. 117–28.

Gallup Organization (1981), 'Religion in America, 1981', *Gallup Opinion Index* 184, pp. 1–76.

Gerbner, George *et al.* (1984), *Religion and Television* (Philadelphia, Pa: Annenberg School of Communications, University of Pennsylvania).

Hadden, Jeffrey K., and Swann, Charles E. (1981), *Prime Time Preachers: The Rising Power of Televangelism* (Reading, Mass.: Addison-Wesley).

Head, Sydney (1979), 'Islam and Christianity', in Head, Sydney W. (ed.), *Broadcasting in Africa* (Philadelphia, Pa: Temple University Press), pp. 201–4.

Hughes, Robert Don (1989), 'Models of Christian missionary broadcasting', *Review and Expositor*, 81, 1, pp. 31–42.

Mainland, Edward, Pomar, Mark, and Carlson, Kurt (1986), 'The voice present and future: VOA, The USSR and Communist Europe', in Short, K. R. M. (ed.), *Western Broadcasting over the Iron Curtain* (London: Croom Helm), pp. 113–36.

Mumper, Sharon E. (1986), 'The missionary that needs no visa', *Christianity Today*, 30, 3, pp. 24–6.

Powell, David E. (1975), *Antireligious Propaganda in the Soviet Union: A Study of Mass Persuasion* (Cambridge, Mass.: MIT Press).

Spring, Beth (1983), 'The Voice of America beams religion to the Soviet bloc', *Christianity Today*, 27, 14, pp. 35–6.

Stacey, William, and Shupe, Anson (1982), 'Correlates of support for the electronic church', *Journal for the Scientific Study of Religion*, 21, 4, pp. 291–303.

Tweedie, Stephen W. (1978), 'Viewing the Bible Belt', *Journal of Popular Culture*, 11, 4, pp. 865–76.

Van Der Puy, Abe C. (1984), 'Evangelism through radio', in Douglas, J. D. (ed.), *The Work of an Evangelist* (Minneapolis, Minn.: World Wide Publications), pp. 854–5.

World Radio TV Handbook, 1987 (1987) (New York: Billboard Publications).

18 Communications and information flows in the commonwealth Caribbean: a force for integration?

ANNE LYEW-AYEE

Introduction

The territories commonly referred to as the Commonwealth Caribbean comprise a number of island-states and two mainland countries which were or are colonies of the United Kingdom in the region. In this paper, the focus is on the two mainland countries (Belize and Guyana), and on Jamaica, Trinidad and Tobago, Barbados, and the Windward and Leeward Islands in the eastern Caribbean (Figure 18.1). The Cayman Islands, the Turks and Caicos Islands and the British Virgin Islands are very small colonies which are excluded from discussion in this paper because of their

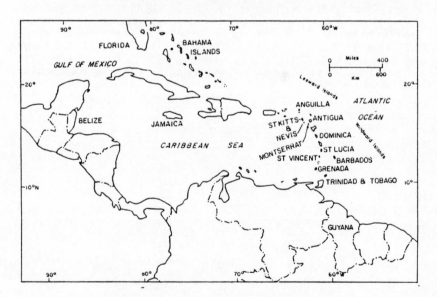

Figure 18.1 The Commonwealth Caribbean

Table 18.1 The Commonwealth Caribbean: some relevant statistics.

Country	Area (km²)	Population 1982	Per capita National income 1983 (US $)	No. of radios per 1000 population 1982	No. of telephones/ 1000 population, 1982
Barbados	430	250,400	3,990	542	267
Belize	22,963	152,000	870	483	43
Guyana	215,000	920,000	560	383	28
Jamaica	10,991	2,250,000	1,270	328	55
Leeward Islands:					
Anguilla★	91	6,500	800	n.a.	90
Antigua & Barbuda	440	77,000	1,690	227	43
Monserrat★	102	11,700	2,400	520	224
St Kitts & Nevis	269	45,000	1,320	450	73
Trinidad & Tobago	5,128	1,200,000	6,830	255	69
Windward Islands					
Dominica	751	74,400	970	473	13
Grenada	344	106,000	830	342	52
St Lucia	616	124,000	1,050	660	64
St Vincent & the Grenadines	389	100,000	840	290	58
Total	257,514	5,317,000			

★Still British colonies.   n.a.   Not available.
Source: *The World in Figures* (London: *The Economist*, 1984).

markedly different economic circumstances. The Bahamas group is also excluded because of its extra-Caribbean location and its distinctly different physical and historical backgrounds (the lack of a sugarcane plantation economy, for example).

The territories which are the focus of this paper are strung out over an expanse of more than 3 million km², but they actually have a combined land area of only 258,000 km², and the total population is only 5.3 million (Table 18.1). Guyana and Belize, on the American mainland, have relatively large areas, but small populations. Among the islands, Jamaica is the largest in both area and population. Anguilla and Montserrat are only about 100 km² each, with estimated populations in 1982 of 6,500 and 11,700, respectively.

The British colonial heritage dates back to the early seventeenth century, and one aspect of this heritage is the marked similarity in the general development of the territories. The sugarcane plantation economy, with its dependence on slave labor imported from Africa until well into the nineteenth century, has left its mark on the region. The overwhelming majority of the population is of African descent. Export agriculture remains the mainstay of the region's economy, although there has been diversification into such other crops as bananas and coconuts. Jamaica and Trinidad and Tobago have the only significant mining sectors, producing bauxite and petroleum respectively. Most of the islands have also developed important tourism sectors, vying with one another for visitors from North America in particular.

The other major aspect of the colonial heritage which is a factor providing a common thread linking the territories is the English language. The term 'English-speaking Caribbean' aptly sets the countries apart from their mainly Spanish-speaking neighbors in the region.

This paper examines the patterns of communications and information flows which have developed among the countries of the Commonwealth Caribbean and attempts to assess the potential of current and future developments in communication technology for promoting closer links among the territories.

Internal communication

The common language imposed by the colonial rulers for more than 300 years did a great deal to facilitate administration from and communication with London, but among the West Indian territories physical separation led to the establishment of small newspapers serving only readers on individual islands. Newspapers appeared as early as 1718 in Jamaica and 1731 in Barbados, and new daily or weekly newspapers continued to make their appearance quite regularly, up to the present time (Table 18.2). However, most of these newspapers were shortlived, and only two newspapers in the

Table 18.2   Number of Commonwealth newspapers, 1700–1971.

Country	1700–49	1750–99	1800–49	1850–99	1900–71	Total
Barbados	1	1	12	13	22	49
Jamaica	2	13	43	34	64	156
Leeward Islands:						
Anguilla	0	0	0	0	2	2
Antigua	0	4	5	4	9	22
Monserrat	0	0	0	2	4	6
St Kitts & Nevis	1	4	1	14	11	31
Trinidad & Tobago	0	2	10	19	18	49
Windward Islands:						
Dominica	0	10	5	9	16	40
Grenada	0	3	5	9	9	26
St Lucia	0	1	8	8	11	28
St Vincent	0	0	5	8	6	19
Total	4	38	94	120	172	428

Source: Lent, 1977, p. 44.

region have a publication record dating continuously back to the nineteenth century (one each in Jamaica and St Lucia) (Lent, 1977).

Today, a large part of the total circulation of daily newspapers in the region is accounted for by three newspapers in Trinidad and Tobago, two in Jamaica, and one in Barbados – and in each of these countries the circulation is almost entirely internal. Especially in the larger countries, it is difficult and expensive to distribute a daily newspaper to all the communities within the country. Low incomes and low literacy rates in less accessible areas combine to make it even less likely that publishers would bother with the effort of having their newspapers reach these areas daily.

Many of the countries are faced with severe foreign exchange problems which are reflected in, for instance, the widespread use of obsolete equipment and techniques for the production of newspapers. Only the largest-circulation newspapers, such as the *Gleaner* in Jamaica, can afford the computerized typesetting and production processes which are the norm in the developed countries. Foreign-exchange crises also lead to periodic shortages of newsprint. The withholding by governments of import permits and of the necessary foreign exchange for newsprint imports may also work against total freedom of the press, most notably in Guyana.

There were slight improvements in the communication links between the territories with the growth of trade with Britain in the nineteenth century, and the telegraph, which came into general use in the 1870s, was important in linking the territories not only internally but also with the rest of the world.

The other major means of internal communication until the advent of the electronic media in the twentieth century was the postal service. The village post office, which is also the receiving and distribution point for telegrams, plays an important role in keeping inhabitants of rural areas in touch with friends and families. In rural areas, where mail is generally not delivered to individual homes, the post office also has a social role as a point where people meet when they gather to pick up their mail.

Interestingly, it was the West Indian interest in cricket, itself another aspect of the British colonial heritage, which spurred the introduction of radio broadcasting in the region. In the 1930s, radio commentaries on cricket matches were being avidly followed in Guyana, Barbados, Trinidad and Jamaica. The fact that the system was not wireless until the middle of the twentieth century meant that relatively few people had access to this means of communication.

As elsewhere in the Third World, the transistor radio had a dramatic impact on the ability of people living in remote or physically inaccessible areas to be brought abreast of events in the larger world. Many rural areas which still lack electricity are able to keep in touch with at least the local radio station(s) on their island by means of battery-powered radios. High figures for radio ownership (Table 18.1) reflect the relatively low cost of a radio set, as well as the inability of television to reach many remote areas. According to Schramm (1964), among the yardsticks suggested by the United Nations Educational, Scientific and Cultural Organization (UNESCO) for measuring the sufficiency or insufficiency of mass communication facilities in the developing countries is a figure of five radio receivers for every hundred inhabitants; all the countries in the Commonwealth Caribbean would have achieved more than this 'minimum level'.

The radio is a powerful medium for reaching large numbers of people in each territory, especially in situations where not only do the people not have access to a daily newspaper but many are illiterate. The oral information tradition is rooted in the plantation/slavery era. Great interest in radio soap operas is exploited by Jamaica's National Family Planning Board, which uses the format of a fifteen-minute weekly tale following the daily lives and travails of a fictional community called 'Naseberry Street' to push the need for birth control and family planning in the country.

Politicians also make use of the medium to get their message across. In virtually every one of the countries the government has an information service which uses the various media (press, radio, television, billboards, etc.) to inform the public on government policies and programs, for instance, but also to serve as a taxpayer-funded propaganda unit. This, in conjunction with government ownership and operation of radio and/or television stations, gives rise to charges of partisan political manipulation of the media, especially in periods leading up to elections. 'Paid political broadcasts' are a regular part of Sunday afternoon radio fare in Jamaica, for example, in or out of election years.

Television lags behind radio as a means of internal communication. It obviously cannot reach those areas which do not have electricity supply. A television set is also considerably more expensive than a small transistor radio. In addition, the difficulties of mountainous terrain add to the high cost of ensuring that television signals can reach remote areas. In Jamaica, television transmission originates at the studios of the sole station in Kingston, and the signal is picked up and retransmitted by no fewer than eight other repeaters scattered strategically around the island. Establishment and maintenance costs for such a system are high, and there are still repeated complaints about poor television reception in many rural areas.

Television was not introduced into the Commonwealth Caribbean until independence had been achieved by such countries as Jamaica, Trinidad and Tobago, and Barbados from the early 1960s. It was to be almost twenty years before the advent of color television in the region. Some of the smaller and/or poorer territories (e.g., Dominica and Guyana) remained without a local television station throughout this period. Financial problems as well as an express political decision kept Guyana without a local television service until 1 January 1988, when the newly established local station initiated a one-hour-per-day transmission.

Intraregional communications: patterns and problems

Mail, telegraph and telephone services are well established as means of communication among the territories in the region. Such services have improved considerably in the past two decades, especially with the setting up of satellite tracking stations in Trinidad, Barbados, Jamaica and Guyana in the early 1970s. In 1977 Cable and Wireless significantly improved telecommunication links in the region with its 1,300 km Caribbean microwave system linking twelve islands.

One of the factors deterring the greater use of telecommunication services is the high cost of such services, even over relatively short distances. The smaller media houses, in particular, cannot afford to lease telecommunication circuits and carry more news items about neighboring islands. Cuthbert (1982, pp. 131-2) points out that 'it is cheaper for Reuters to file news from the French Caribbean islands of Martinique and Guadeloupe to Paris than to file from those islands to neighboring Barbados, because the French have a press rate while the Caribbean does not', and that 'a news link from Barbados to nearby Venezuela in 1979 cost about three thousand dollars per month while a link betwen New York and Venezuela cost only one thousand dollars per month.'

The University of the West Indies, itself an example of an institution funded by the governments of the English-speaking territories in the region, initiated in 1983 an experimental program in distance teaching. The University of the West Indies Distance Teaching Experiment (UWIDITE) links the three campuses (in Jamaica, Barbados and Trinidad) and extramural centers in Antigua, Dominica, Grenada, St Lucia and St Vincent by means of a leased telecommunication circuit (Jamaica and Trinidad are linked by INTELSAT satellite). In addition, within Jamaica, local centers in Mandeville, Montego Bay and Port Antonio are linked by microwave to the base in Kingston. Conferences involving persons at all these locations, as well as a number of courses, have underscored the advantages of such a modern facility in promoting regional links. In particular, medical practitioners in the smaller territories, for example, make use of the system to confer with their colleagues at the main teaching hospital on the Jamaica campus. Persons may continue to be employed in non-campus territories and enrol for courses via their local UWIDITE facility without having to be physically present at one of the campuses.

Still at the experimental stage, this program in distance teaching faces severe problems of financing in keeping the program going as well as in expanding the network. Not quite yet the 'open university' concept so successful in such countries as the United Kingdom and Australia, the distance teaching idea is a potentially powerful force for expanding the availability of tertiary education to larger numbers of people in the region.

Newspapers have not moved beyond their insular readerships. No West Indian newspaper can claim a regional circulation, although some, such as the Barbados *Advocate*, are available on nearby islands. Physical separation adds significantly to the costs of distributing daily newspapers, especially where, in some cases, there are no daily air links between countries.

On the other hand, at least the Sunday editions of some major newspapers from the United States and the United Kingdom (especially the New York *Times*, the Miami *Herald* and the *Sunday Times* of London) are 'airspeeded' to several points in the Caribbean, as are the major English-language weekly newsmagazines (*Time*, *Newsweek*, *The Economist*, *US News & World Report*, among others).

As far as the flow of news is concerned, there continues to be little cooperation among the countries. The major international news agencies (Reuters, Associated Press and United Press International) were, until recently, the sources for even regional news, so that a media house in Jamaica, for example, might get news of an event in the eastern Caribbean via one of these agencies, and such an event, to be carried by these agencies, had to be of international interest. It was not until 1968 that Reuters moved its Caribbean desk from London to Barbados.

The mass media began to take on a regional look during the shortlived West Indies Federation (1958–62), when some thirteen British colonies in the region came together in a political union. To foster interest in other federal territories and their people, the Jamaica *Gleaner* and the Barbados *Advocate*, for example, began to carry West Indian news columns. Profiles of the various members of the proposed federation, and coverage of the elections to the federal parliament and, later on, of news from that parliament in Trinidad, were carried in the press and on radio. However, even before the breakup of the federation in 1962, the media houses were feeling the cost of covering regional news, a problem exacerbated by the lack of cooperation among the media in the different territories.

The Caribbean Broadcasting Union (CBU) was launched in 1970 to encourage greater interest in regional affairs. In 1976 the establishment of the Caribbean News Agency (CANA) had a more specific aim: mass media houses in fifteen countries got together to provide an 'indepth' news service for the region. Not only do the media houses provide news items for one another, CANA also maintains desks in such cities as Toronto, New York and Washington, DC, which cover news of special interest to the region. Reuters and AP remain the major sources of extraregional news in the press.

For radio broadcasting, the problem of overcoming the great distances separating the countries is compounded by frequency incompatibilities and the smallness of the stations' operations. At the same time, newscasts and other programs transmitted by the British Broadcasting Corporation (BBC) and by the Voice of America (VOA), for instance, reach large numbers of listeners either directly on their own shortwave receivers or through the hookups provided by their local radio stations.

Cricket is again a unifying force in the region. The world-class West Indies cricket team is made up of players from the various countries, and commentaries on its matches as far away as Australia and New Zealand are carried live on most West Indian radio stations. Regional cricket competition among the territories is also well covered in the media. However, there appears to be little interest in any other sporting activity in another part of the Commonwealth Caribbean.

The costs of local television programming are high compared to those in countries such as the United States. Even international cricket matches are not covered live on local television stations. Indeed, a sports enthusiast in the region is more likely to be able to watch live television coverage of such annual sports events as tennis at Wimbledon or horse racing at the Kentucky Derby, or other major sporting activities (e.g., the Olympics, the World Cup soccer competition), than of local or regional competitions.

The cost of producing programs to fill even a few hours of television time each evening is beyond the financial capability of most stations and local programming is uniformly confined to newscasts, magazine programs, interviews and panel discussions. A detailed study of Jamaican television in 1980 and 1981 (Gould and Lyew-Ayee, 1981) revealed that more than 60 percent of programs aired on the sole station were of foreign origin – primarily the United States. This picture was much the same at the end of the decade, not just for Jamaica but also for Trinidad and Barbados, the two other countries with long-established television stations. Even advertisements on television may be of foreign origin; thus, for example, while an American soft drink may be locally bottled, the television commercial for it shows an American celebrity promoting its consumption.

Perhaps the greatest impact on television in the region has been the availability of receivers capable of picking up satellite television transmissions, especially from the United States. The proliferation of domestic 'dishes', as these receivers are popularly called, all over Jamaica probably reflects many Jamaicans' dissatisfaction with the one-channel government-run station. The Jamaica Broadcasting Corporation (JBC) has itself acquired two large dishes from which it videotapes programs (including newscasts) for later transmission over the airwaves. Authorities prefer to turn a blind eye to the ethics of such 'piracy'. Indeed, much discussion continues in the region and abroad on the question of whether the interception of United States television satellite programming in the Caribbean is illegal (see, for example, Brady, 1987).

Elsewhere, Antigua, Dominica and Montserrat are among the countries where households may opt for access to multiple channels via a cable feed from a central dish. Guyana had two such cable television services long before its government initiated its limited television service in 1988.

Not only is television now available to a larger audience in the region than ever, but the nature of the programming available must lead to more serious questions than before of 'cultural imperialism'. When television programs are beamed directly into homes from satellites in space, and when those programs are overwhelmingly of American origin, the impact of foreign lifestyles and ideas has got to be significant. Even where local stations have a nightly newscast, anchored by their own personnel and leading off with local news stories, the segment dealing with foreign news relies entirely on news clips from the three major American networks or from Cable News Network. The observant viewer will not fail to notice that the foreign news stories are better illustrated, whereas the items of local or regional news may carry no film coverage at all, primarily because of the stations' inability (due to lack of finances, manpower, etc.) to have camera crews everywhere that things are happening.

Nettleford (1987) bemoans the effective 'Dallasization' of consciousness in the Caribbean as images of material opulence in the United States are transmitted via very popular television programs to mass audiences in the region. Deep concern has been expressed about the effects of American television programming on the Caribbean cultural identity. A study has been commissioned by the Caribbean Publishing and Broadcasting Association (CPBA) the report of which was to be published in early 1988. The CPBA was expected to follow up this study by sponsoring a conference to develop a plan to address some of the ills of the media identified in this report.

This kind of concern is being taken up at the highest levels of government. A meeting of regional Ministers of Communication in September 1987 pointed to the obsolete communication technology and the lack of technological and management skills in their countries, and discussed in particular the impact of satellite developments on the region.

This is part of the ongoing concern in Third World countries about the need for a 'New World Information Order' (see, for example, Robinson, 1982 , and Mowlana, 1985). West Indians have long had links, directly and indirectly, with such developed countries as the United States, Canada and the United Kingdom, each of which has large West Indian communities. The idea of a 'West Indian cultural identity' is seen to be even more endangered by the impact of current developments in communication.

However, there may be reason to doubt whether the various territories do in fact have a common cultural identity. The language which is said to be common to all the countries has also produced quite distinct 'patois' forms in different countries. Even the names of common fruits may vary so greatly among the islands as to cause confusion. 'Soursop' does not refer to the same fruit in Jamaica and St Kitts, for example: 'golden apple' is the name of two entirely different fruits in Jamaica and Trinidad.

Music forms have also developed differently, so that calypso is synonymous with the eastern Caribbean, while reggae has put Jamaica on the world's musical map.

Air travel between points in the Commonwealth Caribbean is often indirect, time-consuming and expensive. There is no nonstop commercial air route between Jamaica and its fellow members of the Commonwealth Caribbean. The main regional air carrier, British West Indian Airways, owned by the government of Trinidad and Tobago, provides daily air links between the two ends of the island arc (Jamaica and Trinidad), but with three or more stops en route. The Dutch Antillean carrier, ALM, links Jamaica with Trinidad via Haiti and Curaçao. A Belizean en route to a Caribbean island can reach his destination only by changing planes in Miami, Florida. Air Jamaica initiated a shortlived direct flight between Jamaica and Trinidad in the early 1980s, but at the present time the only Caribbean destination served by the airline from Jamaica is the British colony of Cayman to its northwest.

Complicated regulations governing international flights and the rights of non-regional airlines to carry passengers between points in the Caribbean have led to a situation where those countries with airports capable of accommodating large jet aircraft are well linked to, especially, North American points such as Miami and New York. Thus American Airlines, for example, has daily flights between the United States and Jamaica, Barbados, Trinidad and Antigua, and British Airways links these islands to London.

Most of the smaller islands do not have the airports that would accommodate large commercial jet aircraft. A fleet of small, twin-engined planes serves as the air link for Dominica, St Lucia, St Vincent, etc., to the larger outside world through regular hops to Barbados or Antigua.

Even a direct, nonstop flight from Barbados or Trinidad to Jamaica would take four hours, about the same time as a flight from Jamaica to New York or Toronto. The air fares from Jamaica to the latter cities are also lower than those from Jamaica to the eastern Caribbean.

Prospects for closer integration

The absence of a language problem among the countries of the Commonwealth Caribbean has not facilitated or promoted closer communication or integration in the region.

Some of the main reasons for the failure of the West Indian Federation are rooted in the separateness of the constituent territories. The parochialism engendered by the physical separation of the territories and the widely varying stages of development among them made difficult the pooling of resources for economic development and viability which was an aim of federation. In the mid 1980s, with continuing economic problems and nonattainment of the desired levels of growth, some Commonwealth Caribbean leaders had begun to reassess the prospects of federation once more, but the general view is that the disagreements over politics, finances, trade, etc., which marked the federation years of 1958–62 have not dissipated. The smaller islands, in particular, see advantages in pooling resources, and there is ongoing discussion on whether the seven-member Organization of Eastern Caribbean States (OECS), now a loose, United Nations-style grouping, may get together in a formal political union.

In the federation period there had been calls for a Caribbean news agency, for a regular newspaper service to all the islands, and for a national newspaper. Only one of these objectives in the realm of regional communication has been realized; even so, it was not until 1976 that the Caribbean News Agency (CANA) was actually launched.

The expansion of the regional cable, wireless, telephone and telegraph links in the twentieth century did not significantly increase the flow of news within the region, but did increase the flow of foreign news into the region. This trend has accelerated with the developments in communication technology of the last decade.

Direct dialing on international telephone services is available for most of the countries in the Commonwealth Caribbean. The telegraph service has been supplemented by telex and other more recent, sophisticated systems (e.g. Fax). While these are more expensive to install, the possibility of rapid direct communication with other countries does exist. The technology, however, is entirely foreign in origin, and the use of such systems emphasizes the region's dependence on First World expertise in communication technology.

The effects of satellite television transmission have been discussed. The trend in the region is of more and more countries taking the route of entertaining their populations via such television fare rather than undertaking to put on their own programming. The concern expressed in various circles about the effect of such foreign domination of the medium has to be weighed against the simple economics of the matter.

The countries of the Commonwealth Caribbean already cooperate on such regional communications projects as the Caribbean News Agency, and the Distance Teaching Experiment of the University of the West Indies. The Caribbean Institute of Mass Communication, based at the Jamaica campus of the regional university, is the only such facility in the Commonwealth Caribbean for training media personnel.

However, politics and finances are serious obstacles in the way of greater regional cooperation and integration. However small a territory may seem in area and population, each has a government and institutions in place whose views may be difficult to incorporate harmoniously into a regional consensus. The financial problems of such countries as Jamaica and Guyana, viewed against the relative prosperity of Trinidad and Tobago and Barbados, work against even the operation of a common market for trade within the region.

The difficulties of overcoming sheer physical distance are very real, and have not been markedly ameliorated by late twentieth century developments in communication technology. Indeed, these very developments suggest that the countries of the Commonwealth Caribbean will continue to have close links with institutions in the developed countries rather than among themselves. If 'the communication satellite [has] removed most of the distance barrier from communication' (Schramm, in Preface to Edelstein, 1982, p. 11), that barrier which has been removed as far as the Commonwealth Caribbean is concerned is between the region and the outside world, especially North America; the barriers to communication within the region remain formidable.

References

Brady, H. C. W. (1987), 'The interception of United States satellite programming in the Caribbean', paper presented to the American Bar Association, in association with the International Law Institute, reprinted in the Jamaica *Sunday Gleaner*, 4 January 1988.

Cuthbert, M. (1982), 'Evolving channels for development news in the Caribbean', in Robinson, G. J. (ed.), *Assessing the New World Information Order Debate: Evidence and Proposals*, International Communications Division, Association for Education in Journalism, pp. 122–34.

Edelstein, A. S. (1982), *Comparative Communication Research* (Beverly Hills: Sage Publications).

Fischer, H.-D., and Merrill, J. C. (1976), *International and Multicultural Communication* (New York: Hastings House).

Gerbner, G. (ed.) (1977), *Mass Media Policies in Changing Cultures* (New York: John Wiley).

Gould, P., and Lyew-Ayee, A. (1981), *The Structure of Jamaican Television: A Pilot Study*, University Park, Mona, Jamaica: International Television Flows Project, Discussion Paper no. 13.

——(1983), 'Jamaican television: images and geographic connections', *Caribbean Geography*, 1, 1, pp. 36–50.

——(1985), 'Television in the Third World: a high wind on Jamaica', in Burgess, J., and Gold, J. R. (eds), *Geography, the media and popular culture* (London: Croom Helm), pp. 33–62.

Lent, J. A. (1977), *Third World Mass Media and Their Search for Modernity: The Case of Commonwealth Caribbean, 1717–1976* (London: Associated Universities Press).

Lerner, D., and Schramm, W. (eds) (1969), *Communication and Change in the Developing Countries* (Honolulu: East–West Center Press).

Lowenthal, D. (ed.) (1961), *The West Indies Federation* (Washington, DC: American Geographical Society Research Series, No. 23).

Mowlana, H. (1985), *International Flow of Information: A Global Report and Analysis* (Paris: United Nations Educational, Scientific and Cultural Organization).

Nettleford, R. (1987), 'Cultivating a Caribbean sensibility: media, education and culture', *Caribbean Review*, XV, 3, pp. 4–8.

Richstad, J. (ed.) (1977), *New Perspectives in International Communication* (Honolulu: East–West Communication Institute).

Robinson, G. J. (ed.) (1982), *Assessing the New World Information Order Debate: Evidence and Proposals*. International Communication Division, Association for Education in Journalism.

Schramm, W. (1964), *Mass Media and National Development* (Stanford, Calif.: Stanford University Press).

Index